Deep Learning Techniques for Biomedical and Health Informatics

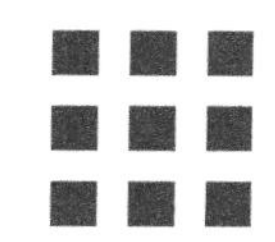

Deep Learning Techniques for Biomedical and Health Informatics

Edited by

Basant Agarwal

Valentina Emilia Balas

Lakhmi C. Jain

Ramesh Chandra Poonia

Manisha

Academic Press is an imprint of Elsevier
125 London Wall, London EC2Y 5AS, United Kingdom
525 B Street, Suite 1650, San Diego, CA 92101, United States
50 Hampshire Street, 5th Floor, Cambridge, MA 02139, United States
The Boulevard, Langford Lane, Kidlington, Oxford OX5 1GB, United Kingdom

Notices

Knowledge and best practice in this field are constantly changing. As new research and experience broaden our understanding, changes in research methods, professional practices, or medical treatment may become necessary.

Practitioners and researchers must always rely on their own experience and knowledge in evaluating and using any information, methods, compounds, or experiments described herein. In using such information or methods they should be mindful of their own safety and the safety of others, including parties for whom they have a professional responsibility.

To the fullest extent of the law, neither the Publisher nor the authors, contributors, or editors, assume any liability for any injury and/or damage to persons or property as a matter of products liability, negligence or otherwise, or from any use or operation of any methods, products, instructions, or ideas contained in the material herein.

Library of Congress Cataloging-in-Publication Data
A catalog record for this book is available from the Library of Congress

British Library Cataloguing-in-Publication Data
A catalogue record for this book is available from the British Library

ISBN: 978-0-12-819061-6

For information on all Academic Press publications
visit our website at https://www.elsevier.com/books-and-journals

Publisher: Mara Conner
Acquisition Editor: Chris Katsaropoulos
Editorial Project Manager: Gabriela D. Capille
Production Project Manager: Punithavathy Govindaradjane
Cover Designer: Mark Rogers

Typeset by SPi Global, India

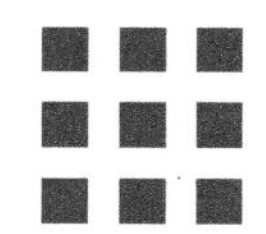

Contents

Contributors

Selam Ahderom Electron Science Research Institute, Edith Cowan University, Joondalup, WA, Australia

Kamal Alameh Electron Science Research Institute, Edith Cowan University, Joondalup, WA, Australia

Salah Alheejawi Department of Electrical and Computer Engineering, University of Alberta, Edmonton, AB, Canada

Ashish Anand Department of Computer Science and Engineering, Indian Institute of Technology Guwahati, Guwahati, India

Rangel Arthur School of Technology (FT), State University of Campinas (UNICAMP), Limeira, Brazil

Muhammad Waseem Ashraf GC University Lahore, Lahore, Pakistan

Valentina Emilia Balas Department of Automation and Applied Informatics, Aurel Vlaicu University of Arad, Arad, Romania

Richard Berendt Cross Cancer Institute, Edmonton, AB, Canada

Animesh Biswas Department of Mathematics, University of Kalyani, Kalyani, India

Mou De Netaji Subhash Engineering College, Kolkata; Computer Innovative Research Society, Howrah, India

Vijaypal Singh Dhaka Manipal University Jaipur, Jaipur, India

Reinaldo Padilha França School of Electrical Engineering and Computing (FEEC), State University of Campinas (UNICAMP), Campinas, Brazil

Bappaditya Ghosh Department of Mathematics, University of Kalyani, Kalyani, India

E.A. Gopalakrishnan Center for Computational Engineering and Networking (CEN), Amrita School of Engineering, Amrita Vishwa Vidyapeetham, Coimbatore, India

P. Gopika Center for Computational Engineering and Networking (CEN), Amrita School of Engineering, Amrita Vishwa Vidyapeetham, Coimbatore, India

Yuzo Iano School of Electrical Engineering and Computing (FEEC), State University of Campinas (UNICAMP), Campinas, Brazil

Vijay Jeyakumar Department of Biomedical Engineering, SSN College of Engineering, Chennai, India

Naresh Jha Cross Cancer Institute, Edmonton, AB, Canada

Anirban Kundu Netaji Subhash Engineering College, Kolkata; Computer Innovative Research Society, Howrah, India

Preethi Kurian Department of Biomedical Engineering, SSN College of Engineering, Chennai, India

Cheng Lu CASE Western Reserve University, Cleveland, OH, United States

Swanirbhar Majumder Department of Information Technology, Tripura University, Agartala, India

Mrinal Mandal Department of Electrical and Computer Engineering, University of Alberta, Edmonton, AB, Canada

Navid Mavaddat Electron Science Research Institute, Edith Cowan University, Joondalup, WA, Australia

D.A. Meedeniya University of Moratuwa, Moratuwa, Sri Lanka

Takhellambam Gautam Meitei Department of Electronics and Communication Engineering, North Eastern Regional Institute of Science and Technology, Nirjuli, India

Ana Carolina Borges Monteiro School of Electrical Engineering and Computing (FEEC), State University of Campinas (UNICAMP), Campinas, Brazil

P. Naga Srinivasu Department of CSE, GIT, GITAM Deemed to be University, Visakhapatnam, India

Ramesh Chandra Poonia Norwegian University of Science and Technology (NTNU), Alesund, Norway

Nitesh Pradhan Manipal University Jaipur, Jaipur, India

Geeta Rani Manipal University Jaipur, Jaipur, India

Nivedita Ray De Sarkar Netaji Subhash Engineering College, Kolkata; Computer Innovative Research Society, Howrah, India

I.D. Rubasinghe University of Moratuwa, Moratuwa, Sri Lanka

Subhashis Sahu Department of Physiology, University of Kalyani, Kalyani, India

Sunil Kumar Sahu Department of Computer Science and Engineering, Indian Institute of Technology Guwahati, Guwahati, India

Sinam Ajitkumar Singh Department of Electronics and Communication Engineering, North Eastern Regional Institute of Science and Technology, Nirjuli, India

K.P. Soman Center for Computational Engineering and Networking (CEN), Amrita School of Engineering, Amrita Vishwa Vidyapeetham, Coimbatore, India

V. Sowmya Center for Computational Engineering and Networking (CEN), Amrita School of Engineering, Amrita Vishwa Vidyapeetham, Coimbatore, India

T. Srinivasa Rao Department of CSE, GIT, GITAM Deemed to be University, Visakhapatnam, India

Muhammad Imran Tariq The superior University, Lahore, Pakistan

Shahzadi Tayyaba The University of Lahore, Lahore, Pakistan

Valentina Tiporlini Electron Science Research Institute, Edith Cowan University, Joondalup, WA, Australia

Hongming Xu Cleveland Clinic, Cleveland, OH, United States

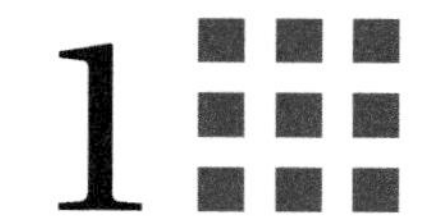

Unified neural architecture for drug, disease, and clinical entity recognition

Sunil Kumar Sahu, Ashish Anand
DEPARTMENT OF COMPUTER SCIENCE AND ENGINEERING, INDIAN INSTITUTE OF TECHNOLOGY GUWAHATI, GUWAHATI, INDIA

1.1 Introduction

Biomedical and clinical named entity recognition (NER) in the text is an important step in several biomedical and clinical information extraction tasks [1–3]. State-of-the-art methods formulate an NER task as a sequence labeling problem where each word is labeled with a tag and, based on the tag sequence, entities of interest are identified. In comparison to the generic domain, recognizing entities in the biomedical and clinical domains are difficult due to several reasons, including the use of nonstandard abbreviations or acronyms, multiple variations of the same entities, etc. [3, 4]. Furthermore, clinical notes often contain shorter, incomplete, and grammatically incorrect sentences [3], thus making it difficult for models to extract rich context. Most widely used models, including conditional random field (CRF), maximum entropy Markov model (MEMM), or support vector machine, use manually designed rules to obtain morphological, syntactic, semantic, and contextual information of a word or piece of text surrounding a word, and then use them as features for identifying correct labels [5–10]. Performance of such models is limited by the choice of explicitly designed features specific to the task and its corresponding domain. For example, Chowdhury and Lavelli [6] explained several reasons why features designed for biological entities such as proteins or genes are not equally important for disease name recognition.

Deep learning-based models have been used to reduce manual efforts for explicit feature design [11]. Here, distributional features are used in place of manually designed features and a multilayer neural network is used in place of a linear model to overcome the needs of task-specific meticulous feature engineering. Although proposed methods outperform several generic domain sequence labeling tasks, it fails to overcome the state of art in a biomedical domain [12]. There are two plausible reasons behind this: First, it learns features only from word-level embedding and, second, it takes into account only a fixed length context of the word. It has been observed that word-level embeddings preserve the

Deep Learning Techniques for Biomedical and Health Informatics. https://doi.org/10.1016/B978-0-12-819061-6.00001-X

syntactic and semantic properties of a word but may fail to preserve morphological information that can also play an important role in NER [6, 13–16]. For instance, drug names *Cefaclor, Cefdinir, Cefixime, Cefprozil,* and *Cephalexin* have a common prefix, and *Doxycycline, Minocycline,* and *Tetracycline* have a common suffix. These common prefixes/suffixes are often sufficient to predict entity types. Furthermore, window-based neural architecture can only consider words that appear within the user-decided window size as context and thus is likely to fail in picking up vital clues lying outside the window.

This work aims to overcome the two previously mentioned issues. To obtain both morphologically as well as syntactically and semantically rich embedding, two bi-directional long short-term memory networks (BLSTMs) are used in a hierarchy. First, BLSTM works on each character of the words and accumulates morphologically rich word embedding. Second, BLSTM works at the word level of a sentence to learn contextually rich feature vectors. To make sure all context lying anywhere in the sentence is utilized, we consider the entire sentence as input and use a first-order linear chain CRF in the final prediction layer. The CRF layer accommodates dependency information about the tags.

We evaluated the proposed model on three standard biomedical entity recognition tasks, namely *Disease NER, Drug NER,* and *Clinical NER*. This study distinguished features compared in several other studies [15, 17–19]. Ma and Hovy [15] focused on sequence labeling tasks in the generic domain and used a convolutional neural network (CNN) for learning character-based embedding in contrast to bi-directional LSTM used in this study. Luo et al. [17] used a similar architecture as ours for the BioCreative V.5.BeCalm Tasks focused on patents. Luo et al. [19] used attention-based BLSTM with CRF for chemical NER, whereas Zeng et al. [18] used a similar architecture only for drug NER. However, our work still has a lot to offer to readers. First, we evaluate a unified model on the different genre of texts (clinical notes vs. biomedical research articles) for multiple entity types. None of the previously mentioned studies evaluate the different genre of texts. Second, extensive analyses are performed to understand the significance of various components of the model architecture, including CRF (sentence-level likelihood for accounting for tag dependency) versus word-level likelihood (treating each tag independently); feature ablation study to understand the importance of each feature type; and the significance of word and character embedding. Lastly, error analysis is also performed to gain insight as to where new models should focus to further improve the performance. We compare the proposed model with the existing state-of-the-art models for each task and show that it outperforms them. Further analysis of the model indicates the importance of using character-based word embedding along with word embedding and CRF layer in the final output layer.

1.2 Method

1.2.1 Bi-directional long short-term memory

Recurrent neural network (RNN) is a variant of neural networks that utilizes sequential information and maintains history through its recurrent connection [20, 21]. RNN can be used for a sequence of any length; however, in practice, it fails to maintain long-term dependency due to vanishing and exploding gradient problems [22, 23]. Long short-term memory (LSTM) network [24] is a variant of RNN that takes care of the issues associated with vanilla RNN by using three gates (input, output, and forget) and a memory cell.

We formally describe the basic equations pertaining to the LSTM model. Let $h^{(t-1)}$ and $c^{(t-1)}$ be hidden and cell states of LSTM, respectively, at time $t-1$, then computation of current hidden state at time t can be given as:

$$
\begin{aligned}
i^{(t)} &= \sigma(U^{(i)}x^{(t)} + W^{(i)}h^{(t-1)} + b^i) \\
f^{(t)} &= \sigma(U^{(f)}x^{(t)} + W^{(f)}h^{(t-1)} + b^f) \\
o^{(t)} &= \sigma(U^{(o)}x^{(t)} + W^{(o)}h^{(t-1)} + b^o) \\
g^{(t)} &= \tanh(U_l^{(g)}x^{(t)} + W^{(g)}h^{(t-1)} + b^g) \\
c^{(t)} &= c^{(t-1)} * f^{(t)} + g^{(t)} * i^{(t)} \\
h^{(t)} &= \tanh(c^{(t)}) * o^{(t)}
\end{aligned}
$$

where σ is sigmoid activation function, $*$ is an element-wise product, $x^{(t)} \in \mathbb{R}^d$ is the input vector at time t, $U^{(i)}$, $U^{(f)}$, $U^{(o)}$, $U^{(g)} \in \mathbb{R}^{N \times d}$, $W^{(i)}$, $W^{(o)}$, $W^{(f)}$, $W^{(g)} \in \mathbb{R}^{N \times N}$, $b^i, b^f, b^o, b^g \in \mathbb{R}^N$, $h^{(0)}, c^{(0)} \in \mathbb{R}^N$ are learning parameters for LSTM. Here, d is dimension of input feature vector, N is hidden layer size, and $h^{(t)}$ is output of LSTM at time step t.

It has become common practice to use LSTM in both forward and backward directions to capture both past and future contexts, respectively. First, LSTM computes its hidden states in the forward direction of the input sequence and then does it in the backward direction. This way of using two LSTMs is referred to as bi-directional LSTM or simply BLSTM. We use bi-directional LSTM in our model. The final output of BLSTM at time t is given as:

$$h^{(t)} = \overrightarrow{h^{(t)}} \oplus \overleftarrow{h^{(t)}} \tag{1.1}$$

where $\oplus$ is concatenation operation and $\overrightarrow{h^{(t)}}$ and $\overleftarrow{h^{(t)}}$ are hidden states of forward and backward LSTM at time t.

1.2.2 Model architecture

Similar to any NER task, we formulate the biomedical entity recognition task as a token-level sequence tagging problem. We use beginning-inside-outside (BIO) tagging scheme in our experiments [25]. Architecture of the proposed model is present in Fig. 1.1. Our model takes the whole sentence as input and computes a label sequence as output. The first layer of the model learns local feature vectors for each word in the sentence. We use concatenation of word embedding, PoS tag embedding, and character-based word embedding as a local feature for every word. Character-based word embedding is learned through applying a BLSTM on the character vectors of a word. We call this layer *Char BLSTM* (Section 1.2.3.1). The subsequent layer, called *Word BLSTM* (Section 1.2.5), incorporates contextual information through a separate BLSTM network. Finally, we use a CRF to encode the correct label sequence on the output of *Word BLSTM* (Section 1.2.5). Now onward, the proposed framework will be referred to as *CWBLSTM*. Entire network

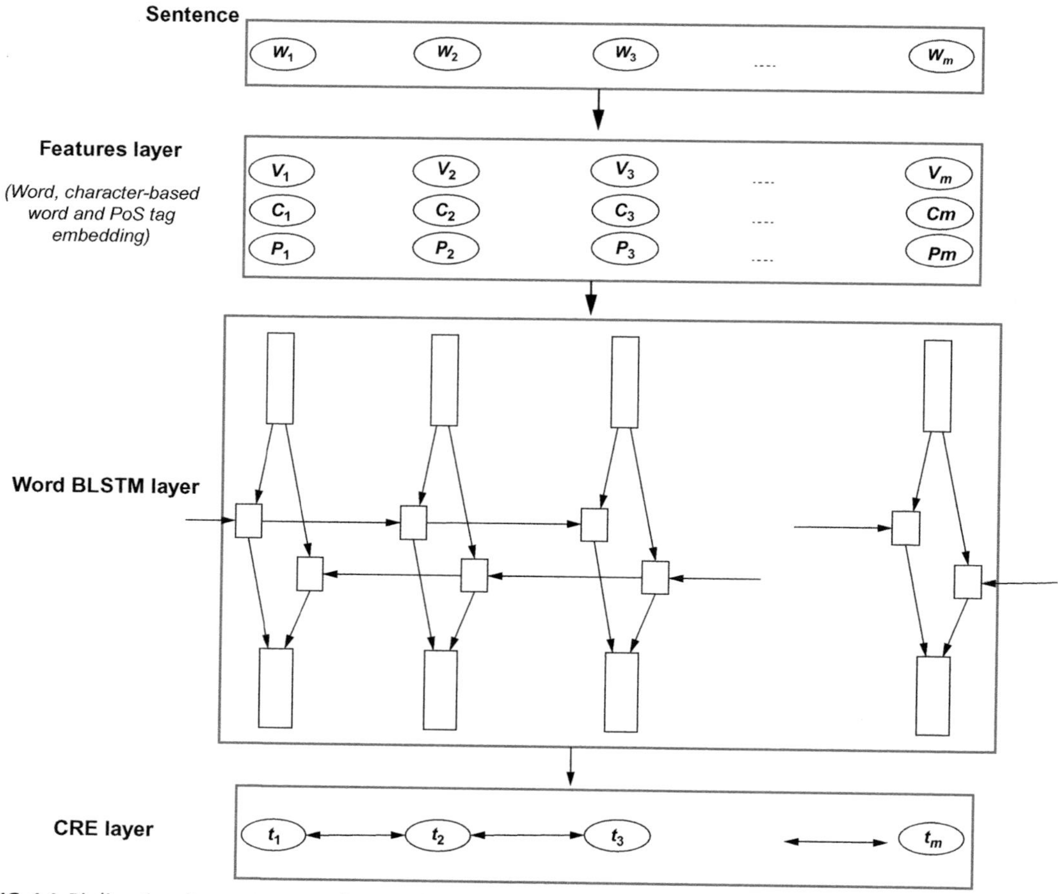

FIG. 1.1 Bi-directional recurrent neural network-based model for biomedical entity recognition. Here, $w_1w_2\ldots w_m$ is the word sequence of the sentence, and $t_1t_2\ldots t_m$ is its computed label sequence, and m represents length of the sentence.

parameters are trained in an end-to-end manner through cross-entropy loss function. We next describe each part of the model in detail.

1.2.3 Features layer

Word embedding or distributed word representation is a compact vector representation of a word that preserves lexico-semantic properties [26]. It is a common practice to initialize word embedding with a pretrained vector representation of words. Apart from word embedding, in this work PoS tag and character-based word embedding are used as features. We use the GENIA[a] tagger to obtain PoS tags in all the datasets. Each PoS tag was

[a]See http://www.nactem.ac.uk/GENIA/tagger/.

initialized randomly and was updated during the training. The output of the feature layer is a sequence of vectors, say $x_1, \ldots, x_m$ for the sentence of length m. Here, $x_i \in \mathbb{R}^d$ is the concatenation of word embedding, PoS tag embedding, and character-based word embedding. We next explain how character-based word embedding is learned.

1.2.3.1 Char BLSTM

Word embedding is a crucial component of all deep learning-based natural language processing (NLP) tasks. Capability to preserve lexico-semantic properties in the vector representation of a word makes it a powerful resource for NLP [11, 27]. In biomedical and clinical entity recognition tasks, apart from semantic information, the morphological structure such as prefix, suffix, or some standard patterns of words also gives important clues [4, 6]. The motivation behind using character-based word embedding is to incorporate morphological information of words in feature vectors.

To learn character-based embeddings, we maintain a vector for every character in an embedding matrix [13, 14, 17]. These vectors are initialized with random values in the beginning. To illustrate, assume *cancer* is a word for which we want to learn an embedding (represented in Fig. 1.2), so we would use a BLSTM on the vector of each character of

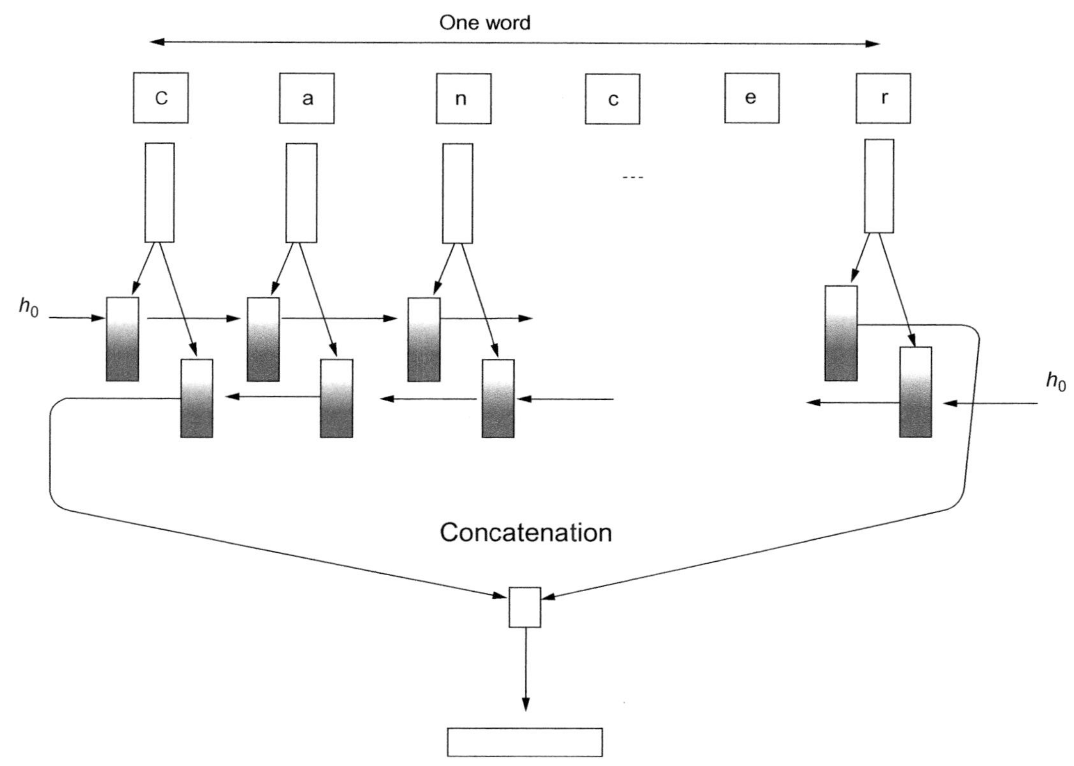

FIG. 1.2 Learning character-based word embedding.

cancer. As mentioned earlier, forward LSTM maintains information about the past in computation of current hidden states, and backward LSTM obtain future contexts, therefore after reading an entire sequence, the last hidden states of both RNNs must have knowledge of the whole word with respect to their directions. The final embedding of a word would be:

$$v_{cw} = \overrightarrow{h^{(m)}} \oplus \overleftarrow{h^{(m)}} \tag{1.2}$$

where $\overrightarrow{h^{(m)}}$ and $\overleftarrow{h^{(m)}}$ are the last hidden states of forward and backward LSTMs, respectively.

1.2.4 Word BLSTM layer

The output of a feature layer is a sequence of vectors for each word of the sentence. These vectors have local or individual information about the words. Although local information plays an important role in identifying entities, a word can have a different meaning in different contexts. Earlier works [6, 11, 12, 16] use a fixed-length window to incorporate contextual information. However, important clues can lie anywhere in the whole sentence. This limits the learned vectors to obtain knowledge about the complete sentence. To overcome this, we use a separate BLSTM network that takes local feature vectors as input and outputs a vector for every word based on both contexts and current feature vectors.

1.2.5 CRF layer

The output of *Word BLSTM* layer is again a sequence of vectors that have contextual as well as local information. One simple way to decode the feature vector of a word into its corresponding tag is to use word-level log likelihood (WLL) [11]. Similar to *MEMMs*, it will map the feature vector of a word to a score vector of each tag by a linear transformation, and every word will get its label based on its scores and independent of labels of other words. One limitation of this way of decoding is that it does not take into account dependency among tags. For instance, in a *BIO tagging* scheme, a word can only be tagged with *I-Entity* (standing for Intermediate-Entity) only after a *B-Entity* (standing for Beginning-Entity). We use CRF [5] on the feature vectors to include dependency information in decoding and then decode the whole sentence together with its tag sequence.

CRF maintains two parameters for decoding, $W_u \in R^{k \times h}$ linear mapping parameter and $T \in R^{h \times h}$ pairwise transition score matrix. Here, k is the size of the feature vector, h is the number of labels present in the task, and $T_{i,j}$ implies pairwise transition score for moving from label i to label j. Let $[v]_1^{|s|}$ be a sequence of feature vectors for a sentence $[w]_1^{|s|}$ and assume $[z]_1^{|s|}$ is the unary potential scores obtained after applying linear transformation on feature vectors (here, $z_i \in R^h$), then CRF decodes this with tag sequence using:

$$P([y]_1^{|s|} | [w]_1^{|s|}) = \underset{t \in Q^{|s|}}{\operatorname{argmax}} \frac{\exp \Psi([z]_1^{|s|}, [t]_1^{|s|})}{\sum_{t^{\psi} \in Q^{|s|}} \exp \Psi([z]_1^{|s|}, [t^{\psi}]_1^{|s|})} \tag{1.3}$$

where

$$\Psi([z]_1^{|s|}, [t]_1^{|s|}) = \sum_{1 \le i \le |s|} (T_{t_{i-1}, t_i} + z_{t_i}) \tag{1.4}$$

Here, $Q^{|s|}$ is a set containing all possible tag sequences of length $|s|$ and t_j is tag for the jth word. Highest probable tag sequence is estimated using Viterbi algorithm [11, 28].

1.2.6 Training and implementation

We train the model for each task separately. We use cross-entropy loss function to train the model. Adam's technique [29] is used to obtain optimized values of model parameters. We use the mini-batch size of 50 in training for all tasks. In all experiments, we use pretrained word embedding of 100 dimensions, which was trained on PubMed corpus using GloVe [30, 31], PoS tag embedding vector of 10 dimensions, character-based word embedding of length 20, and hidden layer size 250. We use l_2 regularization with 0.001 as the corresponding parameter value. These hyperparameters are obtained using the validation set of *Disease NER* task. We considered batch size with values 25, 50, 75, and 100; hidden layer size with values 150, 200, 250, and 300; and 12 regularization with values 0.1, 0.01, 0.001, and 0.0001 for tuning the hyperparameters in a greed search. The corresponding training, validation, and test sets for the *Disease NER* task are available as separate files with NCBI disease corpus. For the other two tasks, we used the same set of hyperparameters as obtained on *Disease NER*. Entire implementation was done in Python language using *TensorFlow*[b] library.

1.3 The benchmark tasks

In this section, we briefly describe the three standard tasks on which we examined the performance of the CWBLSTM model. Statistics of corresponding benchmark datasets are given in Table 1.1.

1.3.1 Disease NER

Identifying disease named entity in the text is crucial for disease-related knowledge extraction [32, 33]. It has been observed that disease is one of the most widely searched entities by users on PubMed [34]. We use *NCBI disease corpus*[c] to investigate the performance of the model on a *Disease NER* task. This dataset was annotated by a team of 12 annotators (2 persons per annotation) on 793 PubMed abstracts [34, 35].

1.3.2 Drug NER

Identifying drug name or pharmacological substance is an important first step for drug-drug interaction extraction and other drug-related knowledge extraction tasks. Keeping

[b]See https://www.tensorflow.org.

[c]See https://www.ncbi.nlm.nih.gov/CBBresearch/Dogan/DISEASE/.

Table 1.1 Statistics of the benchmark datasets corresponding to the three tasks.

Dataset	Corpus	Training set	Test set
Disease NER	Sentences	5661	961
	Disease	5148	961
Drug NER	Sentences	6976	665
	Drug	9369	347
	Brand	1432	59
	Group	3381	154
	Drug_n	504	120
Clinical NER	Sentences	8453	14,529
	Problem	7072	12,592
	Treatment	2841	9344
	Test	4606	9225

this in mind, a challenge for recognition and classification of pharmacological substances in the text was organized as part of SemEval-2013. We used the SemEval-2013 task 9.1 [36] dataset for this task. The dataset shared in this challenge was annotated from two sources: *DrugBank*[d] documents and *MedLine*[e] abstracts. This dataset has four kind of drugs as entities, namely *drug, brand, group,* and *drug_n*. Here, *drug* represents generic drug name, brand is brand name of a drug, *group* is the family name of drugs, and *drug_n* is an active substance not approved for human use [37]. During preprocessing of the dataset, 79 entities (56 *drug*, 18 group, and 5 brand) from the training set and 5 entities (4 drug and 1 group) from the test set were removed. The removed entities of the test set are treated as false negatives in our evaluation scheme.

1.3.3 Clinical NER

For clinical entity recognition, we used a publicly available (under license) i2b2/VA[f] challenge dataset [3]. This dataset is a collection of discharge summaries obtained from Partners Healthcare, Beth Israel Deaconess Medical Center, and the University of Pittsburgh Medical Center. The dataset was annotated for three kinds of entities, namely *problem, treatment,* and *test*. Here, *problems* indicate phrases that contain observations made by patients or clinicians about the patient's body or mind that are thought to be abnormal or caused by a disease. *Treatments* are phrases that describe procedures, interventions, and substances given to a patient to resolve a medical problem. *Tests* are procedures, panels, and measures that are performed on a patient, body fluid, or sample to discover, rule out, or find more information about a medical problem.

[d]See https://www.drugbank.ca/.

[e]See https://www.nlm.nih.gov/bsd/pmresources.html.

[f]See https://www.i2b2.org/NLP/Relations/Main.php.

The downloaded dataset for this task was partially available (only discharge summaries from Partners Healthcare and Beth Israel Deaconess Medical Center) compared to the full dataset originally used in the challenge. We performed our experiments on the currently available partial dataset. The dataset is available in the preprocessed form, where sentence and word segmentation are already done. We removed patient's information from each discharge summary before training and testing, because that never contains entities of interest.

1.4 Results and discussion

1.4.1 Experiment design

We performed separate experiments for each task. We used the *training set* for learning optimal parameters of the model for each dataset and the evaluation is performed on the *test set*. Performance of each trained model is evaluated based on strict matching sense, where the exact boundaries, as well as class, need to be correctly identified for consideration of true positives. For a strict matching evaluation scheme, we used a CoNLL 2004[g] evaluation script to calculate precision, recall, and F1 score in each task. In all our experiments, we trained and tested the models' performance four times with different random initializations of all parameters. Results reported in the paper are the best results obtained among the four different runs. We did this with all baseline methods as well.

1.4.2 Baseline methods

We briefly describe the baseline methods selected for comparison with the proposed models in all the considered tasks. The selected baseline methods were implemented by us, and their corresponding hyperparameters were tuned using the similar strategy as used in the proposed methods.

SENNA: SENNA uses the window-based neural network on the embedding of a word with its context to learn global features [11]. To make inference, it also uses CRF on the output of a window-based neural network. We set the window size five based on hyperparameter tuning using the validation set (20% of the training set), and the rest of all other hyperparameters are set similar to our model.

CharWNN: This model [13] is similar to SENNA but uses the word as well as character-based embedding in the chosen context window [38]. Here, character-based embeddings are learned through convolution neural network and max-pooling scheme.

CharCNN: This method [39] is similar to the proposed model *CWBLSTM* but instead of using BLSTM, it uses convolution neural network for learning character-based embedding.

[g]See http://www.cnts.ua.ac.be/conll2002/ner/bin/conlleval.txt.

1.4.3 Comparison with baseline

Table 1.2 presents a comparison of *CWBLSTM* with different baseline methods on disease, drug, and clinical entity recognition tasks. We can observe that it outperforms all three baselines in each of the three tasks. In particular, when comparing with *CharCNN*, differences are considerable for *Drug NER* and *Disease NER* tasks. The proposed model improved the recall by 5% to gain about 2.5% of relative improvement in F1 score over the second-best method of *CharCNN* for the *Disease NER* task. For the *Drug NER* task, the relative improvement of more than 3% is observed for all three measures—precision, recall, and F1 score—over the *CharCNN* model. The relatively weaker performance on *Clinical NER* task could be attributed to the use of many nonstandard acronyms and abbreviations that makes it difficult for character-based embedding models to learn appropriate representation.

We performed an approximate randomization test [40, 41] to check if the observed differences in performance of the proposed model and baseline methods are statistically significant. We considered $R = 2000$ in an approximate randomization test. Table 1.3 shows the *P*-values of the statistical tests. As the *P*-values indicate, *CWBLSTM* has significantly

Table 1.2 Performance comparison of the proposed model *CWBLSTM* with baseline models on the test set of different datasets.

Tasks	Models	Accuracy	Precision	Recall	F1 score
Disease NER	*SENNA*	97.26	77.93	76.80	77.36
	CharWNN	97.24	78.34	78.67	78.50
	CharCNN	97.61	84.26	78.56	81.31
	CWBLSTM	**97.77**	**84.42**	**82.31**	**83.35**
Drug NER	*SENNA*	96.71	66.93	62.70	64.75
	CharWNN	97.07	69.16	69.16	69.16
	CharCNN	97.09	70.34	72.10	71.21
	CWBLSTM	**97.46**	**72.57**	**74.60**	**73.57**
Clinical NER	*SENNA*	91.56	80.30	78.85	79.56
	CharWNN	91.42	79.96	78.12	79.03
	CharCNN	93.02	83.65	**83.25**	83.45
	CWBLSTM	**93.19**	**84.17**	83.20	**83.68**

Note: Accuracy represents token-level accuracy in tagging. *Bold font* represents the highest performance in the task.

Table 1.3 *P*-values of the statistical significance test comparing performance of *CWBLSTM model* with other models.

Model	*CharCNN*	*CharWNN*	*SENNA*
Disease NER	0.043	0.49×10^{-3}	0.48×10^{-3}
Drug NER	0.090	0.21×10^{-2}	0.49×10^{-3}
Clinical NER	0.088	0.49×10^{-3}	0.21×10^{-3}

outperformed *CharWNN* and *SENNA* in all three tasks (significance level: 0.05). However, *CWBLSTM* can outperform *CharCNN* only in Disease NER task.

One can also observe that, even though *Drug NER* has sufficiently enough training datasets, all models gave a relatively poor performance compared to the performance in the other two tasks. One reason for the poor performance could be the nature of the dataset. As discussed, *Drug NER* dataset constitutes texts from two sources, *DrugBank* and *MedLine*. Sentences from *DrugBank* are shorter and are comprehensive as written by medical practitioners, whereas *MedLine* sentences are from research articles that tend to be longer. Furthermore, the *training set* constitutes 5675 sentences from *DrugBank* and 1301 from *MedLine*, whereas this distribution is reversed in the *test set*, that is, more sentences are from *MedLine* (520 in comparison to 145 sentences from *DrugBank*). The smaller set of training instances from *MedLine* sentences do not give sufficient examples to the model to learn.

1.4.4 Comparison with other methods

In this section, we compare our results with other existing methods present in the literature. We do not compare results on *Clinical NER* as the complete dataset (as was available in the i2b2 challenge) is not available and the results in the literature are for the full dataset.

1.4.4.1 Disease NER

Table 1.4 displays a performance comparison of different existing methods with *CWBLSTM* on NCBI disease corpus. *CWBLSTM* improved the performance of BANNER by 1.89% in terms of F1 score. BANNER is a CRF-based method that primarily uses orthographic, morphological, and shallow syntactic features [16]. Many of these features are specially designed for biomedical entity recognition tasks. The proposed model also gave a better performance than another BLSTM-based model [39] by improving recall by around 12%. The BLSTM model [39] uses a BLSTM network with word embedding, whereas the proposed model makes use of PoS as well as character-based word embeddings as extra features.

1.4.4.2 Drug NER

Table 1.5 reports performance comparison on the *Drug NER* task with submitted results in the SemEval-2013 Drug Named Recognition Challenge [36]. *CWBLSTM* outperforms the

Table 1.4 Performance comparison of *CWBLSTM* with other existing models on *Disease NER* task.

Model	Features	Precision	Recall	F1 score
CWBLSTM	Word, PoS, and character embedding	84.42	**82.31**	**83.35**
BANNER [34]	Orthographic, morphological, syntactic	–	–	81.8
BLSTM+We [39]	Word embedding	**84.87**	74.11	79.13

Bold font represents the highest score.

Table 1.5 Performance comparison of *CWBLSTM* with other existing models submitted in SemEval-2013 *Drug NER* task.

Model	Features	Precision	Recall	F1 score
CWBLSTM	Word, PoS, and character embedding	72.57	**74.05**	**73.30**
WBI [8]	ChemSpot and ontologies	73.40	69.80	71.5
LASIGE [43]	Ontology and morphological	69.60	62.10	65.6
UTurku [9]	Syntactic and contextual	**73.70**	57.90	64.8

best result obtained in the challenge (WBI-NER [8]) by a margin of 1.8%. *WBI-NER* is the extension of the ChemSpot chemical NER [42] system, which is a hybrid method for chemical entity recognition. ChemSpot primarily uses features from a dictionary to make a sequence classifier using CRF. Apart from that, WBI-NER also uses features obtained from different domain-dependent ontologies. Performance of the proposed model is better than LASIGE [43] as well as UTurku [9] systems by a considerable margin. LASIGE is a CRF-based method, whereas UTurku uses Turku Event Extraction System, which is a kernel-based model for entity and relation extraction tasks.

1.4.5 Feature ablation study

We analyzed the importance of each feature type by performing feature ablation analysis. The corresponding results are presented in Table 1.6. The first row in the table presents the performance of the proposed model using all feature types in the three tasks and second, third, and fourth rows show performance when character-based word embedding, PoS tag embedding, and pretrained word embedding are removed from the model subsequently. Removal of pretrained word embedding implies the use of random vectors in place of pretrained vectors.

We can observe that the removal of character-based word embedding led to 3.6%, 5.8%, and 1.1% relative decrements in F1 score on *Disease NER*, *Drug NER*, and *Clinical NER* tasks, respectively. The relative decrement in the performance demonstrates the importance of character-based embedding. As mentioned earlier, character-based word embedding helps our model in two ways: First, it gives morphologically rich vector representation and, second, through character-based word embedding we can also get

Table 1.6 Performance of *CWBLSTM* model under different feature settings.

	Disease NER			*Drug NER*			*Clinical NER*		
Model	**P**	**R**	**F**	**P**	**R**	**F**	**P**	**R**	**F**
CWBLSTM	84.42	82.31	83.35	72.57	74.60	73.57	84.17	83.20	83.68
–CE	80.86	80.02	80.44	64.29	75.62	69.50	83.76	81.74	82.74
–(CE+PE)	82.72	77.73	80.15	65.96	73.42	69.49	83.31	80.51	81.89
–(CE+PE+WE)	79.66	73.78	76.61	65.40	55.80	60.22	79.53	78.28	78.90

Notes: In every blocks, P, R, and F implies precision, recall, and F1 score, respectively. Here in row 4, the model uses random vector in place of pretrained word vectors for word embedding.

Table 1.7 Statistics of unique and OoV words present in different datasets but missing in the dictionary of pretrained word embedding.

Dataset	Unique words	OoV	Percent
Disease NER	8270	819	9.90
Drug NER	9447	1309	13.85
Clinical NER	13,000	2617	20.13

Notes: *OoV* indicates the number of words not found in pretrained word embedding and *percent* indicates their percentage in overall vocabulary of the corresponding datasets.

vector representation for out of vocabulary (OoV) words. OoV words are 9.9%, 13.85%, and 20.13% in *Drug NER*, *Disease NER*, and *Clinical NER* datasets, respectively (shown in Table 1.7). As discussed earlier, these decrements are less in *Clinical NER* because of the presence of acronyms and abbreviations in high frequency, which does not allow the model to take advantage of character-based word embedding. Results (third row) indicate that the PoS tag embedding may not be a crucial feature for any of the three tasks.

In contrast to PoS tag embedding, we observe that use of pretrained word embedding is one of the important feature types in our model for each task. Pretrained word embedding improves the model performance by utilizing large unlabeled corpus. It would help the model to recognize entities in two ways. First, in the training dataset, there are words that appear very few times (in our case, approximately 10% of words in all datasets have appeared only once in the dataset). During training, their randomly initialized vectors would not get sufficiently updated. On the other hand, as mentioned earlier, pretrained word embeddings are obtained through training a model on a large unlabeled corpus. Through pretrained word embedding, we can get appropriate vectors for the words that are rare in the training dataset. Second, word embedding preserves lexical and semantic properties in its embedding, which implies that similar words get similar vectors [26] and similar words would have a similar label for entity recognition. We can observe from Table 1.8 that most of the words in the nearest neighbor set would also have the same labels as the main words.

1.4.6 Effects of CRF and BLSTM

We performed two analyses to understand the effect of using different loss functions in the output layer (CRF vs. WLL) and the effect of using bi-directional or uni-directional

Table 1.8 List of words and its five nearest neighbors obtained through pretrained word embedding used in the model.

Main word	Nearest neighbor words
Cough	Coughing, breathlessness, dyspnea, wheezing, coughs
Tumor	Tumor, tumoral, tumoural, melanoma, tumors
Surgery	Operation, decompression, dissection, resection, parathyroidectomy

Table 1.9 Effect of using CRF and WLL in output layer on the performance of the proposed model on different datasets.

	Disease NER			*Drug NER*			*Clinical NER*		
Model	**P**	**R**	**F**	**P**	**R**	**F**	**P**	**R**	**F**
CWBLSTM	84.42	82.31	83.35	72.57	74.60	73.57	84.17	83.20	83.68
BLSTM+WLL	76.04	78.25	77.13	71.81	70.34	71.07	77.35	80.91	79.09
LSTM+WLL	64.72	77.32	70.46	68.41	69.02	68.71	58.32	68.11	62.83

Note: In every block P, R, and F implies precision, recall, and F1 score, respectively.

(forward) LSTM. For these analyses, we modified our framework and named model variants as follows: Bi-directional LSTM with WLL output layer is called *BLSTM+WLL* and uni-directional or regular LSTM with WLL layer is called *LSTM+WLL*. In other words, *BLSTM+WLL* model uses all the features of the proposed framework except it uses WLL in place of CRF. Similarly, *LSTM+WLL* also uses all features along with uni-directional or regular LSTM instead of bi-directional LSTM and WLL in place of CRF. Results are presented in Table 1.9. A relative decrement of 7.5%, 3.4%, and 5.5% in obtained F1 score on *Disease NER*, *Drug NER*, and *Clinical NER*, respectively, by *BLSTM+WLL* compared to the proposed model demonstrates the importance of using a CRF layer. The deterioration in the performance suggests that an independent identifying tag is not favored by the model, and it is better to utilize the implicit tag dependency. Further observation of average token length of an entity in three tasks indicates the plausible reason for the difference in performance in the three tasks. Average token length is 1.2 for drug entities, 2.1 for clinical, and 2.2 for the disease named entities. The longer the average length of entities, better the performance of a model utilizing tag dependency. Similarly, relative improvements of 12.89%, 4.86%, and 20.83% in F1 score on *Disease NER*, *Drug NER*, and *Clinical NER* tasks, respectively, are observed when compared with *LSTM+WLL*. This clearly indicates that the use of bi-directional LSTM is always advantageous.

1.4.7 Effects of using fixed word embedding

Furthermore, we analyzed the effect of tuning word embedding during training. We retrained the CWBLSTM model by keeping word embedding parameters fixed for all three tasks and referred to it as *CWBLSTM+FixedWE*. Table 1.10 presents the performance comparison of *CWBLSTM* with *CWBLSTM+FixedWE*. Results imply that tuning of word

Table 1.10 Effects of keeping word embedding fixed during training on different datasets.

	Disease NER			*Drug NER*			*Clinical NER*		
Model	**P**	**R**	**F**	**P**	**R**	**F**	**P**	**R**	**F**
CWBLSTM	84.42	82.31	83.35	72.57	74.60	73.57	84.17	83.20	83.68
CWBLSTM+FixedWE	83.84	83.66	83.75	71.43	74.16	72.77	84.65	82.69	83.66

Note: In every block P, R, and F imply precision, recall, and F1 score, respectively.

embedding parameters does not bring any performance enhancement for any of the three tasks under our experimental settings.

1.4.8 Effect of size of the training data

We analyzed the performance of the CWBLSTM model on different training dataset sizes. Fig. 1.3 represents the learning curve of the CWBLSTM model in all three tasks. The graph shows that, although the performance is gradually improving with increasing size of the training datasets, the percentage improvement after 60% of the full dataset is marginal.

1.4.9 Analysis of learned word embeddings

Next, we analyzed characteristics of learned word embeddings after training of the proposed model. As mentioned earlier, we are learning two different representations of each word, one through its characters and other through distributional contexts. We expect that the embedding obtained through character embeddings will be able to capture morphological aspects, whereas distributional word embedding will focus on semantic and syntactic contexts.

We obtained character-based word embedding for each word of the *Drug NER* dataset after training. We randomly picked five words from the vocabulary list of the test set and

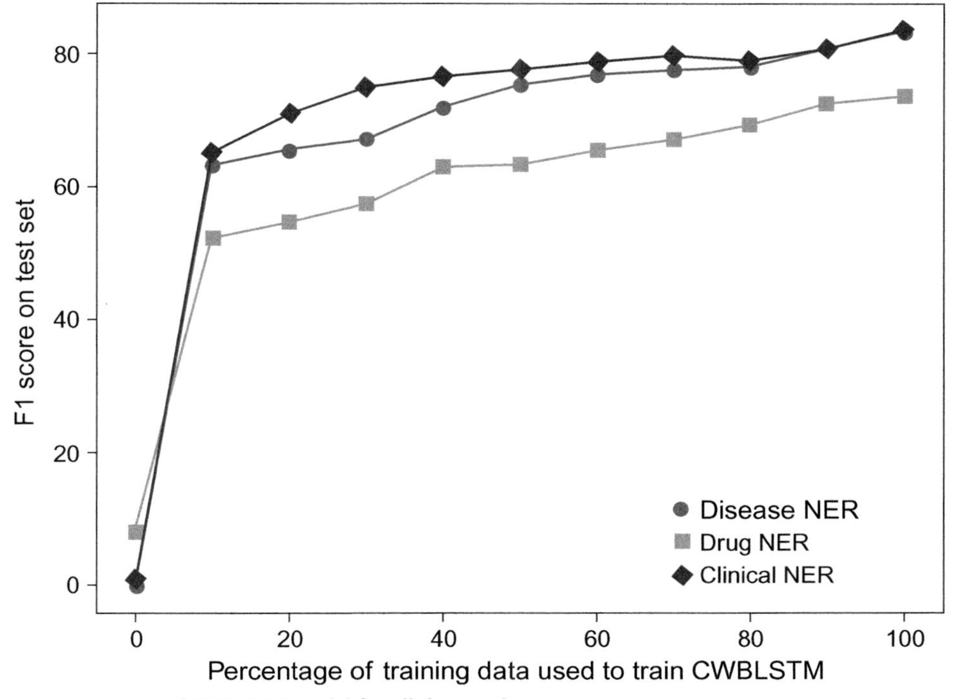

FIG. 1.3 Learning curve of CWBLSTM model for all three tasks.

Table 1.11 Word and its five nearest neighbors (from left to right in increasing order of Euclidean distance) learned by character-level word embedding of our model on *Drug NER* corpus.

Word	Char BLSTM	GloVe
2C19	2C9, 2C8/9, 29, 28.9, 2.9z	NA
Synergistic	Septic, symptomatic, synaptic, serotonergic, synthetic	Synergism, synergy, antagonistic, dose-dependent, exerts
Dysfunction	dysregulation, desensitization, dissolution, addition, administration	Impairment, impaired, disturbances, deterioration, insufficiency
False-positive	False-negative, facultative, five, folate, facilitate	False, falsely, erroneous, detecting, unreliable
Micrograms/mL	Microg/mL, micromol/L, micrograms/ml, mg/mL, mimicked	NA

Note: NA implies word is not present in the list.

observed its five nearest neighbors in the vocabulary list of the training set. The nearest neighbors are selected using both word embeddings, and results are shown in Table 1.11. We can observe that the character-based word embedding primarily focuses on morphologically similar words, whereas distributional word embeddings preserve semantic properties. This analysis suggests that it is advisable to use the complementary nature of the two embeddings.

1.4.10 Error analysis

We performed an analysis to find patterns in incorrect classification by the proposed model. We observed that a fair proportion of errors are due to the use of acronyms and the presence of rare entities. Examples of a few such cases are "CD," "HNPCC," and "SCA1." Due to their short forms, even character-based embedding was not able to learn good embedding for them. However, use of an acronym solver may help in mitigating this issue. We left this for future work. Another major proportion of errors are due to difficulty in recognizing nested forms of entities names. For example, the proposed model can only identify a part of the phrase *ovarian cancer* in *hereditary forms of ovarian cancer* as an instance of disease name and fails to detect the entire phrase as an instance of disease name.

1.5 Conclusion

In this research, we present a unified model for drug, disease, and clinical entity recognition tasks. The proposed model CWBLSTM uses BLSTMs in the hierarchy to learn better feature representation and CRF to infer correct labels for each word in the sentence simultaneously. CWBLSTM outperforms task-specific as well as task-independent baselines in all three tasks. Through various analyses, we demonstrated the importance of each feature type used by CWBLSTM. Our analyses suggest that pretrained word embeddings and character-based word embedding play complementary roles and, along with the use of

CRF layer as the output layer, forms important ingredients for improving the performance of NER tasks in biomedical and clinical domains.

References

[1] B. Rosario, M.A. Hearst, Classifying semantic relations in bioscience texts, in: Proceedings of the 42nd Annual Meeting on Association for Computational Linguistics, Association for Computational Linguistics, 2004, p. 430.

[2] I. Segura-Bedmar, V. Suárez-Paniagua, P. Martínez, Exploring word embedding for drug name recognition, in: Proceedings of the Sixth International Workshop on Health Text Mining and Information Analysis, 2015, pp. 64–72.

[3] Ö. Uzuner, B.R. South, S. Shen, S.L. DuVall, 2010 i2b2/VA challenge on concepts, assertions, and relations in clinical text, J. Am. Med. Inform. Assoc. 18 (5) (2011) 552–556.

[4] R. Leaman, C. Miller, G. Gonzalez, Enabling recognition of diseases in biomedical text with machine learning: corpus and benchmark, in: Proceedings of the 2009 Symposium on Languages in Biology and Medicine, vol. 82, 2009.

[5] J.D. Lafferty, A. McCallum, F.C.N. Pereira, Conditional random fields: probabilistic models for segmenting and labeling sequence data, in: Proceedings of the Eighteenth International Conference on Machine Learning, Morgan Kaufmann Publishers Inc., San Francisco, CA, 2001, pp. 282–289.

[6] M.F.M. Chowdhury, A. Lavelli, Disease mention recognition with specific features, in: Proceedings of the 2010 Workshop on Biomedical Natural Language Processing, BioNLP '10, Association for Computational Linguistics, Stroudsburg, PA, 2010, pp. 83–90.

[7] M. Jiang, Y. Chen, M. Liu, S.T. Rosenbloom, S. Mani, J.C. Denny, H. Xu, A study of machine-learning-based approaches to extract clinical entities and their assertions from discharge summaries, J. Am. Med. Inform. Assoc. 18 (5) (2011) 601–606.

[8] T. Rocktäschel, T. Huber, M. Weidlich, U. Leser, WBI-NER: the impact of domain-specific features on the performance of identifying and classifying mentions of drugs, in: Second Joint Conference on Lexical and Computational Semantics (*SEM), vol. 2, 2013, pp. 356–363.

[9] J. Björne, S. Kaewphan, T. Salakoski, UTurku: drug named entity recognition and drug-drug interaction extraction using SVM classification and domain knowledge, in: Second Joint Conference on Lexical and Computational Semantics (*SEM), vol. 2, 2013, pp. 651–659.

[10] B. Tang, H. Cao, X. Wang, Q. Chen, H. Xu, Evaluating word representation features in biomedical named entity recognition tasks, BioMed. Res. Int. 2014 (2014) 1–6 .

[11] R. Collobert, J. Weston, L. Bottou, M. Karlen, K. Kavukcuoglu, P. Kuksa, Natural language processing (almost) from scratch, J. Mach. Learn. Res. 12 (2011) 2493–2537.

[12] L. Yao, H. Liu, Y. Liu, X. Li, M.W. Anwars, Biomedical named entity recognition based on deep neutral network, Int. J. Hybrid Inform. Technol. 8 (2015) 279–288.

[13] C.N. dos Santos, B. Zadrozny, Learning character-level representations for part-of-speech tagging, in: International Conference on Machine Learning (ICML), vol. 32, JMLR W&CP, 2014, pp. 1818–1826.

[14] G. Lample, M. Ballesteros, S. Subramanian, K. Kawakami, C. Dyer, Neural architectures for named entity recognition, ArXiv preprint arXiv:1603.01360 (2016).

[15] X. Ma, E. Hovy, End-to-end sequence labeling via bi-directional LSTM-CNNs-CRF, in: Proceedings of the 54th Annual Meeting of the Association for Computational Linguistics (vol. 1: Long Papers), Association for Computational Linguistics, Berlin, Germany, 2016, pp. 1064–1074.

[16] R. Leaman, G. Gonzalez, BANNER: an executable survey of advances in biomedical named entity recognition, in: R.B. Altman, A.K. Dunker, L. Hunter, T. Murray, T.E. Klein (Eds.), Pacific Symposium on Biocomputing, World Scientific, 2008, pp. 652–663.

[17] L. Luo, P. Yang, Z. Yang, H. Lin, J. Wang, DUTIR at the BioCreative V. 5. BeCalm tasks: a BLSTM-CRF approach for biomedical entity recognition in patents, 2017.

[18] D. Zeng, C. Sun, L. Lin, B. Liu, LSTM-CRF for drug-named entity recognition, Entropy 19 (6) (2017) 283.

[19] L. Luo, Z. Yang, P. Yang, Y. Zhang, L. Wang, H. Lin, J. Wang, An attention-based BiLSTM-CRF approach to document-level chemical named entity recognition, Bioinformatics 1 (2017) 8.

[20] A. Graves, M. Liwicki, S. Fernández, R. Bertolami, H. Bunke, J. Schmidhuber, A novel connectionist system for unconstrained handwriting recognition, IEEE Trans. Pattern Anal. Mach. Intell. 31 (5) (2009) 855–868.

[21] A. Graves, Generating sequences with recurrent neural networks, ArXiv preprint arXiv:1308.0850 (2013).

[22] R. Pascanu, T. Mikolov, Y. Bengio, Understanding the exploding gradient problem, CoRR abs/1211.5063 (2012).

[23] Y. Bengio, N. Boulanger-Lewandowski, R. Pascanu, Advances in optimizing recurrent networks, in: 2013 IEEE International Conference on Acoustics, Speech and Signal Processing (ICASSP), IEEE, 2013, pp. 8624–8628.

[24] S. Hochreiter, J. Schmidhuber, Long short-term memory, Neural Comput. 9 (8) (1997) 1735–1780.

[25] B. Settles, Biomedical named entity recognition using conditional random fields and rich feature sets, in: Proceedings of the International Joint Workshop on Natural Language Processing in Biomedicine and Its Applications, Association for Computational Linguistics, 2004, pp. 104–107.

[26] Y. Bengio, R. Ducharme, P. Vincent, C. Janvin, A neural probabilistic language model, J. Mach. Learn. Res. 3 (2003) 1137–1155.

[27] J. Turian, L. Ratinov, Y. Bengio, Word representations: a simple and general method for semi-supervised learning, in: Proceedings of the 48th Annual Meeting of the Association for Computational Linguistics, ACL '10, Association for Computational Linguistics, Stroudsburg, PA, 2010, pp. 384–394.

[28] L.R. Rabiner, A tutorial on hidden Markov models and selected applications in speech recognition, Proc. IEEE 77 (2) (1989) 257–286.

[29] D. Kingma, J. Ba, Adam: a method for stochastic optimization, ArXiv preprint arXiv:1412.6980 (2014).

[30] J. Pennington, R. Socher, C.D. Manning, GloVe: global vectors for word representation, in: Proceedings of EMNLP, 2014.

[31] Th. Muneeb, S. Sahu, A. Anand, Evaluating distributed word representations for capturing semantics of biomedical concepts, in: Proceedings of BioNLP 15, Association for Computational Linguistics, Beijing, China, 2015, pp. 158–163.

[32] M. Bundschus, M. Dejori, M. Stetter, V. Tresp, H.-P. Kriegel, Extraction of semantic biomedical relations from text using conditional random fields, BMC Bioinform. 9 (1) (2008) 1.

[33] P. Agarwal, D.B. Searls, Literature mining in support of drug discovery, Brief. Bioinform. 9 (6) (2008) 479–492.

[34] R.I. Doğan, Z. Lu, An improved corpus of disease mentions in PubMed citations, in: Proceedings of the 2012 Workshop on Biomedical Natural Language Processing, BioNLP '12, Association for Computational Linguistics, Stroudsburg, PA, 2012, pp. 91–99.

[35] R.I. Dogan, R. Leaman, Z. Lu, NCBI disease corpus: a resource for disease name recognition and concept normalization, J. Biomed. Inform. 47 (2014) 1–10.

[36] I.S. Bedmar, P. Martínez, M.H. Zazo, SemEval-2013 task 9: extraction of drug-drug interactions from biomedical texts (DDIExtraction 2013), Association for Computational Linguistics, 2013.

[37] I.S. Bedmar, P. Martinez, D.S. Cisneros, The 1st DDIExtraction-2011 challenge task: extraction of drug-drug interactions from biomedical texts, Proceedings of the 1st Challenge Task on Drug-Drug Interaction Extraction, 2011.

[38] C.N. dos Santos, B. Xiang, B. Zhou, Classifying relations by ranking with convolutional neural networks, in: Proceedings of the 53rd Annual Meeting of the Association for Computational Linguistics and the 7th International Joint Conference on Natural Language Processingvol. 1, 2015, pp. 626–634.

[39] S.K. Sahu, A. Anand, Recurrent neural network models for disease name recognition using domain invariant features, in: Proceedings of the 54th Annual Meeting of the Association for Computational Linguistics, ACL 2016, August 7–12, 2016, Berlin, Germany, vol. 1: Long Papers, 2016.

[40] W. Hoeffding, The large-sample power of tests based on permutations of observations, Ann. Math. Stat. 23 (1952) 169–192.

[41] P. Koehn, Statistical significance tests for machine translation evaluation, in: Proceedings of the 2004 Conference on Empirical Methods in Natural Language Processing, 2004.

[42] T. Rocktäschel, M. Weidlich, U. Leser, ChemSpot: a hybrid system for chemical named entity recognition, Bioinformatics 28 (12) (2012) 1633–1640.

[43] T. Grego, F. Pinto, F.M. Couto, LASIGE: using conditional random fields and ChEBI ontology, in: Proceedings of the 7th International Workshop on Semantic Evaluation, 2013, pp. 660–666.

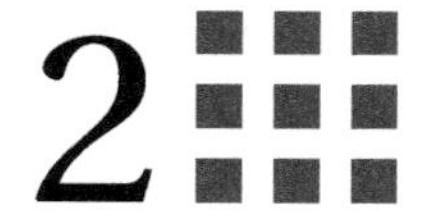

Simulation on real time monitoring for user healthcare information

Mou De[a,b], Anirban Kundu[a,b], Nivedita Ray De Sarkar[a,b]

[a] NETAJI SUBHASH ENGINEERING COLLEGE, KOLKATA, INDIA
[b] COMPUTER INNOVATIVE RESEARCH SOCIETY, HOWRAH, INDIA

2.1 Introduction

2.1.1 Background

Since last decade, social media has played an important role in human life. People are virtually connected with the entire world. People habituate in complex lifestyles, which cause several health issues. Social media performs a key role in social and personal life by collecting users' personalized and communal data. Social media involvement grows faster in daily life. Social networks [1] have been used to collect data to check human behavior, and for monitoring and surveillance. Social network [2] analysis has been done on a large scale to real-time user data to process and store on a daily basis. A large quantity of informative data from social network analysis, social media, and sensors are focused on human surveillance and monitoring their behaviors [3] within the community and with others.

Social network [4] surveillance [5, 6] is a complicated task due to the network's structure [7], along with a huge amount of real-time data. Users' privacy [8, 9] could be hampered, and their social and personal lives become open to a group. Social network provides a platform for everyone to mutually exchange private data and monitor each other. Organizations, community, government, and others use social network to monitor [10] and help remote users. Typically, perception regarding surveillance is negative as it might give certain advantages and benefits to a specific group. Surveillance means collecting specific information from a user to protect him or her from harm, violence, kidnapping, natural disasters, or sudden health issues. Health monitoring [11] systems are useful for remote users such as elders, patients, and children who would receive instant help during an emergency.

Human behaviors [12, 13] and their activities have been modeled in social media. Human behavior [14] could be different depending on individual choice, age, lifestyle, or gender. Everyone has to share opinions and their personal details to the outside world. New technology chooses a social media platform to satisfy its users. These lead people toward green and efficient awareness for both the society and environment.

Deep Learning Techniques for Biomedical and Health Informatics. https://doi.org/10.1016/B978-0-12-819061-6.00002-1

The social cloud [15] provides an interface on social networks and in social media. Users can share their information through a social network. The user connects with other users via the social network to exchange data for communication purposes. All users' aggregated data are then accumulated in the social cloud [16, 17] to maintain faster data access and relationship among users with virtual world. The social cloud offers reliable and sustainable resource sharing with the social network to avoid complex system architecture.

2.1.2 Challenges

Our approach has a challenge to monitor and surveil users in a real-time basis, analyzing user data in case of an emergency situation. Fast rescue service with a proper categorization of alert types is a big challenge to identify and reach the user in an adverse situation.

2.1.3 Objectives

The aim of our approach is to provide emergency services (within a predefined area) to the user equipped with a sensor-based device using typical surveillance using real-time data transferred to server-side framework for analysis within a particular time interval.

2.1.4 Scope

Server-based monitoring and surveillance system generates an automated alert in an emergency situation using sensor-based, real-time user data for assisting users with abnormal health/behavioral statistics.

2.1.5 Motivation

We have designed a server-based surveillance system to safeguard users in critical circumstances on the client side by detecting and/or minimizing human trafficking and health risk factors.

2.1.6 Organization

The rest of this chapter is organized as follows: past research are furnished in a literature review section; proposed work consists of design framework, flowchart, algorithm, and analytical discussions; experimental observations section shows results, analysis, and real-time observations; and we finally conclude that a successful server-based monitoring and surveillance framework can be achieved.

2.2 Literature review

2.2.1 Past researches

Real-time user monitoring and surveillance [18, 19] in a closed environment using sensor devices generate a huge number of sensor data [20] for storing and processing in emergencies. A cloud-based approach [21] is used to make available large-scale [22] heterogeneous data on an on-demand basis. A cloud-based approach provides faster data access in a social network and connects an enormous number of devices and services. A surveillance [23] based system provides security among users from intruders by tracking movements using typical global system for mobile (GSM) sensors. Most monitoring and surveillance systems have been designed using an embedded system [24] with the help of various types of sensors for tracking users' movements using pyroelectric infrared (PIR) sensors [25, 26], health sensors, etc. PIR sensors [27] are typically used to accumulate user behaviors based on the surrounding environment, which helps control different types of automated systems. The main purpose of various sensor network systems is to collect users' data in a periodic manner to avoid intrusions.

Cloud computing offers scalability, efficiency, and the fastest application services, which increases its popularity among users. As a result, high energy expenditure is raised to maintain cloud computing. Drainage of a huge amount of energy makes it a threat toward present environment and society. A cloud-computing approach [28, 29] has been introduced to enhance cloud computing with limited energy resources. Lesser power consumption, energy-efficient resource utilization [30], and less usage of hardware and software are the advantages of cloud computing for a growing society [31].

An automated system is used for remote monitoring [32] of a user for safety and security, surveillance, and other aspects such as healthcare data and environmental data. A wireless sensor network [33] consisting of many sensor devices [34] is used to handle an automated system for collecting location information such as latitude and longitude using a geographical location-based application programming interface (API). Wireless sensor [35] devices are popular for easy access and/or mobility with modular designs. Wireless sensor devices have abilities for processing users' data with local and/or global interactions.

Health monitoring [36, 37] based rescue systems [38, 39] are specially implemented to save humans in remote locations. Healthcare systems and rescue systems are developed using social media for faster and easier response to the user. Automated rescue systems are developed to rescue users via accident detection and hospitalization, to control traffic light signals for faster ambulance services, for automatic disease detection system, and so on. Specific types of health monitoring [40] systems have been developed to serve specific requisites of individuals based on their requirements.

2.3 Proposed model development

2.3.1 Design framework

In our proposed approach, an application-based system is built to collect user information from an enclosed area to monitor the user. A sensor device is attached to a user to check his movement, blood pressure, and pulse rate. The sensor device tracks the user's current position by calculating latitude and longitude values of a particular location.

"Acharya Jagadish Chandra Bose Indian Botanic Garden" is considered for the case study in this regard.

Two different sensors have been used for collecting location data and health data of users, maintaining users' privacy on the client side. Sensor devices were not used to record any audio and/or video data of users. Users have safe and hazard-free devices that track the movement and internal health conditions of the users.

A user's blood pressure and heart pulse rate have been monitored using sensors. A typical range of value is stored within the system for measuring a user's health status. An alert would be generated if data does not match a predefined range of values.

A proposed server-side system framework is used for surveillance depending on the user's movement in a timely manner. If a user stays at a particular location for a long time, an alert would be generated in the monitoring system. The locational and personal monitoring are done on particular time intervals, as per request of the client in an emergency basis.

A cloud-based approach has been developed using small and handy sensor devices to collect user data, which is energy efficient and consumes less power. Wi-fi connection is used to connect sensor devices with the network to store data in the data cloud. The data cloud provides faster communication between sensor devices and Web servers via social media. The data cloud reduces resource utilization to enhance energy efficiency and power consumption. In case of emergencies, service cloud is initiated through social media to minimize time and network overhead.

A proposed approach shows some benefits among users having a harmless and protected environment within an enclosed area without breaking users' privacy. A social network gives a user a friendly and simplified accessible platform for connecting various service communities that minimize network overhead and increase services. A victim and his/her relatives could easily be informed during an emergency through social media using an automated service.

The system sends direct or indirect help according to the specific problem. Direct help is provided by a security rescue team that helps users to rescue them from antisocial activities, extortion, kidnapping, and so on. In case of indirect help, an automated system alert is generated for a specific help service like calling an ambulance, hospital, police station, or fire station for the users, and a rescue team is sent to the victim's location for required services. After generating the alert, a security rescue team is sent to the particular area where the incident happened. If any unusual and/or abnormal activity is seen or monitored over the network, a rescue team is immediately sent to the particular location. In this scenario, the proposed device could be damaged or lost due to unavoidable or unwanted situations. In a health emergency, preliminary treatment is executed before sending a user to the hospital.

In the proposed system (Fig. 2.1), a user's authentication is performed by checking his user ID and password. The data cloud stores each user's login information irrespective of new and old users. Heart pulse rate, blood pressure, and body temperature in typical

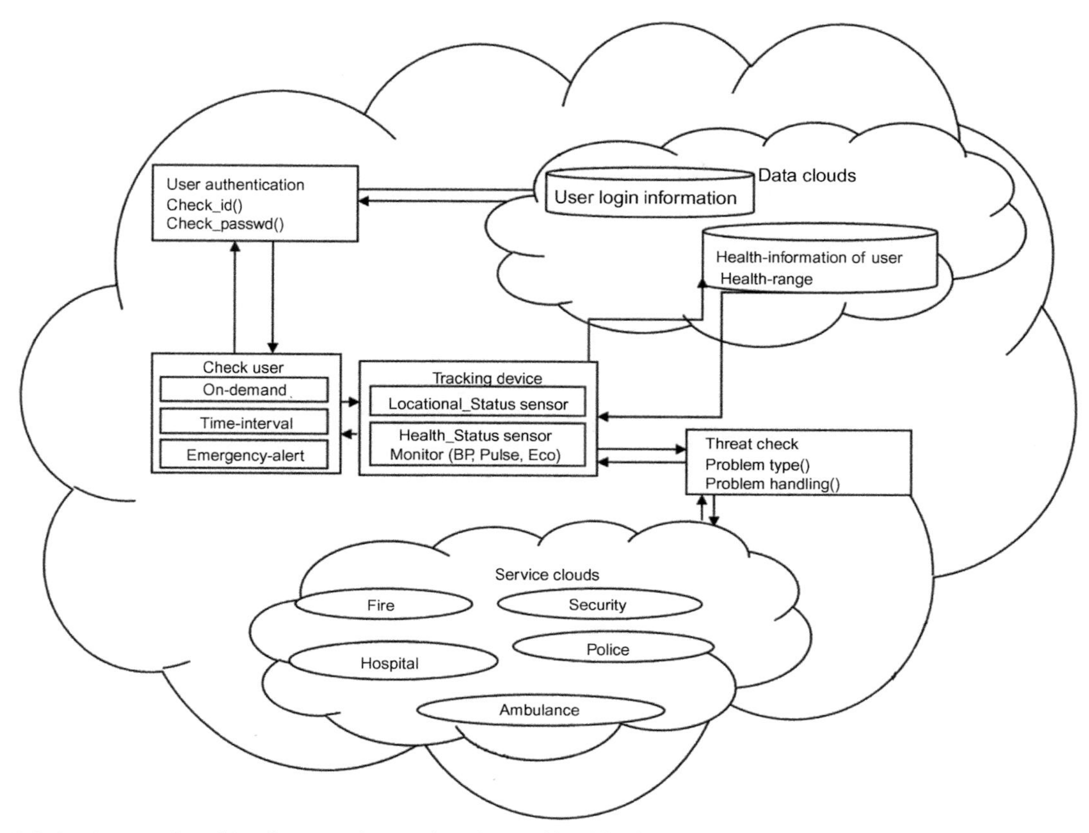

FIG. 2.1 Proposed tracking framework using location and health of users.

circumstances is stored. In an abnormal situation, a user has health issues that will typically cross the threshold limit. Then data cloud is used to store a large amount of user data within a time interval on an on-demand basis. This type of remote monitoring has been done by the tracking device, which tracks user's physical and locational parameters. A sensor device senses real-time user health data, and sends it to the data cloud. The sensor also tracks a user's locational data by GPS tracking sensor for safety. A threat check module takes the data from the tracking device and analyzes the data to categorize the types of problems that occurred with a user. The proposed system generates an alert and takes the necessary steps to solve the problem based on its nature. An automated message is sent for community assistance, such as a police station, fire brigade, or hospital using the social network.

In Fig. 2.2, we have explained the entire flow of data from sensor tracking device to system monitoring with the help of the Web server. We have monitored users from a

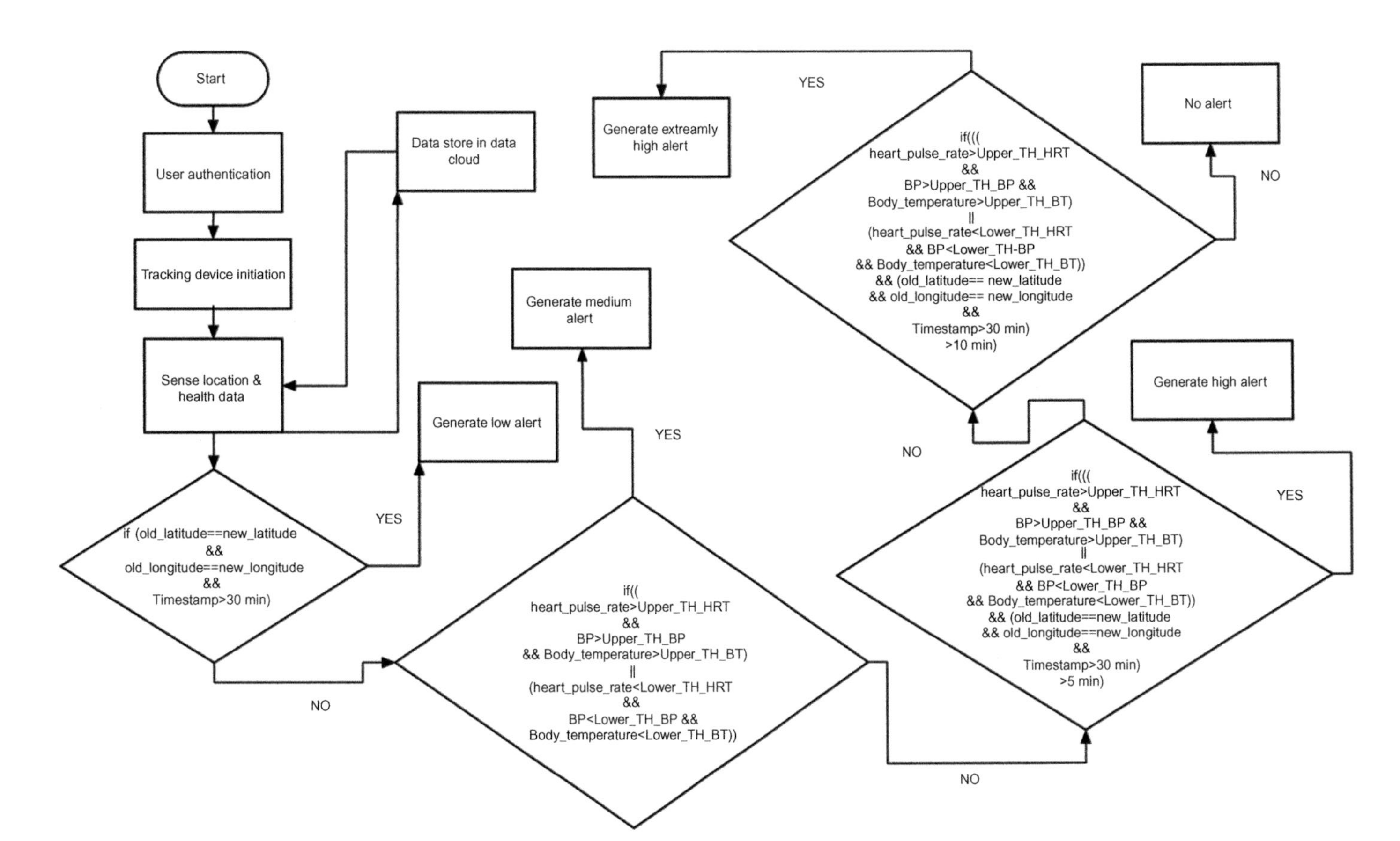

FIG. 2.2 Detailed data flow diagram for distinct alert generations for user safety.

selected location. We have also provided a device to wear by each individual. After that, they validated themselves with their personal information. Each device has an unique device ID number for identification. A device would be activated with a user login ID and device ID. After authorization of the user, the device is activated to track a user's location and health conditions. All user data is sent to the data cloud maintaining a time interval. Data is immediately stored in case of emergency. GPS tracking is measured by a sensor device, which collects latitude and longitude values. If locational values change within a minimum time interval, it indicates the user is alive and in good condition. However, if the latitude and longitude values of a device are fixed for a long period of time, then the sensors collect the user's health parameters on an urgent basis. Health data is sent to the data cloud where health parameters are used to compare real-time values with preloaded health care data while maintaining proper thresholds.

The system initiates an alert when a result is found above the threshold values. In a real-time situation, it might happen that the health status is normal even though the GPS tracker has not changed its position, then the system again issues an alert. An automated alert is also generated when a user loses the particular device.

A user's health data such as blood pressure, body temperature, and pulse rate are recorded as initial data.

Continuous monitoring of a person's location (latitude and longitude) is stored within the server-side database.

In Fig. 2.3, the proposed alert types are shown for managing user health conditions using latitude and longitude patterns. Proposed alerts are defined as follows:

Low alert—Generates a low alert when a user's two consecutive positions (latitude and longitude) have not changed for more than 30 min.

Medium alert—Generates a medium alert when user's initial health pattern exceeds either the lower threshold or upper threshold of concerned health data (blood pressure, body temperature, pulse rate).

High alert—Generates a high alert when a location pattern (latitude and longitude) has not changed for more than 30 min, and also health patterns (blood pressure, body temperature, pulse rate) have moved beyond threshold values for more than 5 min after the initial 30-min duration.

Extremely high alert—Generates a extremely high alert when a location pattern (latitude and longitude) has not changed for more than 30 min, and also health patterns (blood pressure, body temperature, pulse rate) have beyond threshold values for more than 10 min after the initial 30-min duration.

2.3.2 Procedures

Assumption of Algorithm 1: Real-time user health data are stored in the proposed system. Individual user data (location and health) are stored for monitoring each user.

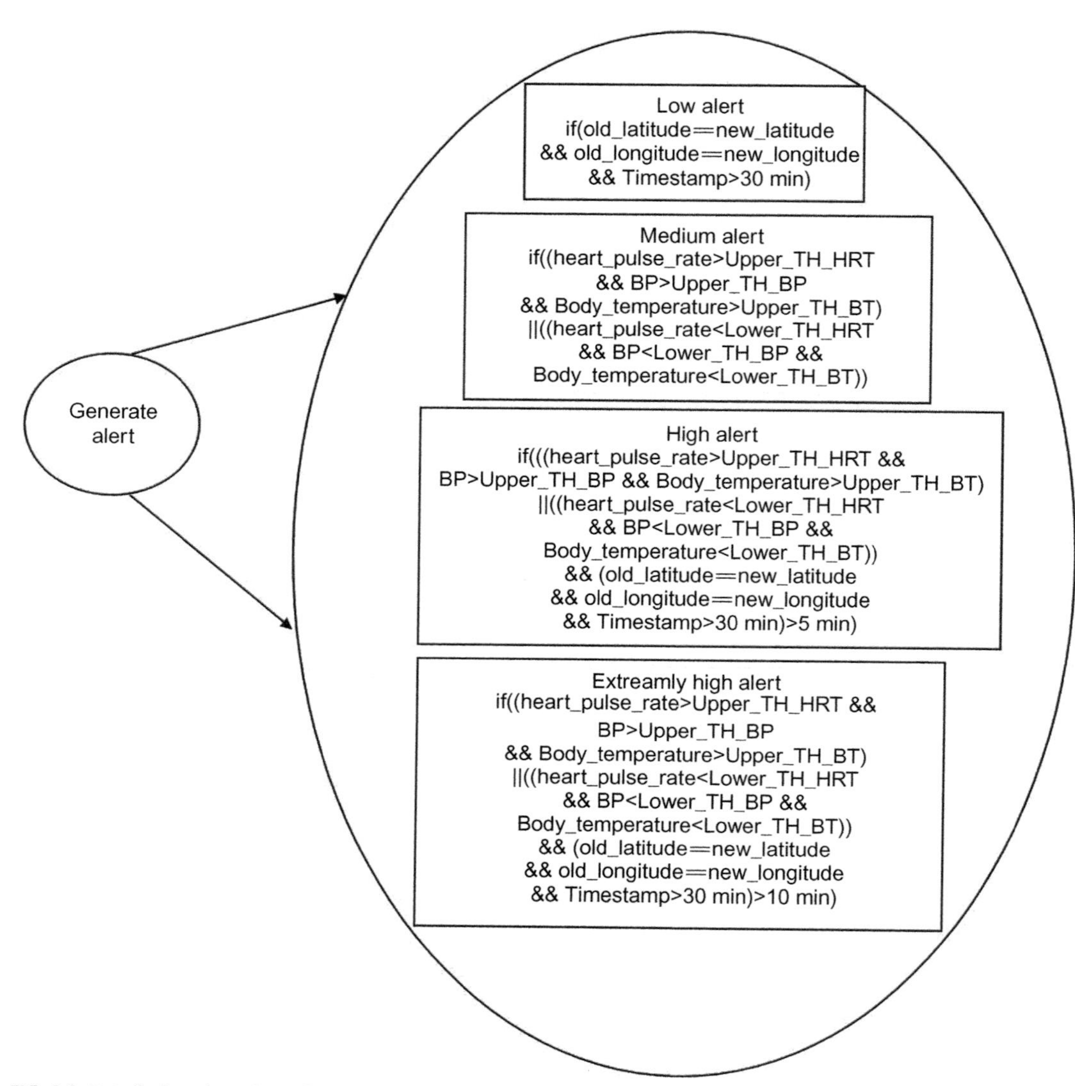

FIG. 2.3 Detailed explanation of alert types.

Algorithm 1

User_Health_Validate

```
Input: User_id, password, device_id, pulse_rate, blood_pressure,
body_temp
Output: User data validation
Void Validate_Device(intdevice_id,intuser_id, String Password,
intuser_phno, intu_pulserate, intu_bodytemp, intu_BP)
 {
 If(device_id!=Null)
 {
 Check device_id and device_state from database
 If(device_state==active)
 {
  If(pulse_rate==Threshold_pulse && body_temp==Threshold_temp &&
blood_pressure==Threshold_bp)
    {
    If (user_id!=Null && user_password!=Null && user_phno!=Null)
    {
    send_otp to user_phno;
    device_id validated with user_id;
    Initialize device with pulse_rate, blood_pressure, body_temp;
    }
    Else
   {
    Error_message("User information does not match");
   Exit(1);
   }
  Else
  {
  Error_message("User health condition not satisfactory");
  Exit(1);
  }
 Else
 {
 Error_message("Device inactive");
 Exit(0);
  }
 Else
 {
 Error_message("Device not available");
 }
 }
```

Explanation of Algorithm 1: User validation is accomplished in Algorithm 1 using device ID, user ID, password, phone number, and customized user health data such as user pulse rate, user body temperature, and user blood pressure. Initially, the device status is checked before user validation takes place. User's customized health data are verified with threshold values of pulse rate, body temperature, and blood pressure. Once users are provided with their user ID, password, and a valid phone number, a one-time password (OTP) is sent to the user's phone number, and a device ID is registered with each user ID. The device is initialized to fetch data from the user after registration.

Assumption of Algorithm 2: The device fetches users' locations and health data, and then sends it to the data cloud where it is stored with the time interval or emergency basis. A range of 120-80 is considered as ideal blood pressure. Pulse rate is assumed to be between 60-100 beats/minute in an ideal case. Body temperature is set to 37°C as ideal.

Time complexity of Algorithm 2 is $O(n)$, because processing is proportional to count of records.

Algorithm 2

Location_Health_Information_Sense_Send

```
  Input: latitude, longitude, pulse_rate, blood_pressure, body_temp
  Output: Storage to Data Cloud
  Sensor_device_module (latitude, longitude, pulse_rate,
blood_pressure, body_temp)
  {
  sense(latitude_data, longitude _data );
  latitude=new_latitude;
  longitude=new_longitude;
  send_geolocation_data_to_Data_Cloud(latitude, longitude);
  sense(pulse_rate, blood_pressure, body_temp);
  send_pulse_rate, blood_pressure, body_temp_to_Data_Cloud(pulse_rate,
blood_pressure, body_temp);
  If(old_latitude == new_latitude && old_longitude== new_longitude)
  {
   If(pulse_rate, blood_pressure, body_temp>upper_health_threshold ||
pulse_rate, blood_pressure, body_temp<lower_health_threshold)
      {
          Data_send_to_Generate_alert();
      }
   }
   }
```

Explanation of Algorithm 2: Once the device is initialized, it starts working to sense current location and health data in real time within a specific time interval. All data coming from the device is redirected to the data cloud from the server-based network interface. If any ambiguous data is found after comparison with the existing database in the data cloud, that is, abnormal data beyond threshold limits, then the system generates an alert depending on the selected criteria as referred in Algorithm 2.

Assumption of Algorithm 3: Sensed data are compared with an ideal range of blood pressure, pulse rate, temperature, and location, which are available in the data cloud. In a typical scenario, each health data (blood pressure, pulse rate, temperature) have upper and lower values. If real-time sensed data cross any of the upper/lower range values, then an alert would be generated. In Algorithm 3, it is assumed that an anomaly has been detected.

Time complexity of Algorithm 3 is $O(n)$, because data processing and storage are proportional to the number of records returned and analyzed.

Algorithm 3

Generate_Alert

```
 Input: latitude_data, longitude_data, pulse_rate, blood_pressure,
body_temp
 Output: Automated alert generation
 Generate_Alert (latitude_data, longitude_data, pulse_rate,
blood_pressure, body_temp)
 {
 receive_geolocation_data_from_Data_Cloud(latitude, longitude);
 receive_pulse_rate, blood_pressure, body_temp_from_Data_Cloud
(pulse_rate, blood_pressure, body_temp);
 if((pulse_rate>upper_pulse_threshold || pulse_rate<lower_
pulse_threshold) &&health_error_time>10 minute)
  {
  if((blood_pressure>upper_pres_threshold || blood_pressure<lower_
pres_threshold) && health_error_time > 10 minute)
  {
   if(idle_time>30 minute)
   {
    Generate_Extremly_High_alert();
   }
   }
 }
 Elseif((pulse_rate>upper_pulse_threshold || pulse_rate
<lower_pulse_threshold) &&health_error_time>5 minute)
  {
```

```
   if((blood_pressure>upper_pres_threshold || blood_pressure<lower_
pres_threshold) &&health_error_time>5 minute)
   {
    if(idle_time>30 minute)
    {
     Generate_High_alert();
    }
    }
  }
  Elseif(pulse_rate>upper_pulse_threshold || pulse_rate<lower_
pulse_threshold)
  {
   if(blood_pressure>upper_pres_threshold || blood_presure<lower_
pres_threshold)
   {
    if(body_temp>upper_temp_threshold || body_temp<lower_
temp_threshold)
    {
     Generate_Medium_alert();
    }
    }
   }
   Elseif(old_latitude == new_latitude && old_longitude==
new_longitude)
   {
   if(idle_time>30 minute)
   {
    Generate_Low_alert();
   }
  }
  else
  {
   No_alert();
  }
  }
```

Explanation of Algorithm 3: Algorithm 3 exhibits automated alert generation depending on the type and nature of problem. It takes the location and health data as inputs from "sense_device_module" (see Algorithm 2). The location and health data are compared with values stored in the data cloud. An extremely high alert is generated if health data, like blood pressure and pulse rate, cross threshold values and the abnormal situation persists for more than 10 min, as well as the user hasn't moved for more than 30 min. When the pulse rate and blood pressure are strangely up/down for more than 5 min, and the user has been idle at a particular place more than 30 min, then a high alert is activated. If a user's health conditions such as blood pressure, heart pulse rate, and body temperature abruptly and frequently change, then the proposed system generates a medium alert. A low alert would be generated when the user is idle at a particular place for more than 30 min. No alert is generated in a typical scenario.

In Fig. 2.4A and B, the movement of different users are plotted using their latitude and longitude values, respectively. Fig. 2.4A and B also depicts an abnormal pattern of user1 movement for more than 30 min, which categorized a low alert.

The purpose of this subsection is to exhibit the proposed tracking system in real time to specify the situation of a particular person moving within the area to be considered. We designed the framework to provide real-life user safety and security in an enclosed area using server-side analysis. Respective security personnel are informed immediately.

Example: "Acharya Jagadish Chandra Bose Indian Botanic Garden" is considered a functional area for analyzing the proposed system. This particular area has been marked with a white color.

A marker (pop-up) is shown to indicate a user's position within the enclosed garden with a particular device (see Fig. 2.5). We have also shown different users' positions and a particular user's start-to-end movement in Figs. 2.6 and 2.7, respectively. A manual border marking is used for easy understanding of the proposed functional area in Figs. 2.5–2.7.

In Fig. 2.5, the tracking system tracks each user using a corresponding device ID and movement using a particular location's latitude and longitude values as its basis.

Fig. 2.6 represents the multiuser tracking system, which tracks different users using device ID and relative movement using a particular location's latitude and longitude.

The proposed tracking system tracks a user using corresponding device ID and the movement using a particular location's latitude and longitude values. If we wish to track all the locations of particular user, then it is very useful as shown in Fig. 2.7.

Comparisons of our proposed approach and an existing approach are depicted in terms of storage, application, power, server structure, and type of services in Table 2.1).

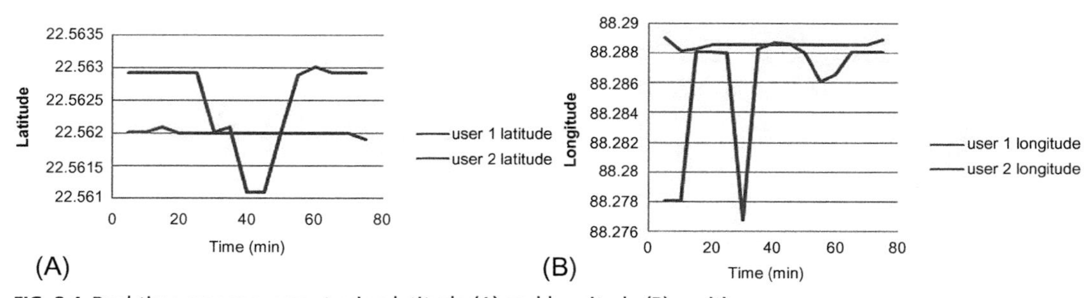

FIG. 2.4 Real-time user movement using latitude (A) and longitude (B) positions.

FIG. 2.5 Tracking user location as latitude and longitude with single device ID.

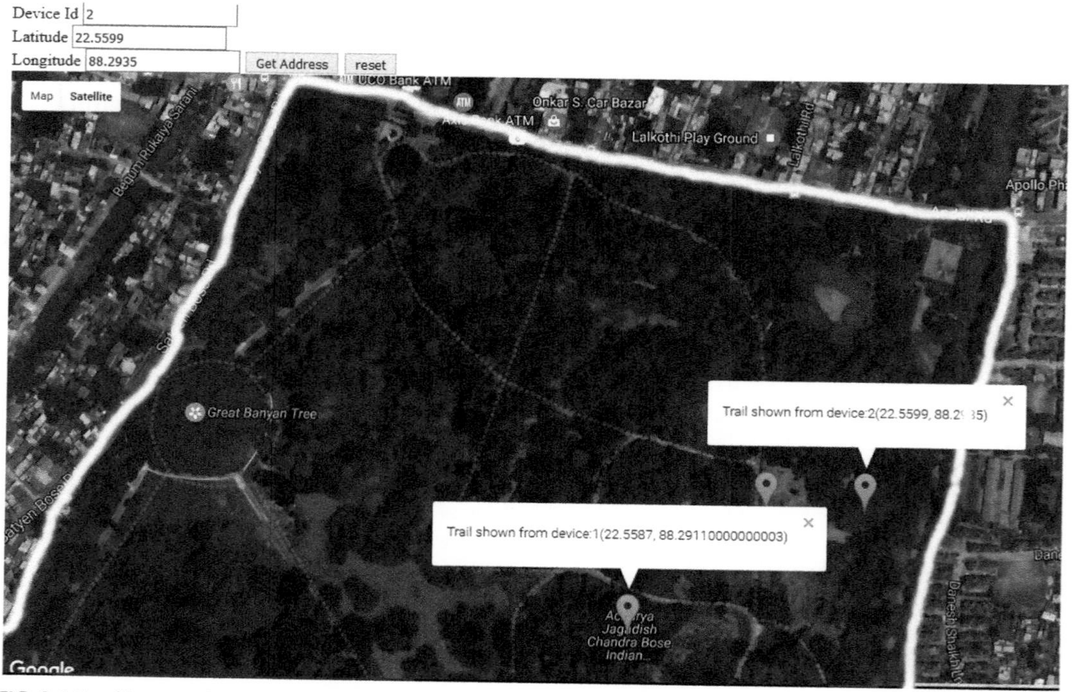

FIG. 2.6 Tracking user location as latitude and longitude with multiple device IDs.

FIG. 2.7 Tracking user locations at different times using particular device IDs.

Table 2.1 Comparison between existing system and proposed system based on distinct input types.

Input type	Existing system	Proposed system
Storage	Multiple DB	Cloud (distributed DB)
Application	Health monitoring, web mining, sensor network	Server-based monitoring and surveillance using social cloud
Power	Huge power and energy consumption	Less power and energy consumption
Server structure	Structured	Distributed
Service type	Specific type	Cost-efficient hardware-based client device is used for health and safety

2.3.3 Performance analysis

Performance ratio is measured using Algorithm 3. System response time for generating a specific alert is considered to derive from our performance-based proposed approach.

Consider, Performance ratio $= PR$;
System response time for $AT_X = S_{Res}$;
Low alert generation time $= AT_L$;
Medium alert generation time $= AT_M$;

High alert generation time $= AT_H$;
Extremly high alert generation time $= AT_{EH}$.

$$\therefore PR = \left[\frac{S_{Res}}{\text{Time Required for}\,(AT_L) + \text{Time Required for}\,(AT_M) + \text{Time Required for}\,(AT_H) + \text{Time Required for}\,(AT_{EH})} \right] \tag{2.1}$$

Fig. 2.8 describes the performance of our proposed system based on distinct alert types. According to the types of emergency, the system responds to generate low, medium, high, and extremely high alerts, which are depicted by orange, maroon, yellow, and red colors, respectively. In the case of an extremely high alert, the proposed system response time is minimal (7 s) compared to low, medium, and high alert response times. Performance ratio is increased during high and extremely high alert where users may be in a severe situation.

2.3.4 Benefits of proposed model

- **(i)** Fast response from the server with linear time complexity.
- **(ii)** Proposed system framework is fully scalable.
- **(iii)** Any number of new users could be connected in our system framework without delay, achieving high performance.
- **(iv)** Proposed system framework is 100% accurate.

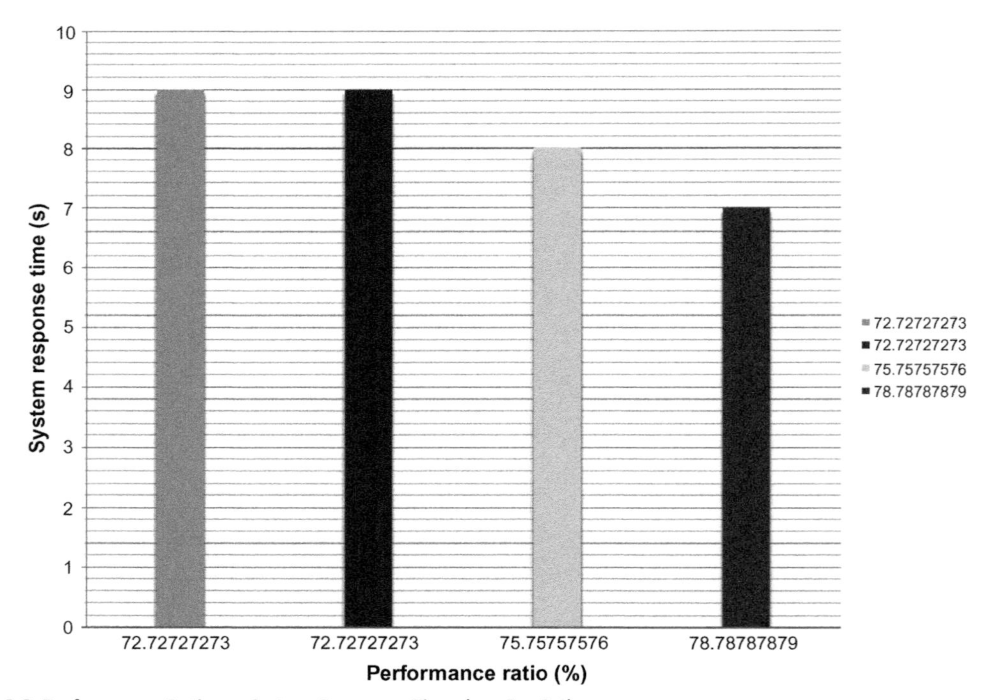

FIG. 2.8 Performance Ratio vs. System Response Time (see Eq. 2.1).

2.4 Experimental observations

"Acharya Jagadish Chandra Bose Indian Botanic Garden" is considered for the experimental study and analysis.

2.4.1 Server-side working environment

A system-level experiment was being executed using "Open Hardware Monitor" and "Jmeter" software in the Windows 7 platform. Real-time observations took place using "Open Hardware Monitor" to check CPU load and memory load. Experimental analysis of the system for load testing was accomplished using "Jmeter."

Number of users considered for real-time experimentation: 500
Area to be considered: 109 ha
Type of participants: Visitors of Acharya Jagadish Chandra Bose Indian Botanic Garden

2.4.2 Privacy policy

It is an outdoor application. Therefore, no privacy-related issues are involved in such cases. Although, sound and picture were not considered for tracking users to maintain their privacy. For our experiment, "Acharya Jagadish Chandra Bose Indian Botanic Garden" of 109 ha was considered. Any person entering the garden received the client device after live registration at the time of purchasing their entry ticket to the garden. Overall, the server-based system monitored and surveilled visitors by measuring their health parameters using sensors as discussed in Section 2.3 (see Figs. 2.1 and 2.2).

2.4.3 Experimental comparisons

In the proposed system, an upper and lower threshold was defined as 120 and 40 mm Hg, respectively. If a user's blood pressure crossed either of the limits, then the system detects an anomaly. An automated alert is generated as a consequence for the service cloud using a social network (see Fig. 2.9).

In the case of the existing system, various blood pressure ranges are categorized. When real-time data matches any of the specified ranges, it determines there is a problem, and a short message service (SMS) is sent to the doctor or ambulance as required (see Fig. 2.10).

Figs. 2.11 and 2.12 are used to describe the functioning of a proposed system for pulse rate detection. The proposed system detects abnormal, normal, high risk, and stressed heart pulse rates. When the system encounters an irregular heart pulse rate, it checks whether other health parameters are within ranges. If all health parameter values exceed predefined ranges, then the system generates an automated alert and initiates required services.

An existing system monitors normal heart rate based on specific ranges. If the range is not matched, then an automated SMS would be generated to the rescue team.

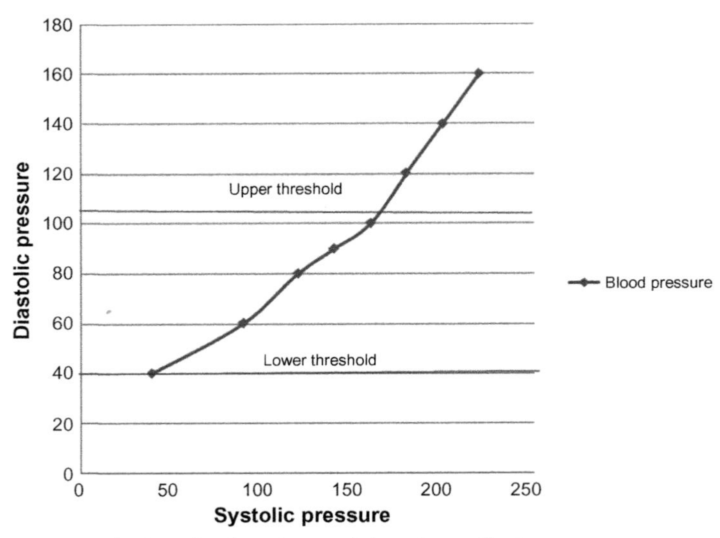

FIG. 2.9 Blood pressure anomaly detection based on real-time data collection.

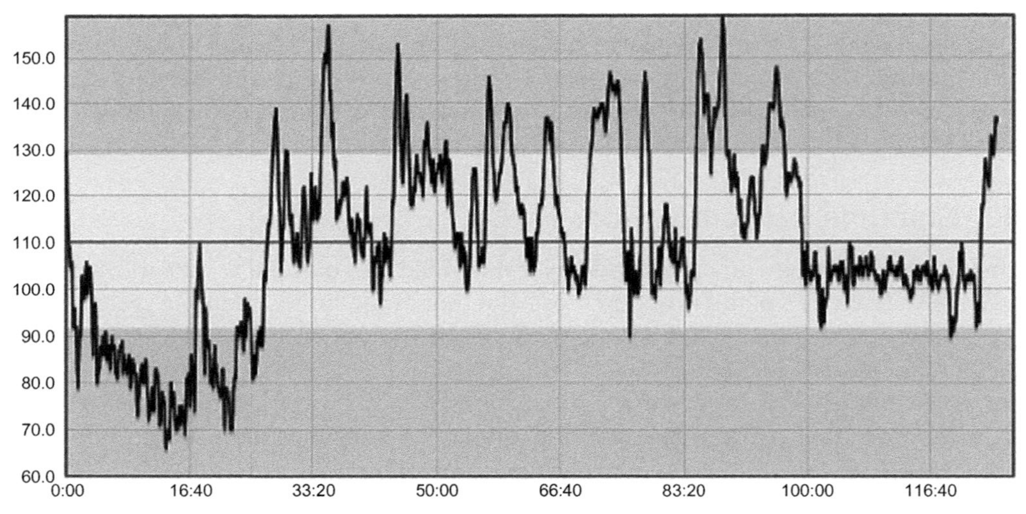

FIG. 2.10 Existing system heart rate abnormality detection [40].

2.4.4 Cost analysis

Weightage and response time have been introduced as two elements for analyzing the cost efficiency of the proposed system. System responses are based on three different predetermined situations, which are considered as related cases.

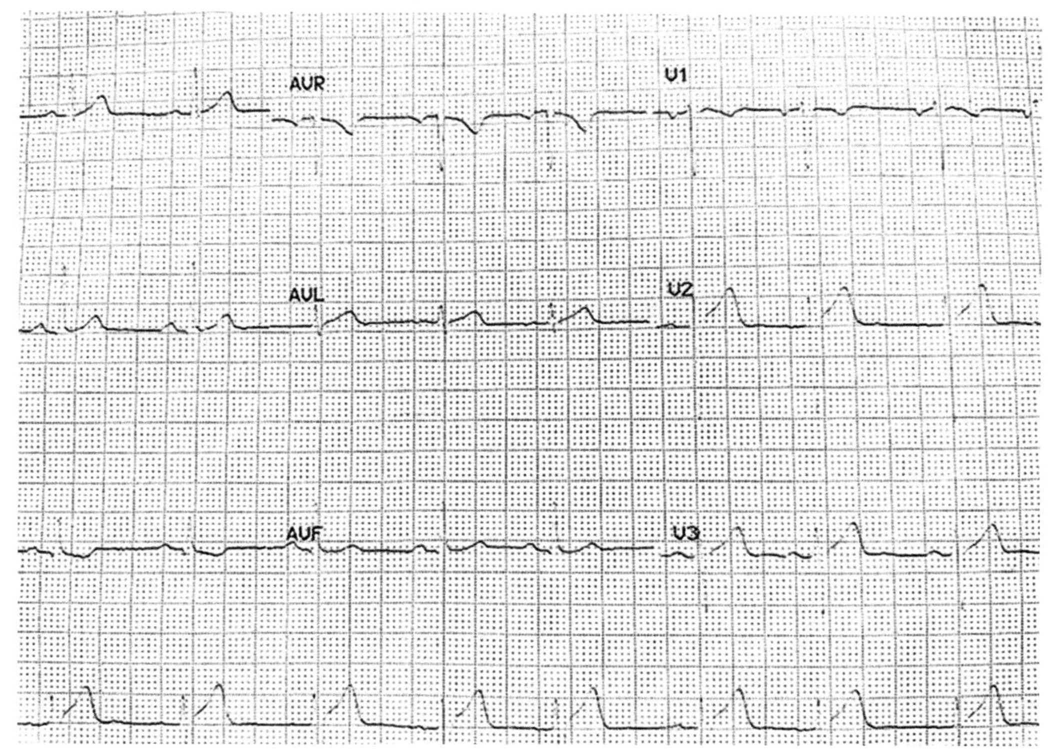

FIG. 2.11 Typical heart pulse rates.

The predetermined situations defined for the proposed system are as follows:

Time interval, on-demand basis, emergency.
Weightage is defined as w.
Response time is defined as T_1.
Delay or waiting time is defined as T_2.
Total execution time, $T = T_1 + T_2$;

Consider,

Case 1 = Time interval; where $w = 0.2$, and $T_2 = 60$ s.
Case 2 = On-demand basis; where $w = 0.3$, and $T_2 = 0$ s.
Case 3 = Emergency; where $w = 0.5$, and $T_2 = 0$ s.

In Case 1, data is transmitted to the data cloud in every 60 s for analysis and storage. If any abnormal situation (value beyond threshold) arises, the system responds in a predetermined manner within 10 s.

In Case 2, the response of the proposed system is prioritized based on arrival frequency of requests sent by the user/system administrator.

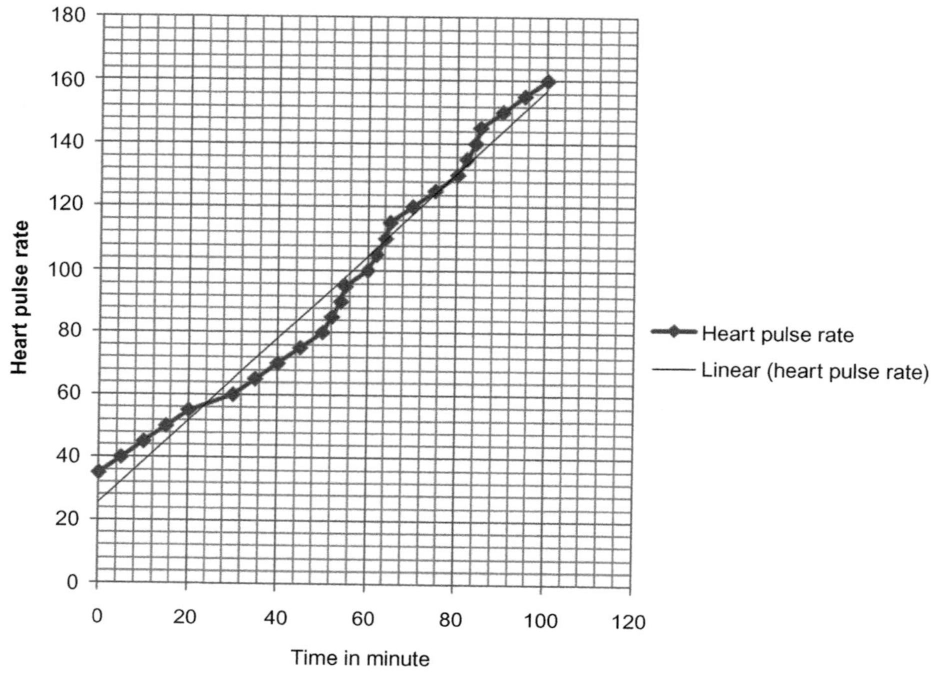

FIG. 2.12 Heart pulse rate detection in the proposed system.

In Case 3, the system reacts based on predefined emergencies.

During an extremely high emergency condition, the system responds within 1–2 s. High emergency conditions are realized within 3–4 s. A medium emergency condition is handled within 5–7 s, and a low emergency situation is realized between 8 and 10 s.

In Fig. 2.13, we exhibit cost efficiency with respect to execution time and weightage.

Cost (C) α Time (T)
Given, $w = \{0.2, 0.3, 0.5\}$
Therefore, $C = wT$; where $w =$ constant.

Execution time is substantially reduced in case of emergency (see Fig. 2.11). Execution time and cost are reduced on an on-demand basis. The value increases as and when high demand arises. During the time interval, cost and execution time increase in a steady manner.

2.4.5 Real-time server-based observations

In this section, server-side interactions are demonstrated using system performance of servers. Servers typically have stored surveillance and tracking data.

In Fig. 2.14, CPU load of the system has been depicted. Web server and other services work continuously. CPU load is high due to full load conditions of web servers. Y axis represents CPU load and X axis represents time in minutes.

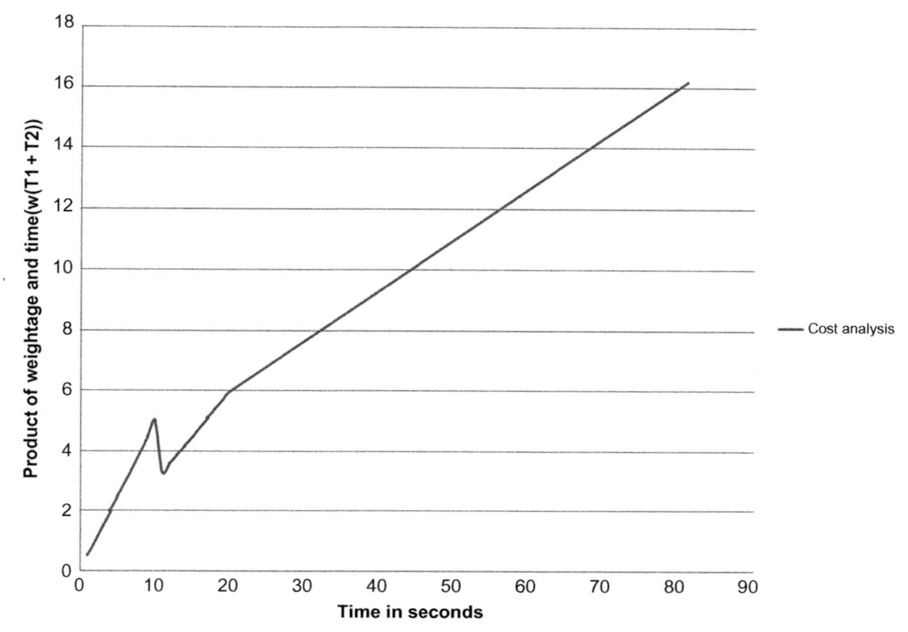

FIG. 2.13 $w(T_1 + T_2)$ vs. time.

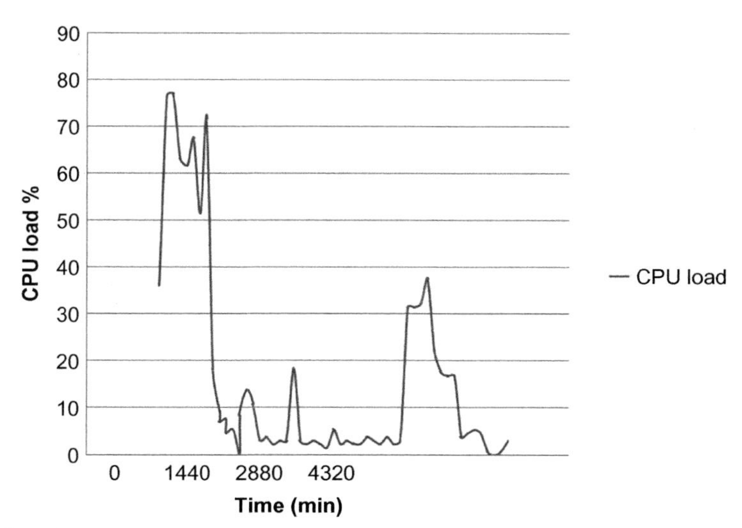

FIG. 2.14 CPU load vs. time.

In Fig. 2.15, CPU core#1 load of the server-side system is shown. When Web servers are in initial load condition, CPU core#1 load is high as the system has internal delay. *Y*-axis represents CPU core#1 load in a percentage, and *X*-axis represents time in minutes. High peaks symbolize heavy load within the server due to handling higher number of users.

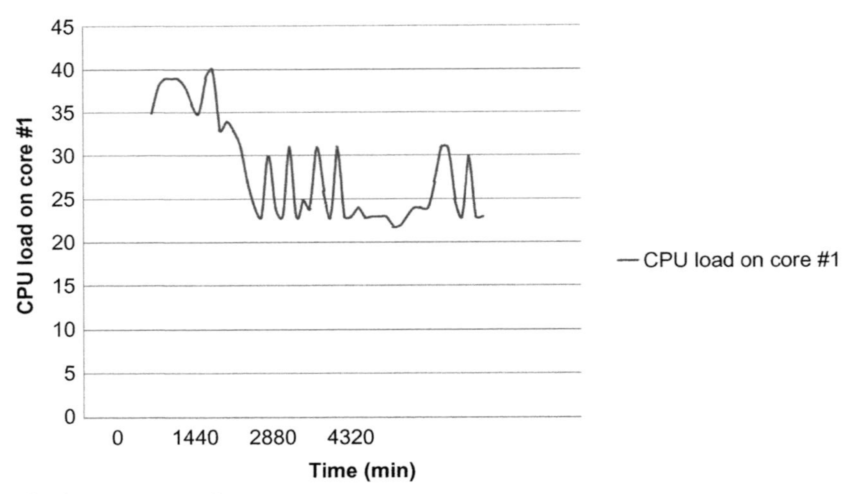

FIG. 2.15 CPU load on Core#1 vs. time.

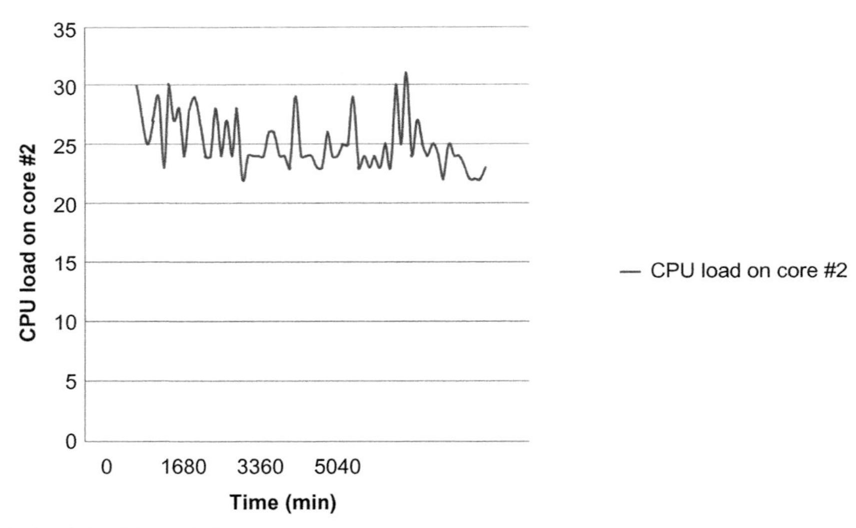

FIG. 2.16 CPU load Core#2 vs. time.

In Fig. 2.16, CPU core#2 load of servers is depicted. CPU core#2 is also important to control the system's behavior with respect to the number of users in real time. *Y*-axis represents CPU core#2 load, and *X*-axis represents time in minutes.

In Fig. 2.17, server temperature for a specific duration is illustrated. Initially, the temperature of the servers is comparatively high. Gradually, the system temperature is controlled and reduced to a certain level. It means the cooling system of the servers

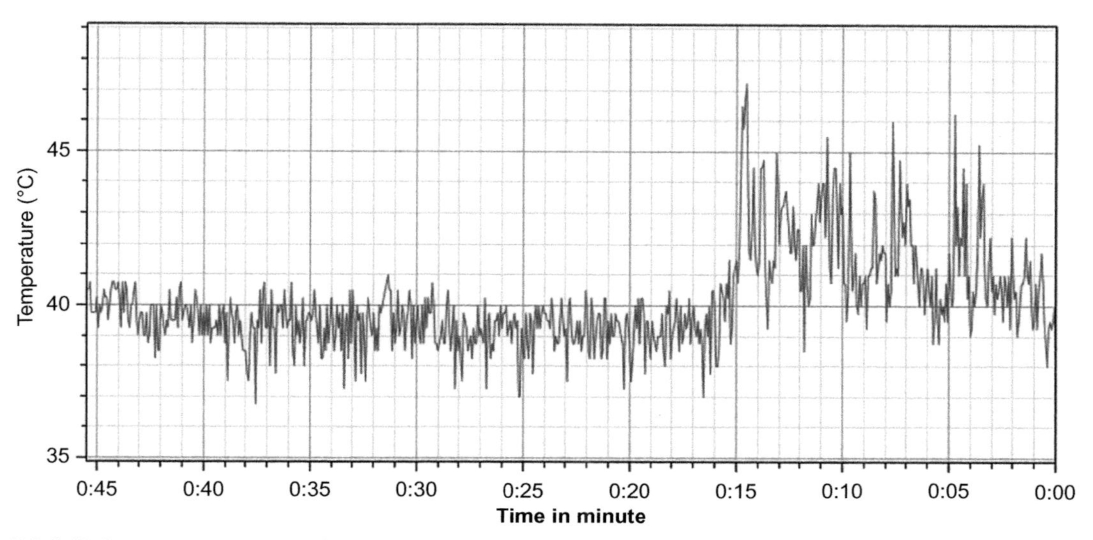

FIG. 2.17 Server temperature vs. time.

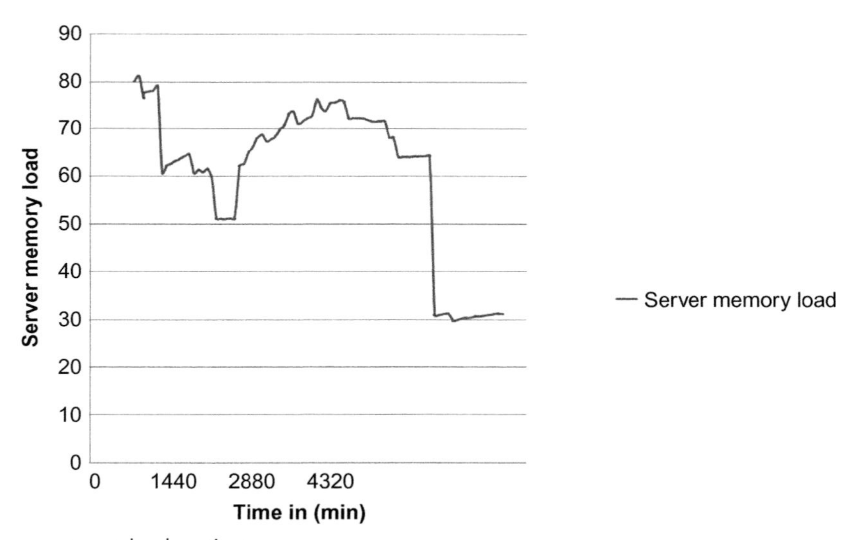

FIG. 2.18 Server memory load vs. time.

is perfectly working. *Y*-axis represents CPU core#1 load, and *X*-axis represents time in minutes.

In Fig. 2.18, the server memory load with respect to time is shown. Initially, memory load is comparatively high, then again it increases due to simultaneous user inputs on the server side. When servers are processing users' requests, the system is in a stable condition. *Y*-axis represents memory load, and *X*-axis represents time in minute.

2.4.6 Scalability analysis

Consider, U_{CA} = No of user connections in an active state.

U_{CI} = No of user connections in an inactive state.

U_C = No of user connections.

$$\because U_C = U_{CI} + U_{CA}$$

$\therefore$ $U_{CA} \subseteq U_C$; and $U_{CI} \subseteq U_C$

Consider, L = Total load in server-side system for handling U_C number of connections from user device.

$$\therefore U_C \propto L \tag{2.2}$$

$\therefore$ $U_C = kL$ where k is the load constant.

Fig. 2.19 shows that the proposed approach has linear scalabilities with respect to user connections and corresponding server load. Load and user connections in the case of distinct load constant (k) values have a similar pattern. Thus, load and user connections always exhibit linear distributions.

2.4.7 Monte Carlo simulation-based analysis

Monte Carlo simulation [41] of deterministic result analysis for scalability is applied to proposed model development to increase efficiency for tracking the number of users with respect to server load.

From Eq. (2.2) (refer to Section 2.4.6), we have derived $U_C \propto L$ (i.e., Number of user connections is proportional to server load).

$\therefore U_C = kL$; where k is the load constant;

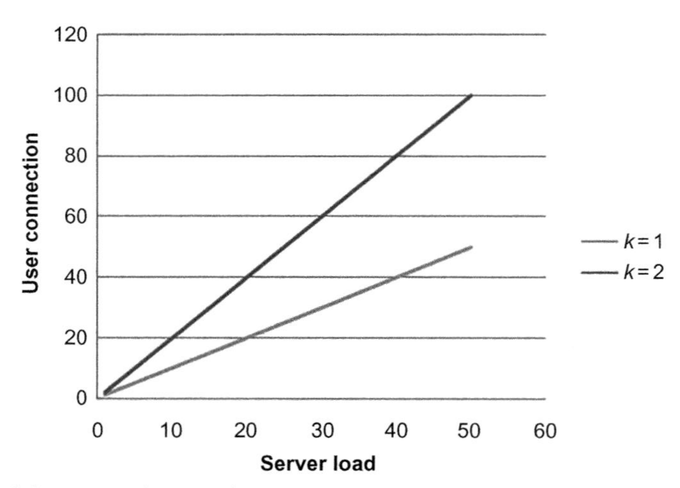

FIG. 2.19 Scalability of the proposed approach.

Consider, U_C is the function J_h, which is similar to $J_h = (\theta, h)$; where, $\theta = I\,\theta + \int$ and I is the integral operator with load constant $k(x', x)$. The k load constant satisfies convergence law of the Neumann series $\{k^n\ 0\} < 1$.

θ = A number of active user connection (i.e., U_{CA});
h = Number of inactive user connections (i.e., U_{CI});
$\{L^n\}$ = Markov-chain of particular load;
π^n = Initial density.
$L\{x', x\}$ = Transition density which is $L(x' \rightarrow x)$

So probability of termination of the load chain (L^n) at value x' is

$$F(x') = 1 - \int L(x', x)dx$$

Consider, N is the random index of last state. A function of user connection of the chain is defined with number of load with expectation J_h. So, the collection estimator is as follows:

$\xi = \sum_{n=0}^{N} \theta n\, h(xn)$; where $UC0 = \frac{\int L_0}{\pi L_0}$; and $UCn = \frac{k(L_{n-1}\, L_n)}{L(x_{n-1}\, x_n)}$;

only if $\{L(x',x),\ k(x',x)\ ,\ \pi(L),\ \int(x)\} \neq 0$

Then, $\xi = \sum_{n=0}^{\alpha} k^n \int, h = (\theta, h) = \int \theta(x) h(x) dx$; where x = Probability of system load;

$\therefore \pi(L) = \frac{\int (L)\theta^*(L)}{\int, \theta^*}$ and, $L(x', x) = \frac{k(x', x)\theta^*(x)}{[k^*\theta^*](x')}$, where $\theta^* = k^* \theta^* + h$

$\therefore D\xi = 0$; and $E\xi = J_h$;

Markov chain simulation is applied with linear data of user connections with load distribution to approximate solution on the basis of following:

$$\theta\,(x) = (\theta, h_x) + \int x \tag{2.3}$$

where $h_x(x') = k(x', x)$.

The Monte Carlo method is invoked to plot the first Eigen values of an integral function by the possibilities of the number of active and inactive user connections by contrary of the number of hit and miss of the server load as follows:

$E[u_{cn} h(x_n)] = (k^n \int, h)$; which extends to a linear algebraic equation of $x + Hx = h$ [42].

From Eq. (2.3),

$$\theta\,(x) = f(\theta,\ \gamma)^* L_F \tag{2.4}$$

where (x) = *Total number of connections.*

L_F = System load with active and nonactive user connection ($0 \leq L_F \leq 1$).

From Eq. (2.4),

$$(\theta, \gamma)\ ^*L_F = \int_0^N [[f(\theta)^* L_F] + [f(\gamma)^* L_F]] \tag{2.5}$$

$$\therefore f_i(\theta) = \omega_i^t \theta + \omega_{i0} \tag{2.6}$$

where $f_i(\theta) = \omega_i$ if $f_i(\theta) > f_j(\theta)$ for all $j \neq i$; ω_{i0} is the bias value; ω_i^t is the weightage factor of active user connection; and t is the time step.

$$\therefore f_k(\gamma) = \omega_k^t \gamma + \omega_{k0} \tag{2.7}$$

where $f_k(\gamma) = \omega_k$ if $f_k(\gamma) > f_l(\gamma)$ for all $k \neq l$; ω_{k0} is the bias value; and ω_k^t is the weightage factor of in-active user connection.

System load with active user connections with respect to total number of connections is plotted on the *X*- and *Y*-axis (see Fig. 2.20 and Eq. 2.6) with a bias value +1.

System load with inactive user connections with respect to total number of connections is shown on *X*- and *Y*-axis (see Fig. 2.21 and Eq. 2.7) with a bias value −1.

2.4.8 Simulation of derived framework for load testing based on real-time data

Jmeter software is being utilized to test data load in real time. Data from 500 users were considered as a sample set of input. The 8080 port was set to interact between client and server during experimentation.

In Fig. 2.22, system performance with respect to time using median values was mentioned. In real time, the system exhibits stable distribution of data points within the graph. A continuous curve was observed in Fig. 2.19 indicating the system stability during high load execution. A median value lies between the lower and upper portions of the curve.

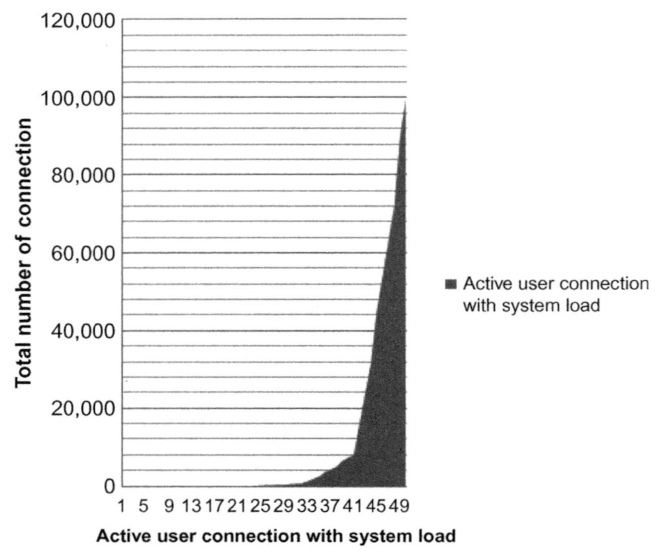

FIG. 2.20 Total number of load vs. active user connections with system load.

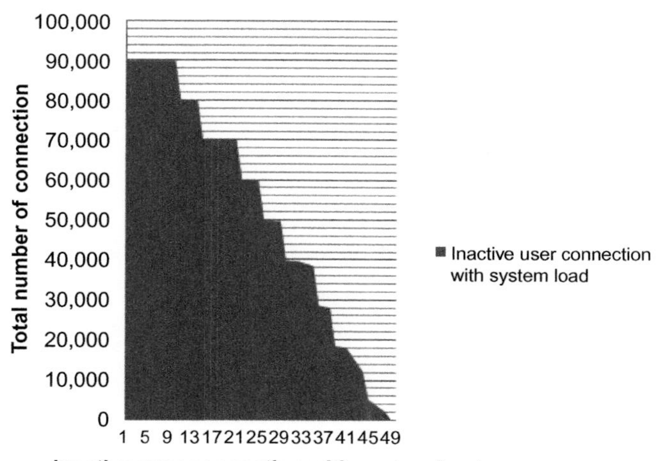

FIG. 2.21 Total number of load vs. inactive user connections with system load.

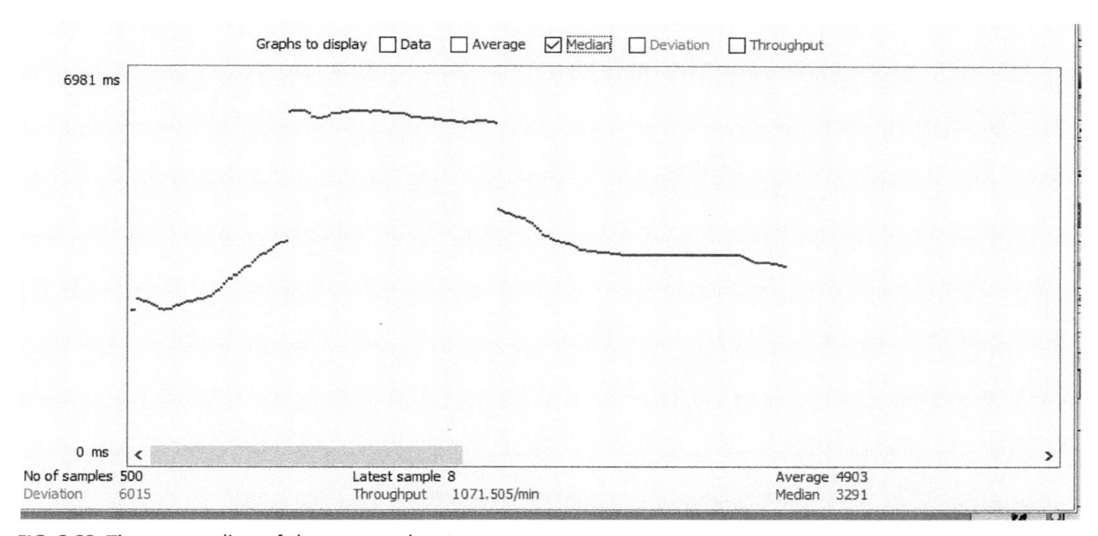

FIG. 2.22 Time vs. median of the proposed system.

Median has a long tail toward the lower side of the curve, which involves better stabilization with respect of performance.

In Fig. 2.23, throughput of the system with respect to time is illustrated. *X*-axis represents throughput values. In case of high load, throughput is also high as shown in the graph. It implies that the load is directly proportional to the throughput. Therefore, the proposed system is energy-efficient and cost-effective.

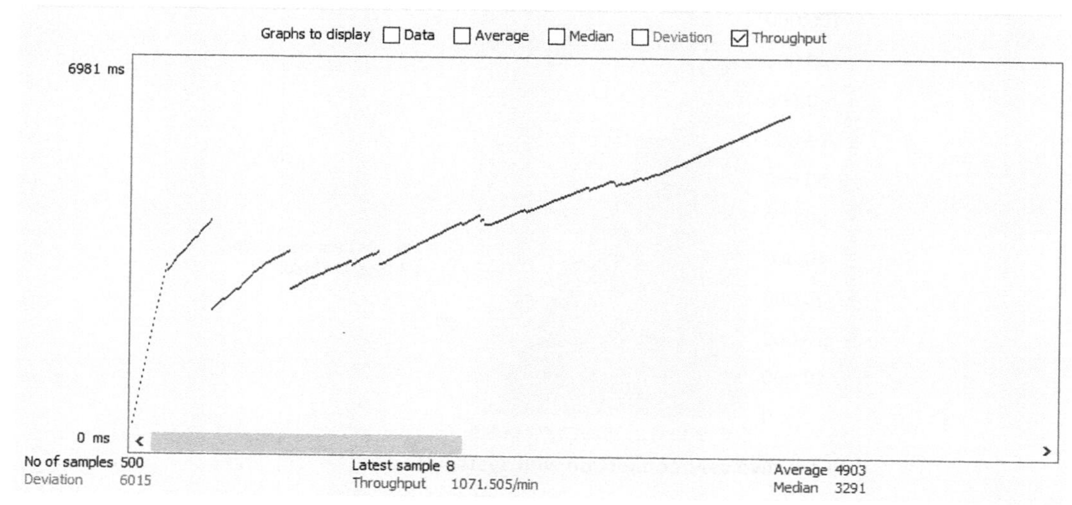

FIG. 2.23 Time vs. throughput of the proposed system.

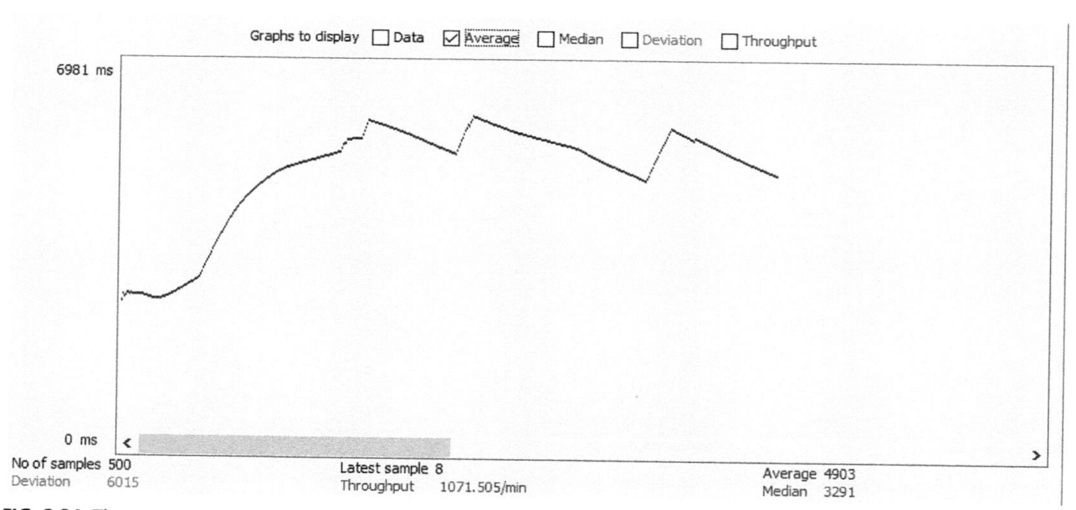

FIG. 2.24 Time vs. Average load of proposed system.

In Fig. 2.24, the average load of the proposed system is depicted with respect of time and is represented by *Y*-axis. *X*-axis represents average values. Web servers are running parallel, and loads are developed simultaneously. In this scenario, real-time system delay takes place. So, the graph takes a pick value, then it falls again.

In Fig. 2.25, deviation of the system with respect to time is presented by the *Y*-axis. *X*-axis represents the deviation values. Web servers are running in parallel, and the

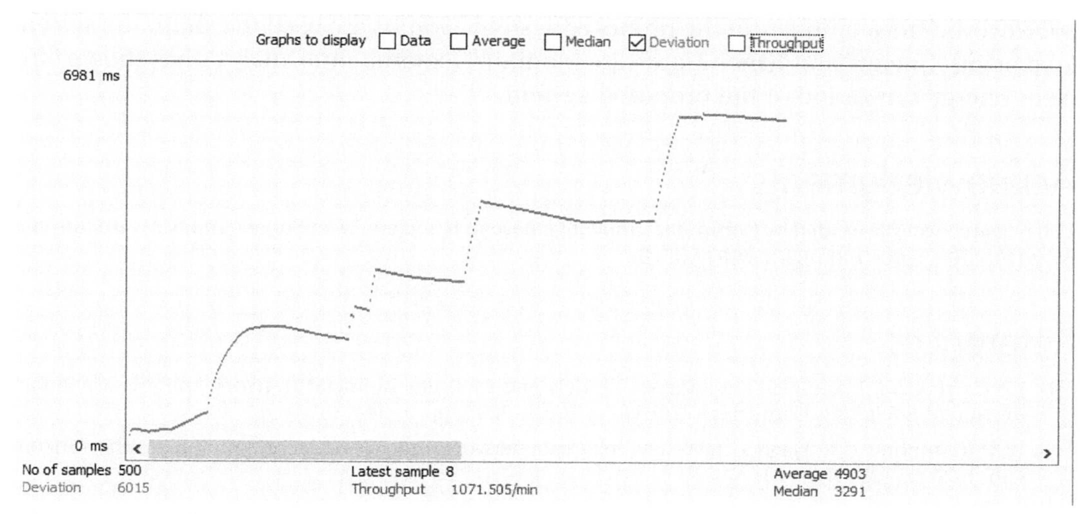

FIG. 2.25 Time vs. deviation of the proposed system.

corresponding load is high for the system. It exhibits a delay occasionally to ramp up the entire procedure and achieve a pick point. As a result, we have a step-up and step-down graph.

2.4.9 Novelty of proposed model

(i) Fully scalable approach
(ii) Cloud-based approach
(iii) Linear time complexity
(iv) 100% accuracy
(v) Successful mapping between geolocation and health statistics

2.5 Conclusion

A real-time server-based monitoring system is achieved for surveillance using sensors at the client's side to rescue users from vulnerable situations. The proposed system has individual sensors to sense users' movements and health conditions to ensure user safety and security without breaking their privacy. Sounds and pictures are not considered to track users for the purpose of protection and privacy. The proposed system ensures a cloud-based approach toward society and environment reducing power, resources, energy consumption by using sensor devices, and cloud and social media. Data accumulation in the cloud is achieved. In case of emergency, an automated alert is activated, and the information is sent to the service cloud, which then takes necessary security measures. Procedures have been used to establish working principles of the entire system. Experimental results have proved the legitimacy of the proposed system. Graphs are used to exhibit the system load using "Jmeter" and "Open Hardware Monitor." Experimental analysis has shown efficient energy utilization of the proposed system.

Acknowledgments

The research work is funded by Computer Innovative Research Society, West Bengal, India. Award number is "2018/CIRS/R&D/1002-05-29/SRTMUHI."

References

[1] J.P. Carrington, J. Scott, S. Wasserman, Models and Methods in Social Network Analysis, Canadian Journal of Sociology, Cambridge University Press, 2005, pp. 1–344.

[2] W.M. Campbell, C.K. Dagli, C.J. Weinstein, Social network analysis with content and graphs, Lincoln Lab. J. 20 (1) (2013) 62–81.

[3] F. Edwards, An investigation of attention-seeking behavior through social media post framing, Athens J. Mass Media Commun. 3 (1) (2017) 25–44.

[4] L. Li, M.F. Goodchild, The role of social networks in emergency management: a research agenda, Int. J. Inform. Syst. Crisis Respon. Manage. 2 (4) (2010) 49–59.

[5] E.A. Marwick, The public domain: surveillance in everyday life, Surveill. Soc. 9 (4) (2012) 378–393.

[6] T. Allmer, C. Fuchs, V. Kreilinger, S. Sevignani, Social networking sites in the surveillance society critical perspectives and empirical findings, Media Surv. Ident. Soc. Perspect. 3 (2) (2014) 6–9.

[7] D. Centola, Failure in complex social networks, J. Math. Sociol. 33 (2009) 64–68.

[8] A. Rust, Participatory (counter-) surveillance and the internet, in: Proceedings of the 21st International Symposium on Electronic Art, Vancouver, Canada, 2015.

[9] C. Fuchs, D. Trottier, Social media surveillance & society, Media Commun. 3 (2) (2015) 6–9.

[10] N. Indumathy, K.K. Patil, Medical alert system for remote health monitoring using sensors and cloud computing, Int. J. Res. Eng. Technol. 3 (4) (2014) 884–888.

[11] M. Shelar, J. Singh, M. Tiwari, Wireless patient health monitoring system, Int. J. Comput. Appl. 62 (6) (2013) 1–5.

[12] A. Acquisti, L. Brandimarte, G. Loewenstein, Privacy and human behavior in the age of information, Science 347 (6221) (2015) 509–514.

[13] A.L. Begun, Human behavior and social environment: the vulnerability, risk and resilience model, J. Soc. Work. Educ. 29 (1) (1993) 26–35.

[14] K. Chard, S. Caton, O. Rana, K. Bubendorfer, Social cloud: cloud computing in social networks, in: The Proceedings of the 3rd International Conference on Cloud Computing (CLOUD), Miami, 2010, pp. 99–106.

[15] A. Archana Lisbon, R. Kavitha, A study on cloud and fog computing security issues and solutions, in: International Journal of Innovative Research in Advanced Engineering (IJIRAE), vol. 4(3), 2017.

[16] H. Sonawane, D. Gupta, A. Jadhav, Social cloud computing using social network, IOSR J. Comput. Eng. 16 (3) (2014) 15–18.

[17] D. Lyon, Surveillance, Power and Everyday Life. A Chapter for Oxford Handbook of Information and Communication Technologies, Business and Management, Technology and Knowledge Management, Social Issues, 2009, pp. 1–37.

[18] F. Stalder, Opinion. Privacy is not the antidote to surveillance, Surveill. Soc. 1 (1) (2002) 120–124.

[19] C. Doukas, I. Maglogiannis, Managing wearable sensor data through cloud computing, in: Third IEEE International Conference on Cloud Computing Technology and Science, 2011, pp. 440–445.

[20] K. Sindhanaiselvan, T. Mekala, A survey on sensor cloud: architecture and applications, Int. J. P2P Netw. Trends Technol. 6 (2014) 1–6.

[21] M. Li, S. Yu, Y. Zheng, K. Ren, W. Lou, Scalable and secure sharing of personal health records in cloud computing using attribute based encryption, IEEE Trans. Parallel Distrib. Syst. 24 (1) (2013) 131–143.

[22] A. Kandhalu, A. Rowe, R. Rajkumar, C. Huang, Y. Chao-Chun, Real-time video surveillance over IEEE 802.11 mesh networks, in: IEEE Conference, 2009, pp. 1–10.

[23] U. Rajendran, A.J. Francis, Anti theft control system design using embedded system, Proced. IEEE 85 (2011) 239–242.

[24] F. Wahl, M. Milenkovic, O. Amft, A distributed PIR-based approach for estimating people count in office environments, in: Proceedings of the IEEE 15th International Conference on Computational Science and Engineering, Paphos, Cyprus, 5–7 December, 2012, pp. 640–647.

[25] J. Yun, M.-H. Song, detecting direction of movement using pyroelectric infrared sensors, IEEE Sensors J. 14 (2014) 1482–1489.

[26] J.-S. Fang, Path-dependent human identification using a pyroelectric infrared sensor and Fresnel lens arrays, Opt. Express 14 (2) (2006) 609–624.

[27] S.P. Ahuja, K. Muthiah, Survey of state-of-art in green cloud computing, Int. J. Green Comput. 7 (1) (2016) 25–36.

[28] A.J. Younge, G.V. Laszewski, L. Wang, S. Lopez-Alarcon, W. Carithers, Efficient resource management for cloud computing environments, in: IEEE, International Conference on Green Computing, Chicago, 2010, pp. 1–8.

[29] Z. Ali, R. Rasool, P. Bloodsworth, Social networking for sharing cloud resources, in: Second International Conference on Cloud and Green Computing, 2012, pp. 160–166.

[30] W.C. Feng, K.W. Cameron, The Green 500 list: encouraging sustainable supercomputing, IEEE Comput. 40 (12) (2007) 50–55.

[31] W.M. Jubadi, S.F.A. Sahak, Heartbeat monitoring alert via SMS, in: 2009 IEEE Symposium on Industrial Electronics and Applications, Kuala Lumpur, Malaysia, 2009, pp. 1–5.

[32] R. Zheng, K. Vu, A. Pendharkar, G. Song, Obstacle discovery in distributed actuator and sensor networks, ACM Trans. Sensor Netw. 7 (3) (September 2010) 22.

[33] Y.C. Chong, P.S. Kumar, Sensor networks: evolution, opportunities, and challenges, Proc. IEEE 91 (8) (2003) 1247–1256.

[34] Y. Zhang, Q. Ji, Efficient sensor selection for active information fusion, IEEE Trans. Syst. Man Cyber. B: Cyber. 40 (3) (June 2010) 719–728.

[35] C. Rainieri, G. Fabbrocino, E. Cosenza, Structural health monitoring systems as a tool for seismic protection, in: The 14th World Conference on Earthquake Engineering, Beijing, China, 2008, pp. 1–8.

[36] B.F. Spencer Jr., M. Ruiz-Sandoval, N. Kurata, Smart sensing technology for structural health monitoring, in: 13th World Conference on Earthquake Engineering Vancouver, B. C., Canada, 2004, pp. 1–13.

[37] P. Bhandari, K. Dalvi, P. Chopade, Intelligent accident-detection and ambulance-rescue system, Int. J. Sci. Technol. Res. 3 (6) (2014) 67–70.

[38] A. Syedul, J. Jalil, I.B.M. Reaz, Accident detection and reporting system using GPS, GPRS and GSM technology, in: 1st International Conference on Informatics, Electronics and Vision, Dhaka, Bangladesh, 2012, pp. 640–643.

[39] M. Hassanalieragh, A. Page, T. Soyata, G. Sharma, M. Aktas, G. Mateos, B. Kantarci, S. Andreescu, Health monitoring and management using internet-of-things (IoT) sensing with cloud-based processing: opportunities and challenges, in: IEEE International Conference on Services Computing, 2015, pp. 285–292.

[40] S. Gayathri, N. Rajkumar, V. Vinothkumar, Human health monitoring system using wearable sensors, Int. Res. J. Eng. Technol. (IRJET) 2 (8) (2015) 122–126.

[41] https://www.encyclopediaofmath.org/index.php/Monte-Carlo_method.

[42] J.H. Curtiss, Monte-Carlo methods for the iteration of linear operators, J. Math. Phys. 32 (4) (1954) 209–232.

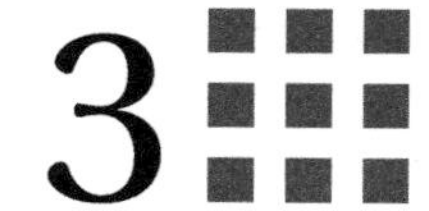

Multimodality medical image retrieval using convolutional neural network

Preethi Kurian, Vijay Jeyakumar

DEPARTMENT OF BIOMEDICAL ENGINEERING, SSN COLLEGE OF ENGINEERING, CHENNAI, INDIA

3.1 Introduction

Advancement in technologies in the medical field has led to the generation of images across different modalities like X-ray, computed tomography (CT), magnetic resonance imaging (MRI), functional-MRI (f-MRI), positron emission tomography (PET), single photon emission computerized tomography (SPECT), endoscopy, ultrasound imaging, and others to inspect and monitor internal organs in the region of interest noninvasively to diagnose and treat diseases. Healthcare centers and medical industries are growing, and the size of healthcare data is expected to grow to an estimated 2314 exabytes by 2020 according to statistics given by international data corporation (IDC) [1]. This increased growth in medical images and modalities will burden medical experts, who lack time and energy, with an error-prone subjective interpretation of disease diagnosis. Machine learning techniques can be used to automate diagnosis by extracting visual features from raw images. There have been many research works carried out in the area of image classification in the field of computer vision and image retrieval for natural images. Deep learning techniques using CNN have shown promising results in image retrieval systems.

3.1.1 Need for medical image retrieval

Diagnosticians can make quick judgments by seeing a medical image report based on their experiences, but the scale of data they have to handle every single day with an increasing number of patients makes it difficult to manage. Automation of data management in hospitals calls for an efficient image retrieval system that operates on a very large database, so as to assist physicians make faster diagnoses by reviewing clinically relevant cases of the same subject or other patients from history via a search query. A text-based approach is the oldest method where text is linked with an image, such as manual annotations, labels, and keywords, which is used to perform image retrieval

Deep Learning Techniques for Biomedical and Health Informatics. https://doi.org/10.1016/B978-0-12-819061-6.00003-3

from an image database [2]. Text query is matched with labels from the image database, and images with similar text are retrieved. Text-based image retrieval systems are fast when the database contains accurately labeled data. However, there are no standards in labeling medical images, so there is no uniformity followed globally. Content-based image retrieval (CBIR) routinely indexes images based on the extraction of their low-level visual features like spatial features, edges, color, and texture determined by image processing such as enhancement and segmentation [3]. Lack of standards in utilizing conventional features, and the semantic differences that arise between the low-level visual features extracted from the imaging devices and high-level human perception of the image, are the challenges in CBIR systems [4]. Some of the search algorithms give results based on text and content with increased mean average precision (MAP) [5], but those systems are semiautomatic.

Accuracy and speed of retrieval are the prime factors that need to be considered. The purpose of the retrieval system is lost if it returns irrelevant search results. A majority of the radiological imaging modalities are in greyscale, and the major challenge is to enable the machine to retrieve similar images by differentiating among multiple modalities; recognizing the organ, tumor, or lesion of interest to help physicians in faster diagnosis and treatment; and helping physicians and researchers in teaching medical students who are less experienced to reach faster decisions by going through similar cases from the repository. Machine learning techniques using convolutional neural network (CNN) have shown promising results [6, 7] for classification of medical images across anatomies and modalities. CNN can be a potential algorithm for image retrieval as it requires very minimal preprocessing, in the sense that the network learns the features unassisted and independently, as an alternative to extracting traditionally handcrafted features, which is tedious. The literature shows that CNN can be used for CBIR [3, 8, 9].

3.1.2 Machine learning-convolutional neural network

Machine learning is a technique that teaches itself by inferring patterns from data over a period of time. The advantage of machine learning is that humans need not give instructions or programs to extract features from the data; rather, the learning is automatically carried out by the machine by fitting the data against the constraints. Neural networks were modeled based on the human brain, which has highly interconnected neurons [10] controlling different actions simultaneously. A recent trend in the field of machine learning is deep learning.

Deep learning employs deep architectures in a neural network to learn intricate features without human intervention and understanding of the domain. CNN is a special neural network designed for image classification and object recognition. Unlike the ordinary multilayer perceptron (MLP), which has fully connected layers, CNN employs convolution filters that reduce the complexity of the model while acquiring local information from an image. The principle layer of the CNN is the convolutional layer, which acts as the feature extractor or the preprocessor.

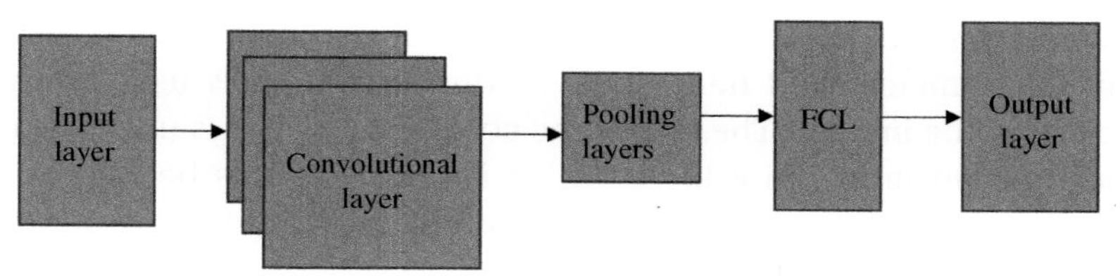

FIG. 3.1 General architecture of CNN.

3.1.3 General architecture of CNN

The basic components of CNN are:

- Input layer
- Convolutional layers
- Pooling layers
- Fully connected layers
- Output layer

The input layer consists of the input image with width, height, and dimension. Convolution layer consist of a number of filters that slide over the input image to generate local feature maps by taking the sum of the products of the filter and the portion of the input image. Pooling layers reduce the dimensionality to minimize the number of training parameters of the network. Max pooling and average pooling are the most commonly used pooling layers in CNN architectures (Fig. 3.1).

As in any artificial neural network (ANN), activation functions are used to introduce nonlinearity in CNN. ReLu layer or rectified-linear units that converge faster than traditional activation functions, like tan sigmoid and log sigmoid, are predominantly used in CNN. It rectifies vanishing gradient problem and enables learning by minimizing error during backpropagation. ReLu is used in hidden layers of CNN. Fully connected layer (FCL) has neurons connected to all activations from the previous layer. Output layer computes the probabilities of classes using Softmax activation and gives the output for the classification problem. CBIR applications on multimodality medical images using CNN are limited, although deep learning has been used for classification of diseases in the recent past.

3.2 Convolutional neural network

CNN is a special neural network, designed specifically for images of multidimensions and video processing. CNNs visualize smaller parts of an input image through filters that slide over the image to collect feature representations of the original image for object recognition problems.

CNN is comprised of different combinations of fully connected layers and convolutional layers, introducing nonlinearity by activations for every neuron in each layer.

Fully connected layers put a huge burden on the memory constraint, increasing the number of training parameters and hence convolution operation is used, which works on smaller patches on the images. The size of the convolution filter is uniform for each neuron as it slides on an image in a layer, which is advantageous because it reduces the required computational cost and improves accuracy for image recognition when compared to ANNs.

3.2.1 Convolutional layer

This is the basic unit of the CNN. The main properties of a convolutional layer are its local connectivity and weight sharing. The convolution layer accepts an input volume of Row $(R1) \times$ Columns $(C1) \times$ Dimension $(Dim1)$. The convolution layer forms the feature maps for object recognition; hence it can be optimized by adjusting the hyperparameters [11].

The four hyperparameters of CNN are:

- Number of filters, K, which are required to produce K feature maps, which will be stacked along the depth for activation for learning.
- Filter size, F is also called receptive field for local connectivity along the depth size of the input volume. The weights in the $F \times F \times D1$ region will be shared.
- Stride size, S is defined as the depth of movement of filter along the pixels of the image. When the stride is 1, then we move the filters one pixel at a time.
- Amount of zero padding, P, which is used to preserve the input data size by adding a border of zeros around the input volume.

The output of the feature map from the convolution layer will be of size $W2 \times H2 \times D2$. The dimensions are calculated by the formulas given in Eqs. (3.1)–(3.3) [12].

$$R2 = \frac{R1 - F + 2P}{S} + 1 \tag{3.1}$$

$$C2 = \frac{C1 - F + 2P}{S} + 1 \tag{3.2}$$

$$Dim2 = F \tag{3.3}$$

where $R1$, $C1$, and $Dim1$ are the row(width), column(height), and dimension of the input image.

$R2$, $C2$, and $Dim2$ are the width, height, and dimension of the feature map after convolution.

The feature maps are obtained by convolving the filters on the input image by dot product. In any neural network, the output is calculated by the formula given in Eq. (3.4).

$$\text{Output} = \sum_{k=0}^{n} (\text{Filter Weight} \times \text{Input}) + \text{bias} \tag{3.4}$$

A visualization of the dot product of input volume of size $(5 \times 5 \times 1)$ with convolution filters of size $(3 \times 3 \times 1)$, zero padding $P = 0$, $S = 1$ to obtain an output volume of $(3 \times 3 \times 1)$ is depicted in Fig. 3.2.

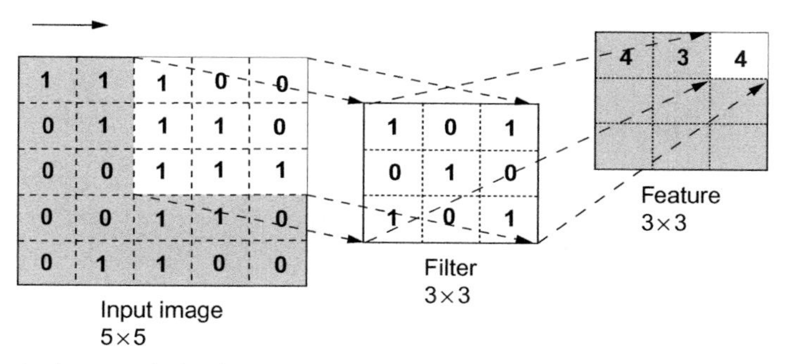

FIG. 3.2 Dot product in the convolution layer.

The output of the third feature by moving the filter across the input data is 4 by the dot product between the input data and the filter as $((1 \times 1) + (0 \times 0) + (0 \times 1) + (1 \times 0) + (1 \times 1) + (0 \times 0) + (1 \times 1) + (1 \times 0) + (1 \times 1)) = 4$.

3.2.2 Pooling layers

A pooling layer is used to minimize the number of parameters used for training, thereby reducing complexity and controlling overfitting [6]. A pooling layer consist of filters with stride of 2 to downsample the height and width of the feature maps. The pooling layer can perform two types of operations: max pooling and average pooling. Max operation takes only the maximum pixel from the extent. Average operation smoothens the feature by averaging the extent area. Examples of the pooling operations are shown in Fig. 3.3.

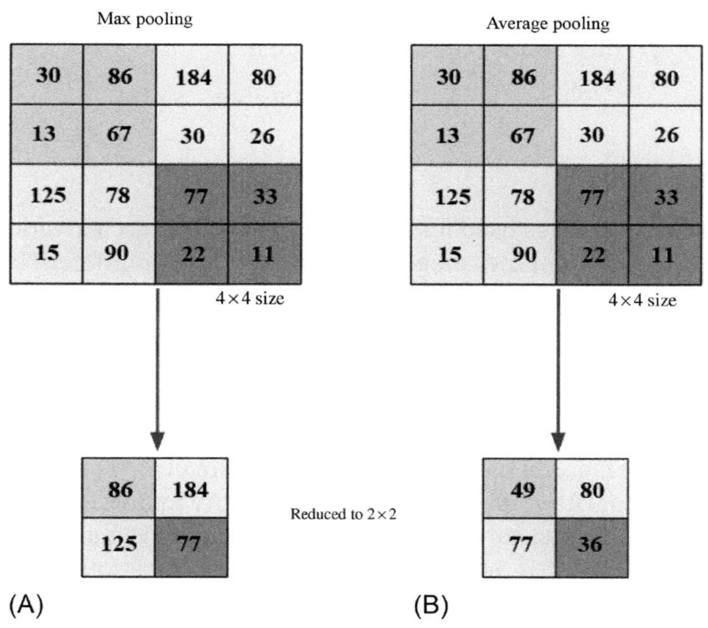

FIG. 3.3 Pooling operations. (A) Max pooling and (B) Average pooling.

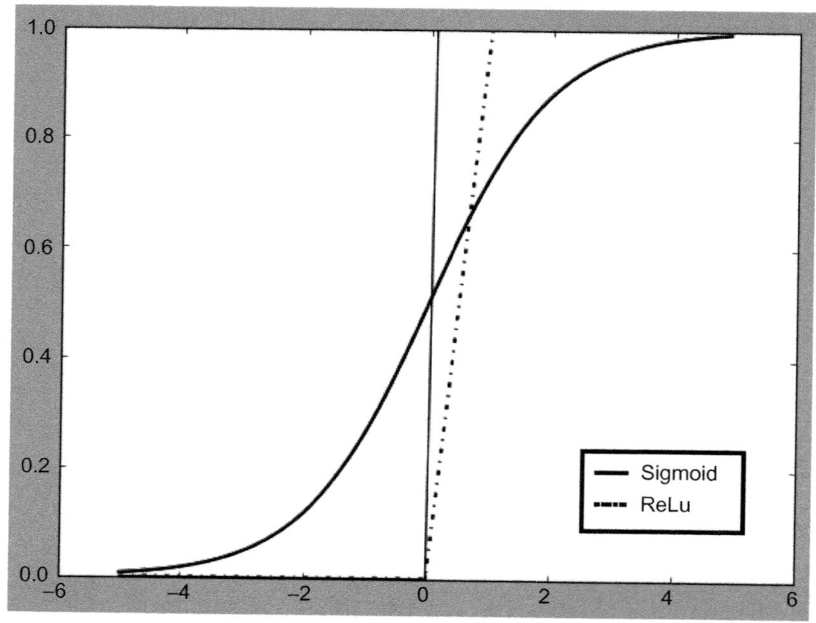

FIG. 3.4 Convergence of sigmoid activation and ReLu activation.

3.2.3 ReLu layer

ReLu layer will apply an elementwise activation function. It does not change the size of the feature map. It converges much faster than the traditional activation function sigmoid as seen in Fig. 3.4. ReLu function is defined by Eq. (3.5).

$$R(z) = \text{Max}\,(0, z), \ \text{for}\, z > 0 \tag{3.5}$$

3.2.4 Softmax layer

The output layer of the CNN is used for classification by Softmax activation function. The CNN can be used to represent the categorical distribution over different labels in terms of probabilities.

3.3 CBMIR methodology

The CBMIR task using CNN deep learning technique is carried out by the following methodology as depicted in Fig. 3.5. It is different from Qayyum et al. [4] as this work includes a large medical database with multianatomy multimodality images. The dataset is divided into train, validation, and test sets. The CNN architectures LeNet and AlexNet are implemented, and the newly created database is trained for classification. The last layers of the models form the feature layer, and the training features are extracted. When the query

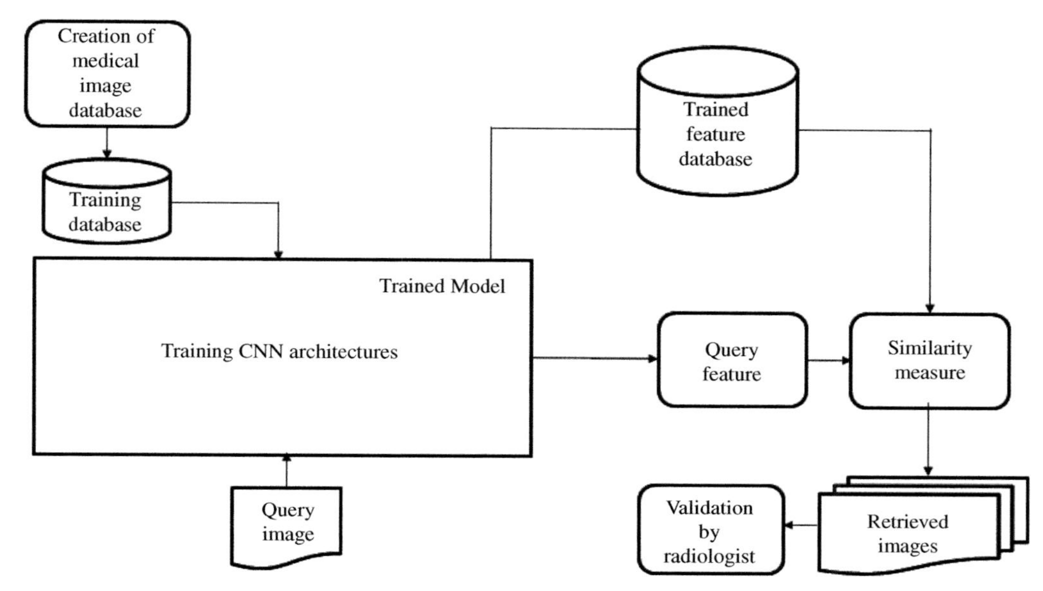

FIG. 3.5 Methodology for CBMIR task using CNN.

image from the testing is given to the trained model, the test feature is extracted, and similarity metric Euclidean distance is used to find the similar images.

3.3.1 Medical image database

The medical image database consists of 49,473 images, which are labeled as 15 classes, acquired from seven modalities with different anatomical information. The image classes were collected from various open-access online databases. The dataset was split into 90% for training, 10% for testing, and 10 images from each class have been considered for validation. The distribution of images under each class is listed in Table 3.1.

Multianatomy X-ray images of different anatomies is obtained from ImageCLEF [22]. MRI and CT volumes from ADNI, Radiopedia, and TCIA were sliced to 2D images, and 10 slices were chosen randomly for each subject for the dataset to reduce redundancy. Skin images have been categorized into clinical and dermatoscope images. All the images are resized to (224,224) and converted to grayscale in PNG format to maintain uniformity. The main advantage of using CNN is that there is no need for further preprocessing except for the input size.

An illustration of images from classes used in the dataset are shown in Fig. 3.6, which shows that there is variation among images of each class, respectively. The images have variations in the plane of orientation for X-ray, CT, and MRI images as shown in Fig. 3.7.

Table 3.1 Medical database containing 15 classes.

Classes	Database	Total	Train	Test
Brain CT	Radiopedia [13]	395	347	38
Brain MRI	ADNI [14], FigShare [15]	3660	3295	365
Breast Mammogram	MiniMammographic database—MIAS [16]	322	291	21
Breast MRI	TCIA [17]	386	338	38
Chest CT	TCIA [17]	501	442	49
Chest Xray	Mendley Data [18]	5791	5203	578
Eye Fundus	Messidor [19]	1200	1071	119
Eye OCT	Mendley Data [18]	10671	9578	1083
Kidney CT	TCIA [17]	348	305	33
Knee Xray	Hospital [20]	995	887	98
Liver CT	TCIA [17] and Radiopedia [13]	308	269	29
Liver MRI	TCIA [21]	147	123	14
Multianatomy X-ray	ImageCLEF2009- IRMA [22]	13076	11760	1306
Skin clinical	HAM10000, Harvard Dataverse [21], ISIC [23]	122	103	9
Skin dermatoscope	HAM10000, Harvard Dataverse [21], ISIC [23]	11542	10383	1149

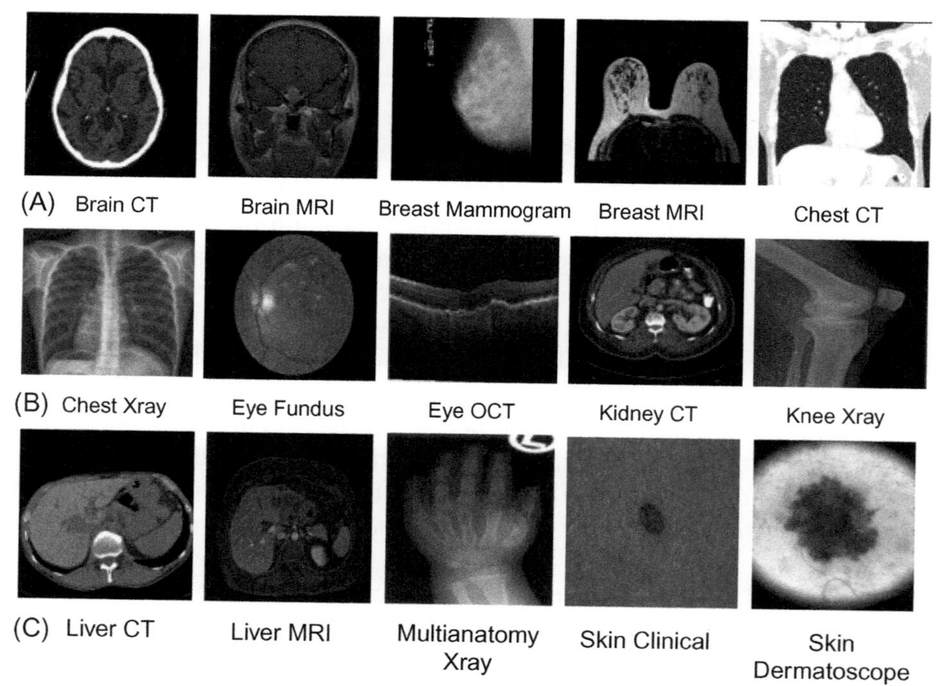

FIG. 3.6 Database containing medical images from 15 classes with multiple modalities. (A) Brain MRI. (B) Chest X-ray. (C) Kidney CT.

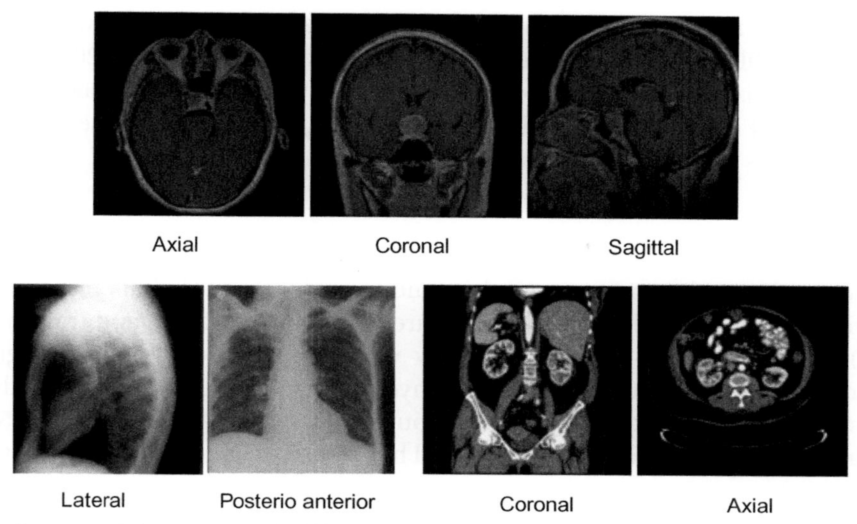

FIG. 3.7 Database containing images with different orientation in each class.

3.3.2 LeNet and AlexNet

3.3.2.1 LeNet-5

LeNet architecture implemented by LeCun [24] was used for handwritten digits on cheques for banking application. It comprises of seven layers with three convolutional layers and two pooling layers after the first and the second convolution layers and two fully connected layers as shown in Fig. 3.8.

The architecture takes a 32 × 32-pixel image as input. First convolution layer C1 produces six features of size 28 × 28. Second layer is the max pooling layer of size 2×2. It reduces the width and height of the feature maps by downsampling at a rate of 2–14×14, but the depth remains unchanged. The third layer is a convolutional layer with filter size 5×5, number of filters 16, thus it results in 16 feature maps of size 10 × 10.

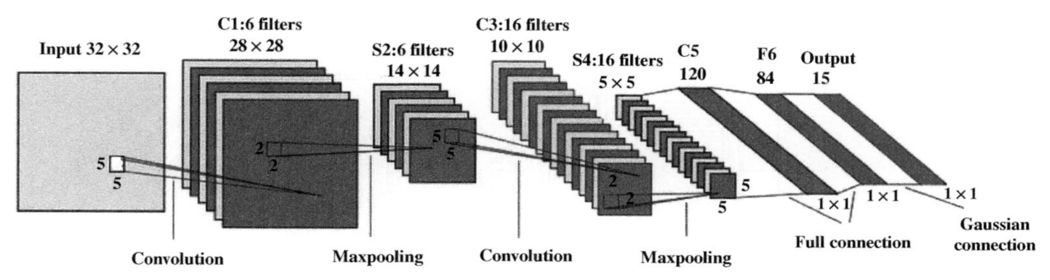

FIG. 3.8 LeNet-5 architecture.

Downsampling of the feature set to 5 × 5 is carried out at the fourth layer with filter size of the max pooling filter being 2 × 2. Fifth layer is a convolution layer with 120 kernels of size 5 × 5, hence the size of the feature set is 1 × 1. Sixth layer is a fully connected layer with 84 neurons [24]. Finally, the output layer classifies the data into its classes. In this work, the final layer was given a Softmax activation for 15 classes.

3.3.2.2 AlexNet

AlexNet [25], designed by Krizhevsky, is one of the deep CNNs designed to handle complex object recognition tasks on Imagenet data and won the ILSVRC challenge in 2012. The main variances between LeNet and AlexNet are (i) number of layers and training parameters, (ii) AlexNet has ReLu as nonlinearity whereas LeNet uses logistic sigmoid for activation, and (iii) AlexNet uses a dropout layer whereas LeNet does not use dropout.

The original implementation was carried out in two pipelines with two GTX 580 GPUs for training. Recent GPUs have good RAM and hence follow the single-line architecture as seen in Fig. 3.9.

AlexNet architecture that consists of eight layers, five convolutional layers and three fully connected layers, accepts RGB images of dimension 224 × 224 as inputs. The first convolution layer takes the input image and extracts 96 features of size 11 × 11 with 4 pixels stride. Nonlinear activation function ReLu is applied after the convolution operation in every convolutional layer. The output is given to the second convolution layer, which generates 256 feature maps of size 5 × 5. After the first two convolution layers, max pooling and normalization is carried out. The pooled output at the second convolutional layer is given to the third layer. The third convolution layer extracts 384 features of size 5 × 5. Fourth convolutional layer also contains 384 filters of size 5 × 5. Convolution layer 5 has 256 filters of size 3 × 3. The two dense layers following the convolutional layers have 4096 neurons, which is flattened. The last fully connected layer forms the output layer with Softmax activation to classify 1000 categories. Dropout layer is added to reduce overfitting. Overfitting is the inability of the trained network to generalize well to a new test data [25].

For implementation in this work, the input images are grayscale instead of RGB. The output layer will have 15 classes for the 15 categories of medical dataset. The last layer before the classification layer is used as the feature layer to find the similarity indices for retrieval.

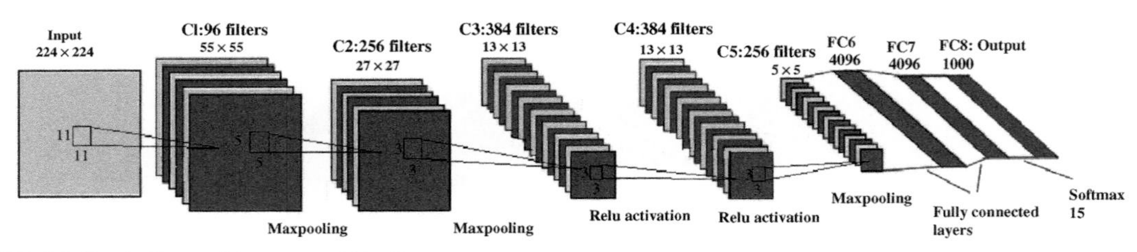

FIG. 3.9 Single-line architecture of AlexNet.

3.3.3 Training of CNN architectures for classification

Training of CNN is carried out by the backpropagation algorithm, which was introduced in the 1990s to train neural networks. It has two parts: (1) Feed forward propagation and (2) backpropagation.

Feed forward propagation:

In forward pass, the net-input to the hidden layer is calculated by the equation:

$$Y_{in_j} = \sum_{i=1}^{N} w_{ji} x_i + \text{Bias} \tag{3.6}$$

where I is the number of neurons from 1 to N, j is the hidden layer number, x is the input vector, w is the weight vector, Y_{in_j} is the net input at the jth layer.

Output of the layer Y_j is calculated by passing through the activation function, generally ReLu in the case of CNN.

$$Y_{j=}f\left(Y_{in_j}\right) \tag{3.7}$$

Similarly, output is calculated through the consecutive layers until it reaches the layer k (Y_k) or classification layer at the end.

Backpropagation:

The error δ_k is calculated from the output layer k:

$$\delta_k = (Target - Y_k) f'(Y_{in_k}) \tag{3.8}$$

where Y_k, Y_{in_k} are output and net input of the output layer k, $f' =$ derivative of activation.

Weight change due to the error correction propagated from kth layer back to jth layer is calculated by:

$$\Delta w_{jk} = -\alpha \delta_k Y_j \tag{3.9}$$

where α is the learning rate.

Thus, the new weights are calculated and updated back through the layers:

$$w_{jk}(new) = w_{jk}(old) + \Delta w_{jk} \tag{3.10}$$

The same process is repeated until it reaches the convergence condition, for example, a certain number of epochs or minimal error/loss.

3.3.3.1 Optimizing the training parameters for CNN learning

The CNN needs to be optimized for better learning by adjusting the learning rate α and weights w according to the loss during backpropagation. Gradient descent method [26] is the standard method for neural networks, which minimizes error δ_k.

Three optimizers were chosen for this analysis.

- **SGD** [26] randomly shuffles the dataset and updates parameter for individual images keeping learning rate constant.
- **AdaGrad** [27] updates learning rate that improves performance on problems with sparse gradients. Disadvantage is that frequently occurring data will be frequently updated and hence the learning rate will be so small that the network fails to learn.
- **Adam** [28] (adaptive moment estimation) makes use of the first and second moments of the gradient for updating the learning rate adapting to each training image.

3.3.4 Training implementation

The methodology is implemented using Anaconda, a worldwide software platform for data science. Keras is a python-based API for neural networks and machine learning compatible to run on top of Tensorflow or Theano. Theano is an efficient compiler and optimizer for evaluating complex networks. CUDA is a software interface that allows parallel computing architecture to work with GPU. Training is accelerated by NVIDIA GeForce GTX 1060 GPU supported on Windows 10 Operating System of 64 GB RAM. Software specifications are listed in Table 3.2.

3.3.4.1 LeNet model training

Train dataset was first trained for classification of 15 categories on the LeNet architecture. Three models were created to be trained for 100 epochs; each model with a different optimizer analyzed which model optimized the network for better learning.

We can observe from Table 3.3 that, even though the training duration was a bit higher, Adam was able to get best validation accuracy of 98.82% and minimal loss of 10.1% at epoch 68 for classification. AdaGrad was able to get good accuracy at epoch 10, but from

Table 3.2 Software specifications.

Anaconda version	V 5.3
Python version	V 3.6.5
Python library	Keras v 2.2.4
Backend library	Theano v 1.0.0
CUDA version	NVIDIA CUDA Toolkit 9.0.176
cuDNN	cuDNN 7

Table 3.3 Optimizer analysis for 100 epochs on LeNet.

Optimizer	Training duration in seconds	Val_acc (%)	Val_loss (%)	Epoch
SGD	347.45	97.97	14.23	85
AdaGrad	376.26	98.50	11.35	10
Adam	406.22	98.82	10.10	68

the graph, we observe that the characteristics of the validation accuracy show large variations.

The losses and accuracy for the validation dataset is unstable and does not show smooth variation in the models using SGD and AdaGrad as seen in the Figs. 3.10 and 3.11, respectively.

The losses and accuracy for validation set is stable over the epochs when the network was optimized with Adam as seen in Fig. 3.12 helps in better learning of CNN at even smaller epochs. Hence Adam is chosen as the best optimizer among the three and will be used for training at higher epochs and for training using AlexNet.

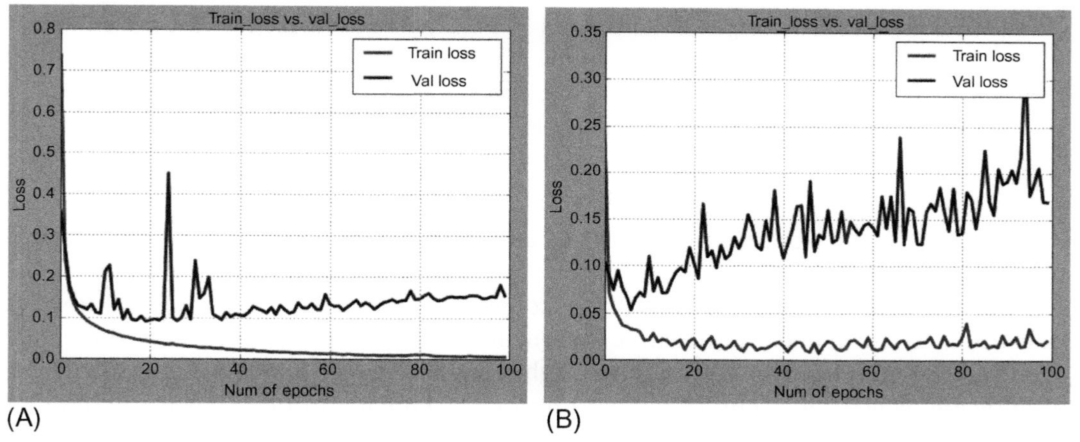

FIG. 3.10 Training loss vs. validation loss for LeNet models using different optimizers for 100 epochs: (A) SGD. (B) AdaGrad.

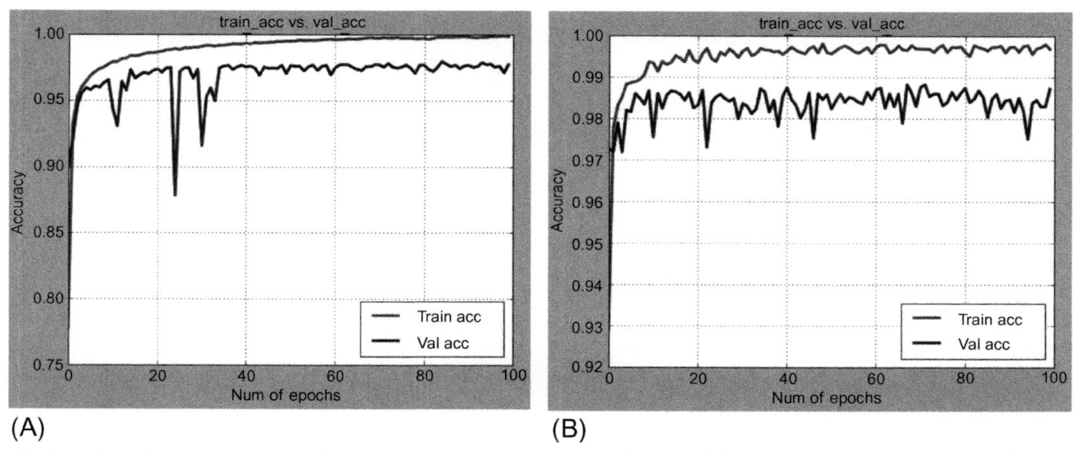

FIG. 3.11 Training accuracy vs. validation accuracy for LeNet models using different optimizers for 100 epochs: (A) SGD. (B) AdaGrad.

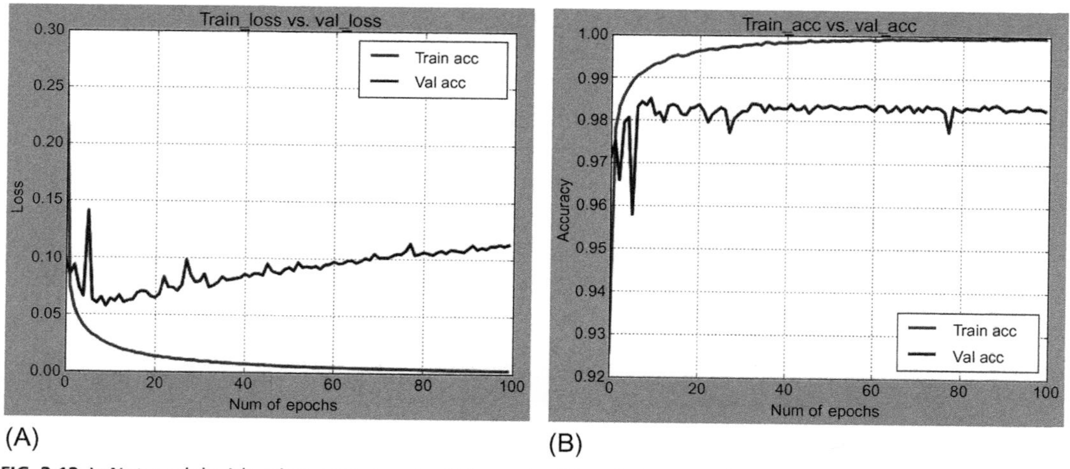

FIG. 3.12 LeNet model with Adam optimizer for 100 epochs: (A) train loss vs. validation loss and (B) train accuracy vs. validation accuracy.

Similarly, the dataset was trained for 500 and 1000 epochs separately with Adam chosen as the optimizer.

From the graphs in Figs. 3.13 and 3.14, a rough variation in convergence was observed. With the increase in epochs, the curve diminishes, and the accuracy decreases due to the overfitting of data. The time to train for higher epochs is very time consuming, and the training and validation accuracy did not improve to a greater extent. Hence, the retrieval process will be carried out with the weights of CNN model with the validation accuracy (98.82%) at the 68th epoch as seen from Table 3.4.

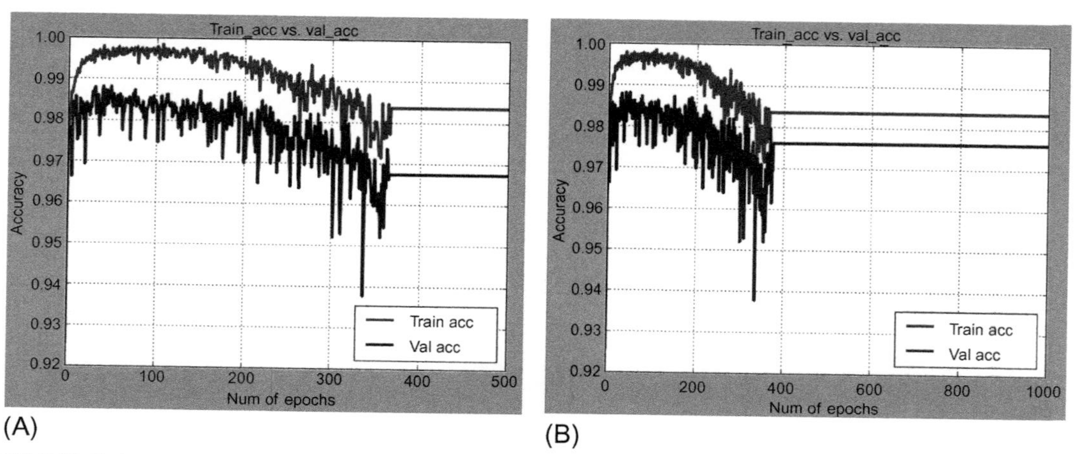

FIG. 3.13 Train accuracy vs. Validation Accuracy plot for different epochs: (A) 500 epochs. (B) 1000 epochs.

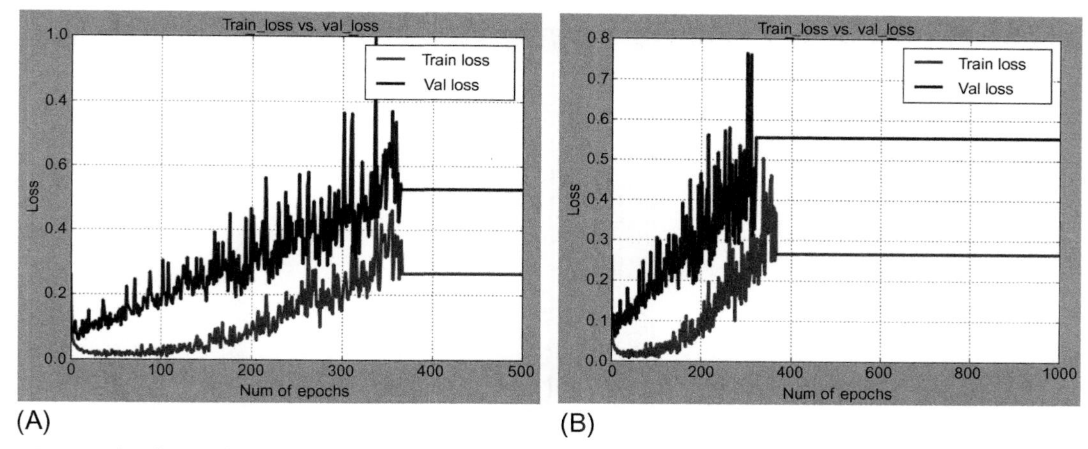

FIG. 3.14 Classifier performance for different epochs: (A) 500 epochs. (B) 1000 epochs.

Table 3.4 Training accuracy of LeNet for different epochs.

Training epochs	Training duration in seconds	Validation accuracy (%)	Converging at epoch
100	406.22	98.82	68
500	2036.93	98.80	41
1000	4097.60	98.66	127

3.3.4.2 AlexNet model training

Three models for AlexNet were created to be trained for 100, 500, and 1000 epochs as depicted in Fig. 3.15. The best validation accuracy 99.29% among three models was observed at epoch 75 as seen in Table 3.5. Therefore, the weights for epoch 75 were used to extract feature maps from the dense layer, i.e., FC8 before the output layer.

Feature layers F6 of LeNet and FC8 of AlexNet will extract 84 and 1000 features from each image on the medical database. The features are stored in the feature database.

3.4 Medical image retrieval results and discussion

3.4.1 Image retrieval by similarity metrics

After feature extraction of the trained dataset, the feature set of the query image should be compared to the trained database features for similarity. The images are then arranged in ascending order of a similarity metric. There are several similarity measures like Minkowski distance, Euclidean distance, and Manhattan distance, which can be used for image retrieval.

Euclidean distance is chosen for this work as the literature [4, 29] has shown improved retrieval performance with the same metric. The Euclidean distance (D) between the

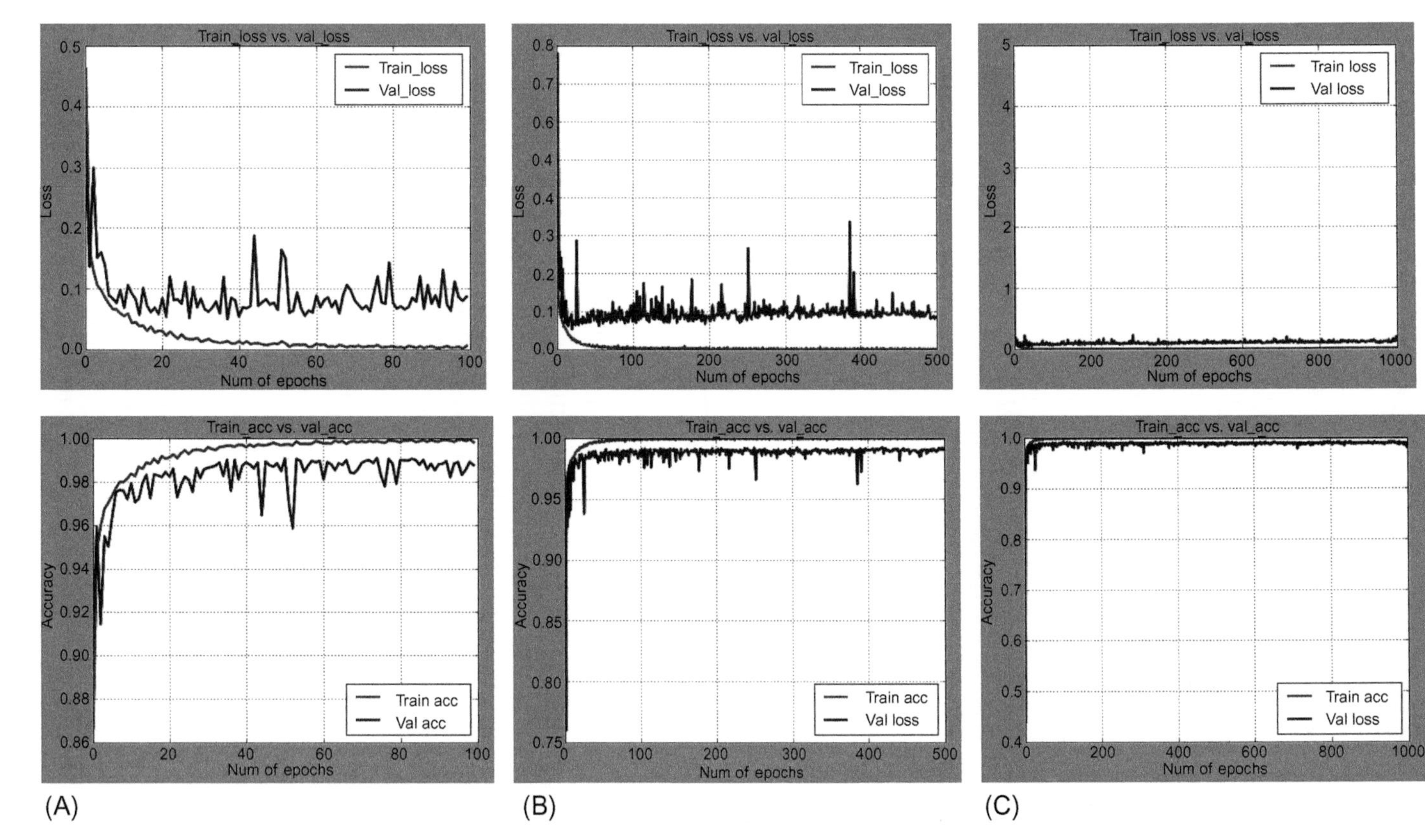

FIG. 3.15 Classifier performance of AlexNet for different epochs: (A) 100 epochs. (B) 500 epochs. (C) 1000 epochs.

Table 3.5 Training accuracy of AlexNet for different epochs.

Train for epochs	Training duration in seconds	Validation accuracy (%)	Converge at epoch
100	6931	99.29	75
500	34,708	99.25	425
1000	68,892	99.21	103

query image feature (y_i) and the database features (x_i) can be calculated using the formula as shown in Eq. (3.11).

$$D = \sqrt{\sum_{i=1}^{k} (x_i - y_i)^2} \tag{3.11}$$

The distances are then ranked in the increasing order, and the indexes of the first k sorted distances are picked from the dataset.

Features were extracted for each image from the medical dataset and stored as feature vectors. Testing is done by giving query images of each class to the trained CNN model. Euclidean distance is determined between the query image features and train dataset features. The index of the database is sorted from smallest to largest as a measure of Euclidean distance. The first 20 images are then retrieved.

3.4.2 Image retrieval performance metrics

The CBMIR system can be evaluated for its performance by using measures like retrieval time, precision, recall, F score and MAP given by Eqs. (3.12)–(3.14) [4] to know how efficiently the system is able to return accurate image search results.

- Retrieval time: Time taken by the system to retrieve k images when a query image is given.
- Precision: Defined as the ratio of the number of images relevant to the query image among the retrieved images to the number of images retrieved to display. This is the most important measurement with respect to image retrieval.

$$\text{Precision} = \frac{\text{Number of images that are relevant among the retrieved}}{\text{Number of images retrieved to display}} \tag{3.12}$$

- Recall: Defined as the ratio of the number of images relevant to the query image among the retrieved images to the total number of images relevant in the database.

$$\text{Recall} = \frac{\text{Number of images that are relevant among the retrieved images}}{\text{Number of relevant images in the database}} \tag{3.13}$$

- $F1$ score: The harmonic mean of precision and recall.

$$F1\ \text{score} = \frac{2 \times \text{Precision} \times \text{Recall}}{\text{Precision} + \text{Recall}} \tag{3.14}$$

- MAP: When the dataset contains multiple classes, precision is averaged with respect to the number of classes.

3.4.3 Retrieval results: LeNet

The retrieval performance metrics were calculated for each of the query images from 15 classes and the MAP is determined. Brain CT, Brain MRI, Chest X-ray, Eye Fundus, Eye OCT, Liver CT and Skin Dermatoscope showed best precision of 100% and the MAP for all classes was found to be 83.9% as seen in Table 3.6.

3.4.4 Retrieval results: AlexNet

The retrieval performance of the AlexNet architecture was slightly improved than that of LeNet architecture but not great enough. We infer that as the depth of the model increases, i.e., the number of layers in the CNN model, the more deeply the model is able to learn the features.

Brain CT, Brain MRI, Chest X-ray, Eye Fundus, Eye OCT, and Liver CT showed 100% precision, whereas Liver MRI and Skin Clinical image did not return any relevant results; the MAP was 86.9% as seen in Table 3.7.

Table 3.6 Image retrieval performance metrics for LeNet.

Query image	Retrieval time (s)	Precision	Recall	*F* score
Brain CT	0.2159	1	0.0576	0.1089
Brain MRI	0.2186	1	0.0060	0.0120
Breast Mammogram	0.2160	0.865	0.0594	0.1112
Breast MRI	0.2171	0.995	0.0588	0.1111
Chest CT	0.2171	0.95	0.0429	0.0822
Chest X-ray	0.2186	1	0.0038	0.0076
Eye Fundus	0.2144	1	0.0186	0.0366
Eye OCT	0.2218	1	0.0020	0.0041
Kidney CT	0.2108	0.935	0.0613	0.1150
Knee X-ray	0.2202	0.91	0.0205	0.0401
Liver CT	0.2093	1	0.0743	0.1384
Liver MRI	0.2171	0.785	0.1276	0.2195
Multianatomy X-ray	0.2173	0.105	0.0001	0.0003
Skin Clinical	0.2183	0.05	0.0097	0.0162
Skin Dermatoscope	0.2184	1	0.0019	0.0038
Mean average	0.2167	0.839	0.0363	0.0671

Table 3.7 Image retrieval performance metrics for AlexNet.

Query image	Retrieval time	Precision	Recall	*F*-score
Brain CT	0.2889	1	0.0576	0.3747
Brain MRI	0.2843	1	0.0060	0.0474
Breast Mammogram	0.2843	0.9	0.0618	0.3735
Breast MRI	0.2827	0.915	0.0541	0.1022
Chest CT	0.2782	0.97	0.0438	0.2812
Chest X-ray	0.2736	1	0.0038	0.0302
Eye Fundus	0.2749	1	0.0186	0.1390
Eye OCT	0.2796	1	0.0020	0.0165
Kidney CT	0.2741	1	0.0655	0.4103
Knee X-ray	0.2796	0.9	0.0202	0.1476
Liver CT	0.2733	1	0.0743	0.4584
Liver MRI	0.2811	0.905	0.1471	0.6758
Multianatomy X-ray	0.2769	0.25	0.0004	0.0008
Skin Clinical	0.2718	0.21	0.0407	0.1836
Skin Dermatoscope	0.2718	0.995	0.0019	0.0152
Mean average	0.2783	0.869	0.0398	0.217

The first 20 retrieved images for query images when given to both the models can be seen in Figs. 3.16–3.24.

Correctly retrieved images for a given query is counted when the class label of the retrieved images matches the query image class labels. AlexNet shows better performance in retrieving accurate images with respect to quality of feature representation. The increase in number of convolutional layers in AlexNet enabled it to learn better.

Analyzing the precision values for all 15 classes for scope size 20 as illustrated in Fig. 3.25, the overall performance of AlexNet is high. Multianatomy X-ray and Skin Clinical classes exhibited poor retrieval precision for both the models.

The recall and F score are observed to be low from Figs. 3.26 and 3.27, respectively. This is because the quantity of images retrieved is very small (scope size, 20), compared to the total number of available images in each of the categories. For instance, Brain CT has 347 images, but the scope size is 20, therefore, the maximum recall value for Brain CT class for 20 relevant images would be $20/347 = 0.0576$. Intuitively, Liver MRI shows the highest recall value and F score as the number of images in that class is low. Maintaining uniformity in the number of images of classes in real-time medical database is not practical. Thus, precision is the best metric to assess the performance of the CNN models.

Retrieval results varying scope sizes (20, 40, 60, 80, 100) is also analyzed. From Fig. 3.28, MAP is found to decrease and the mean recall improved. AlexNet consistently continued to perform better than LeNet as the scope size increased. Although the precision dropped 3.4%, recall improved 15% for AlexNet model compared to a 6.8% drop in precision and 11% improvement in recall for LeNet. Precision for different scope sizes for LeNet and AlexNet is shown in Table 3.8.

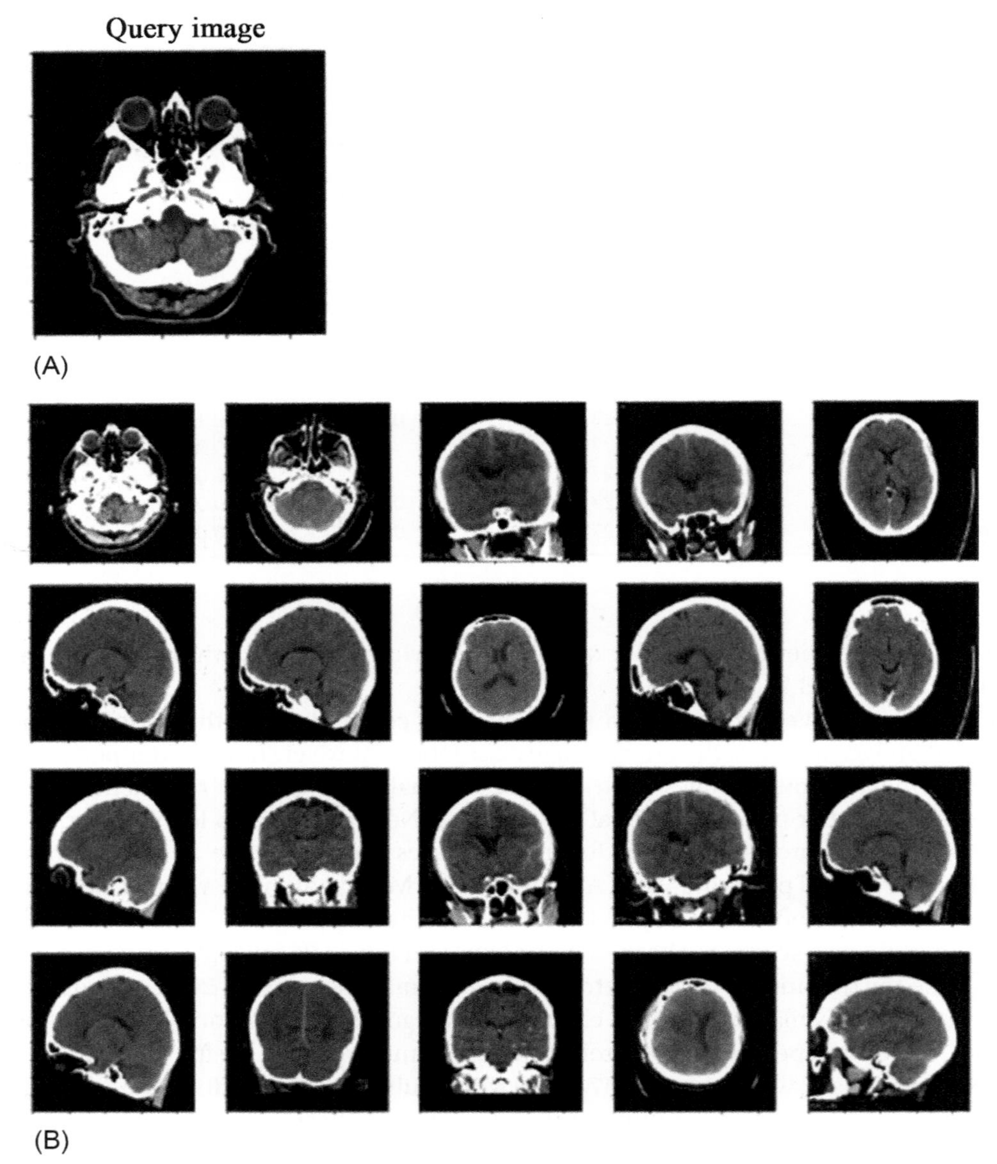

FIG. 3.16 Images retrieved for Brain CT class: Axial (A) query image, (B) retrieved images with LeNet, and

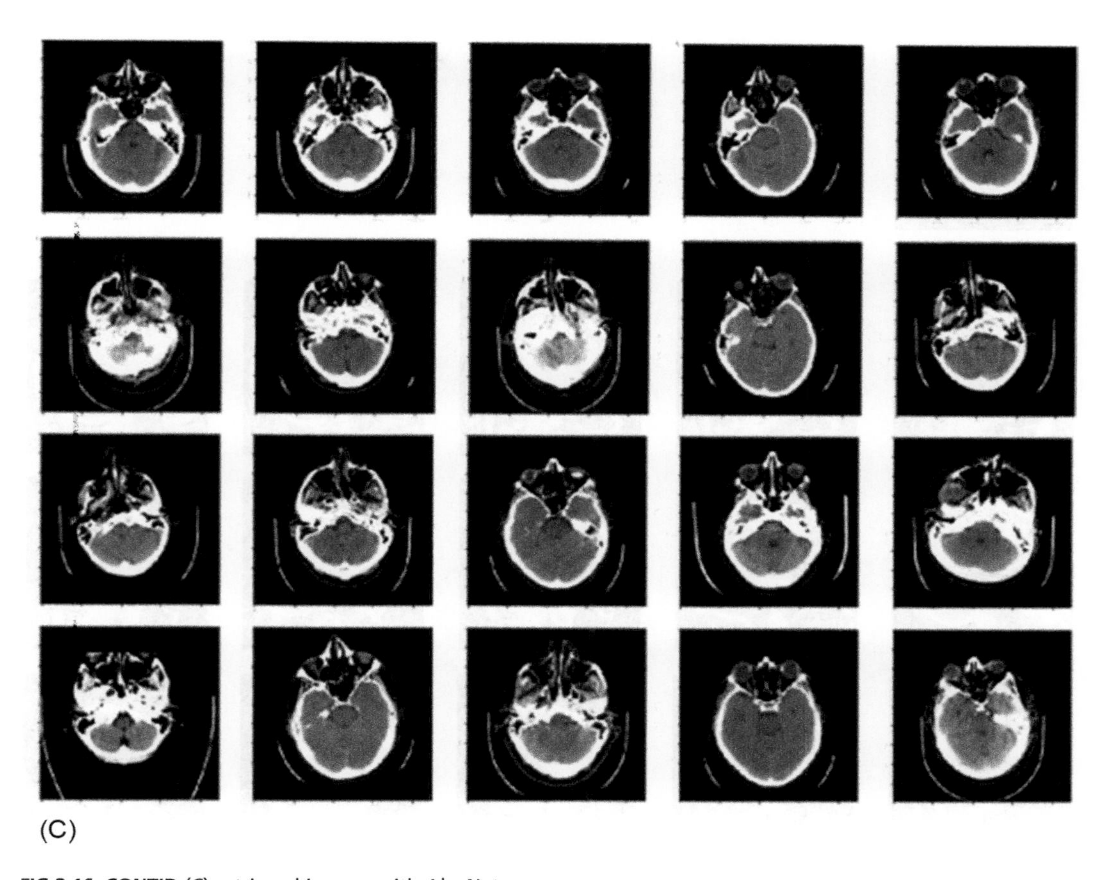

(C)

FIG.3.16, CONT'D (C) retrieved images with AlexNet.

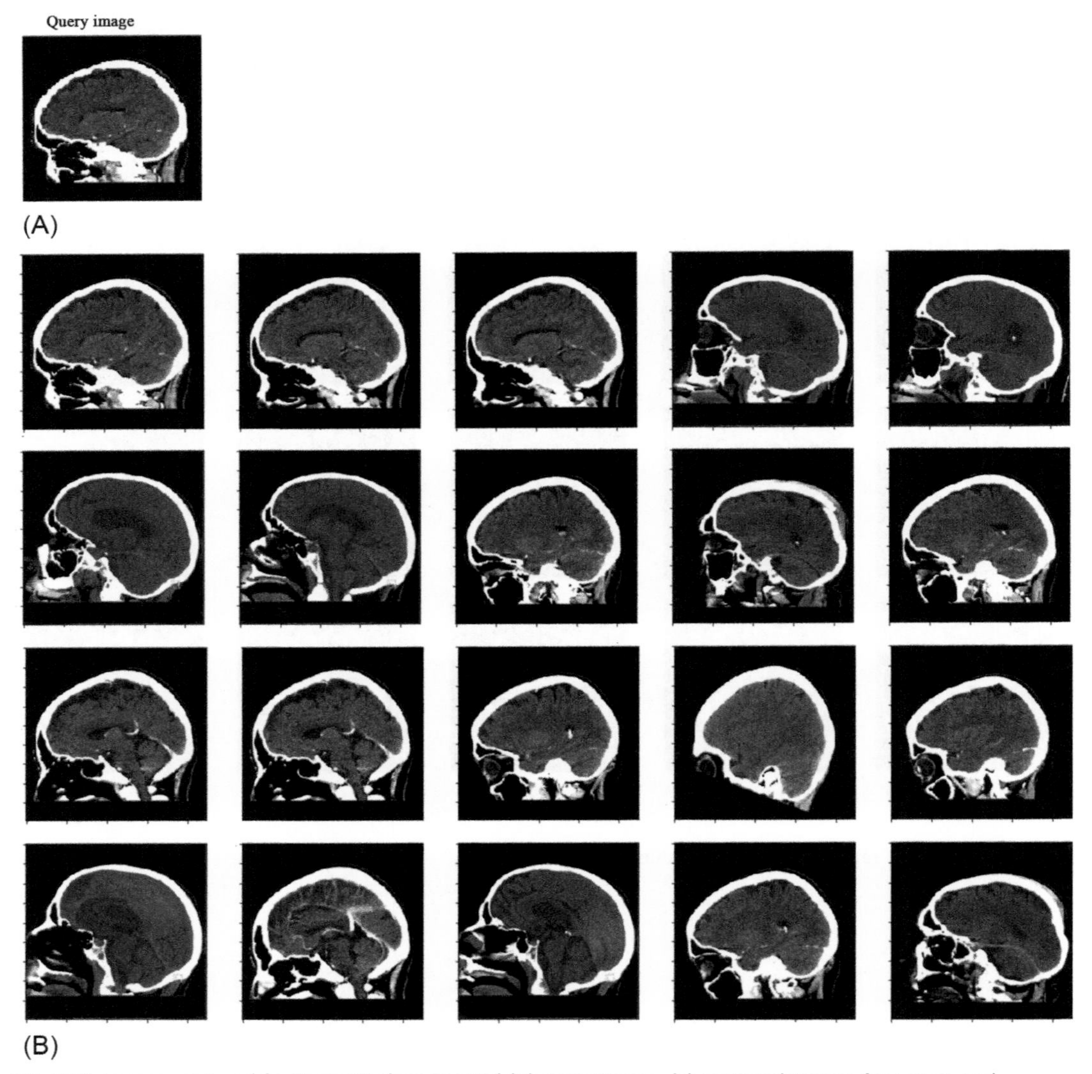

FIG. 3.17 Images retrieved for Brain CT class: Sagittal (A) query image, (B) retrieved images for LeNet, and

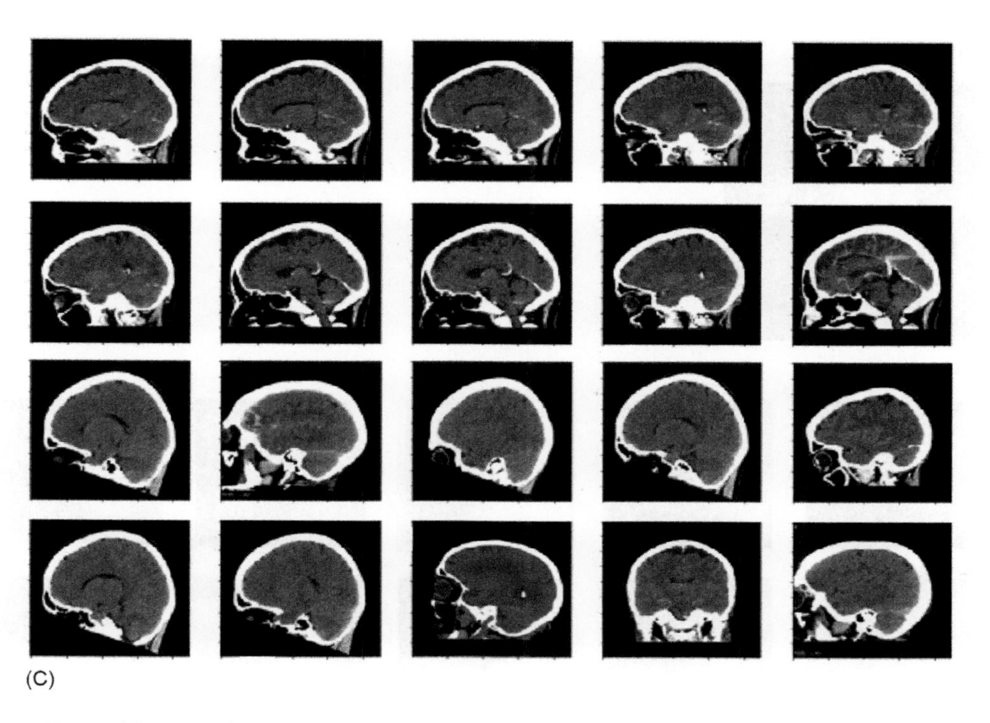

FIG.3.17, CONT'D (C) retrieved images for AlexNet.

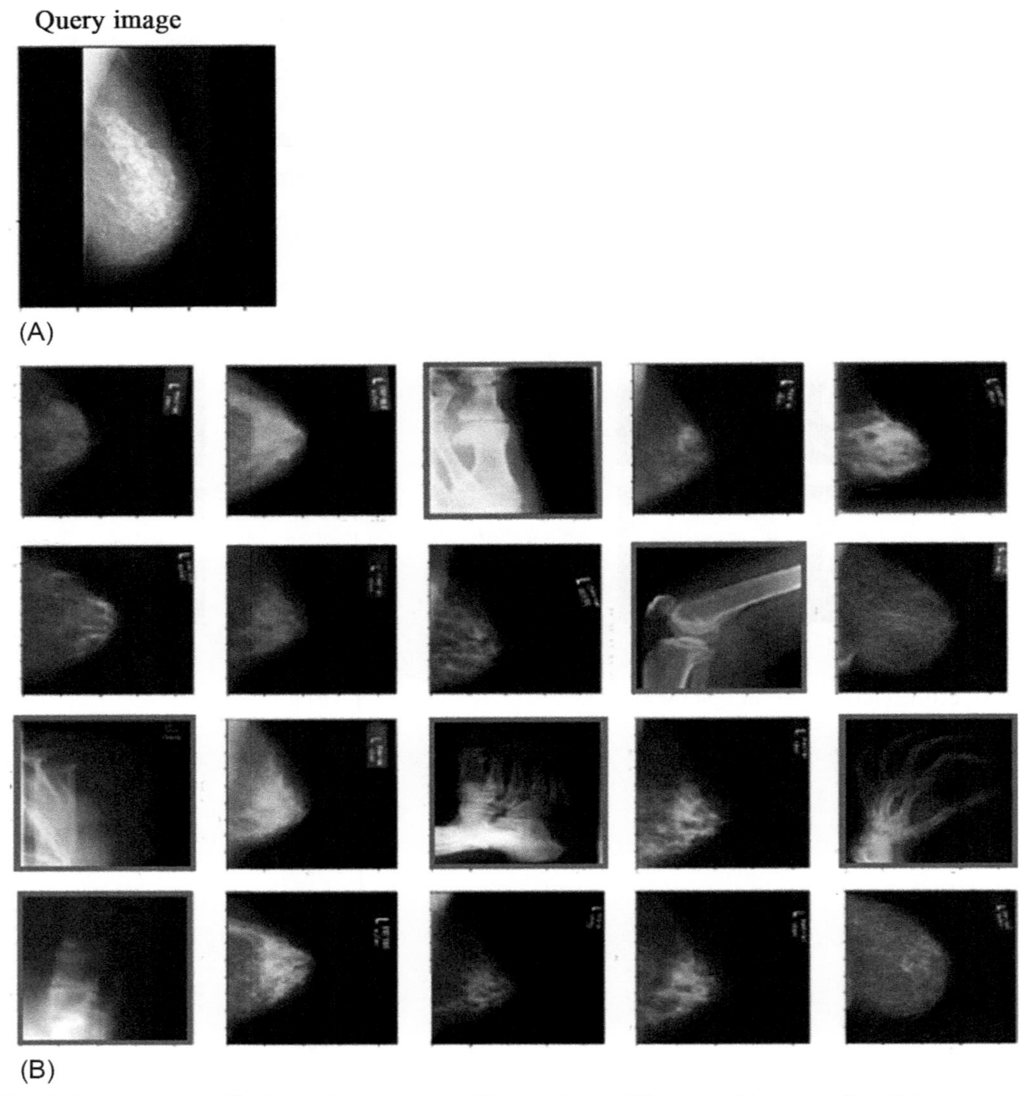

FIG. 3.18 Images retrieved for Breast Mammogram: (A) query image, (B) retrieved images with LeNet, and

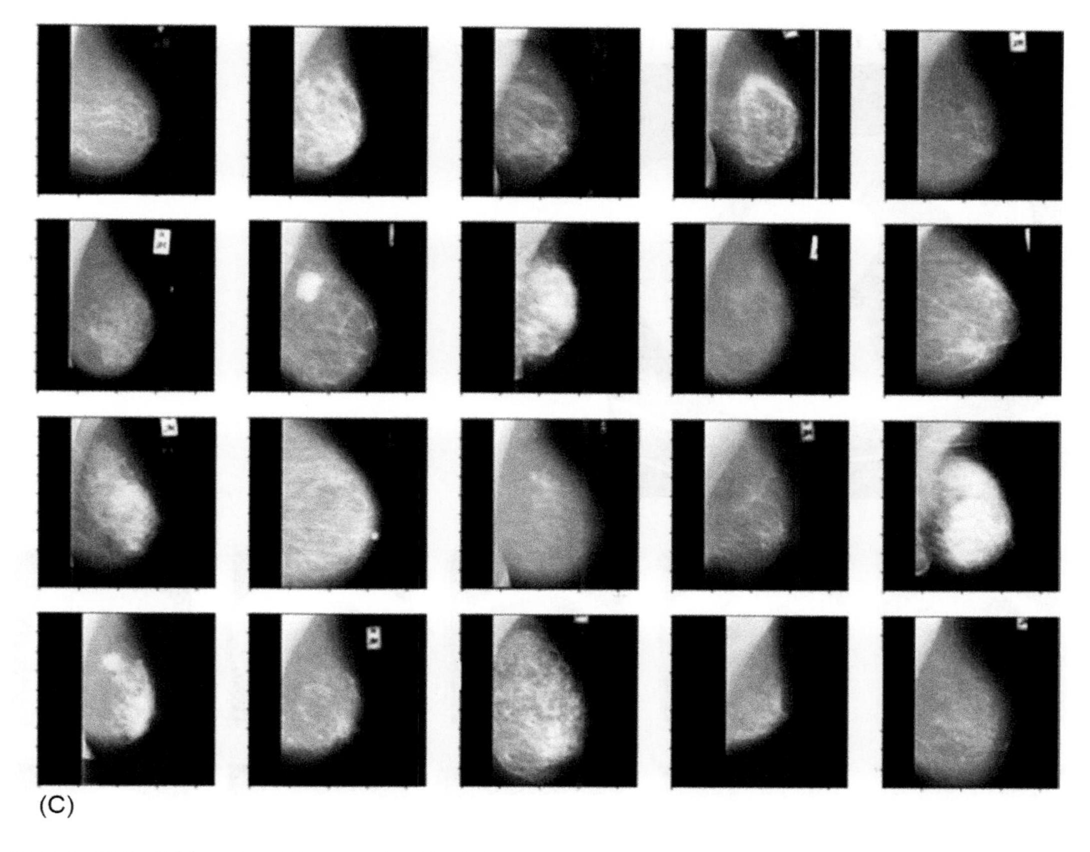

FIG.3.18, CONT'D (C) retrieved images with AlexNet.

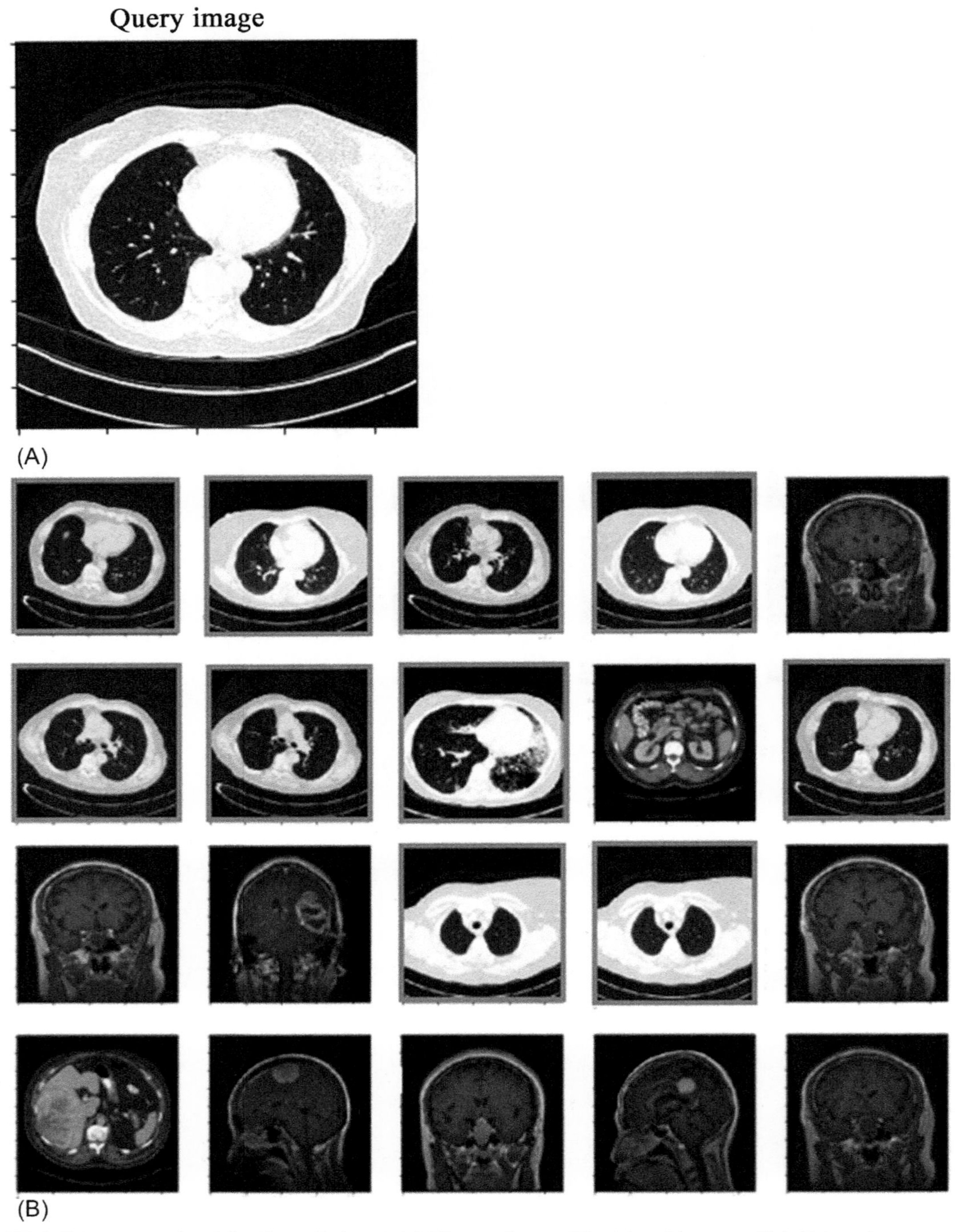

FIG. 3.19 Images retrieved for Chest CT class: Axial (A) query image, (B) retrieved images with LeNet, and

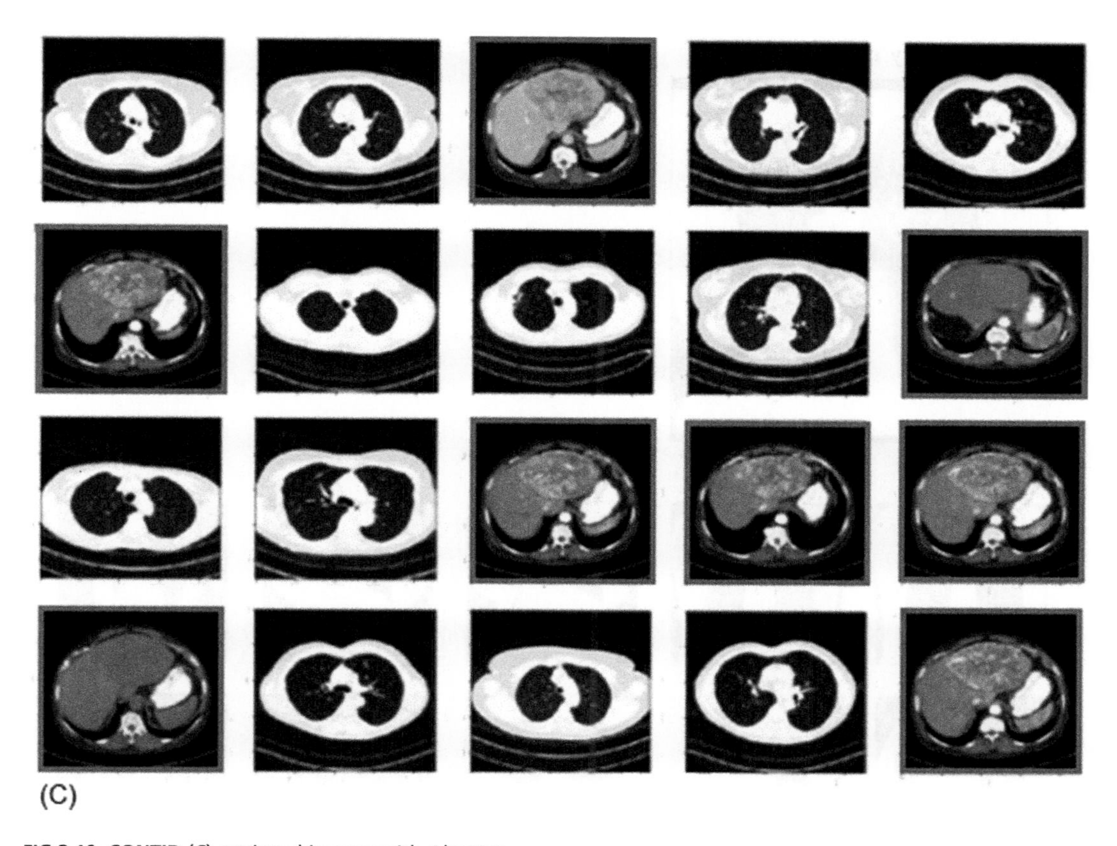

FIG.3.19, CONT'D (C) retrieved images with AlexNet.

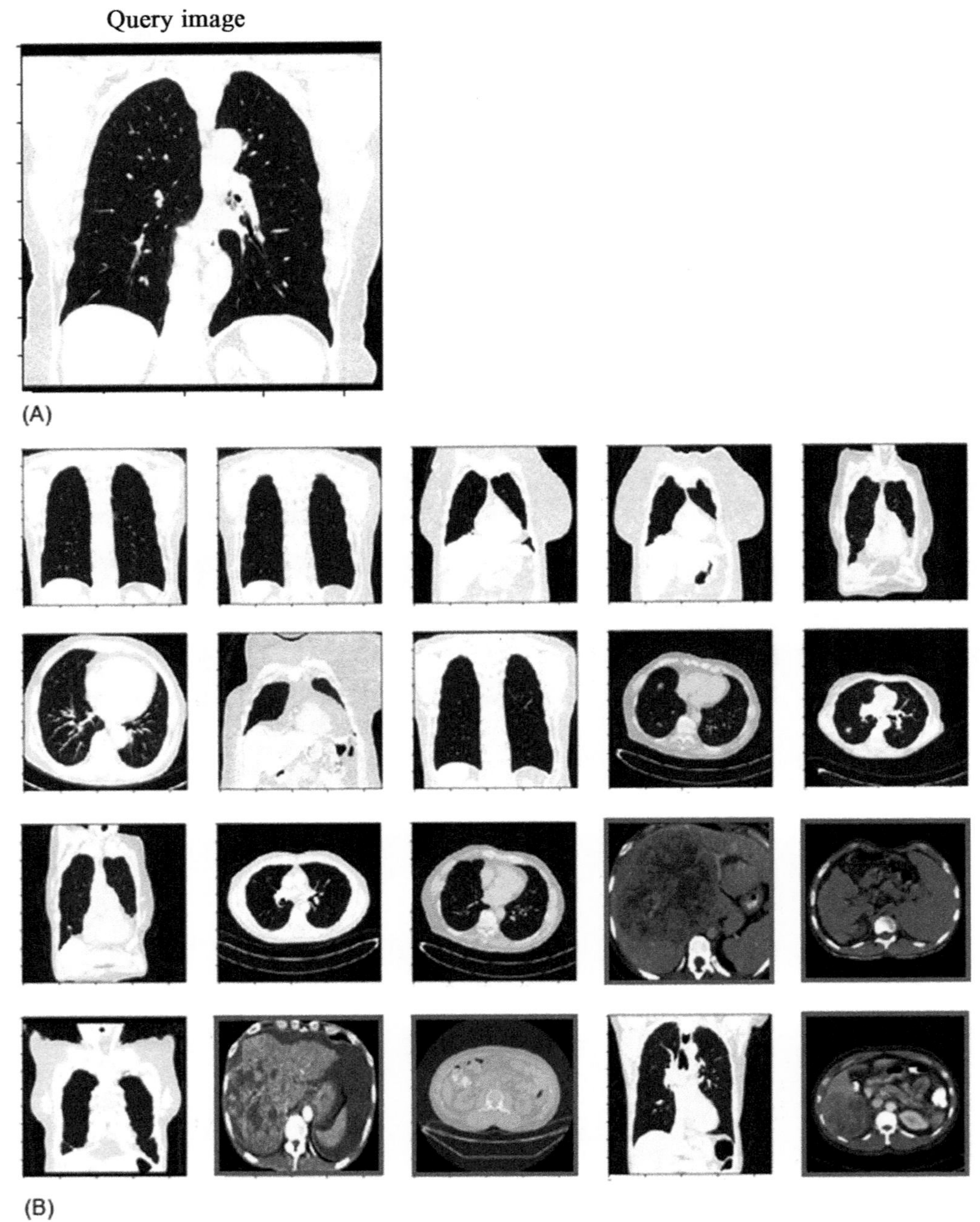

FIG. 3.20 Images retrieved for Chest CT class: Coronal (A) query image, (B) retrieved images with LeNet, and

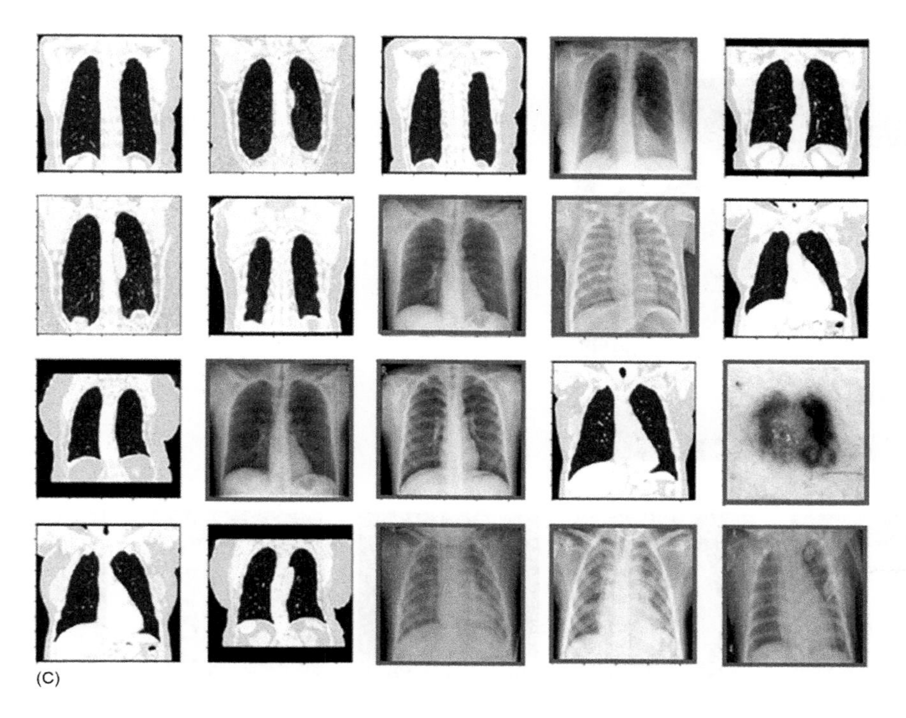

FIG.3.20, CONT'D (C) retrieved images with AlexNet.

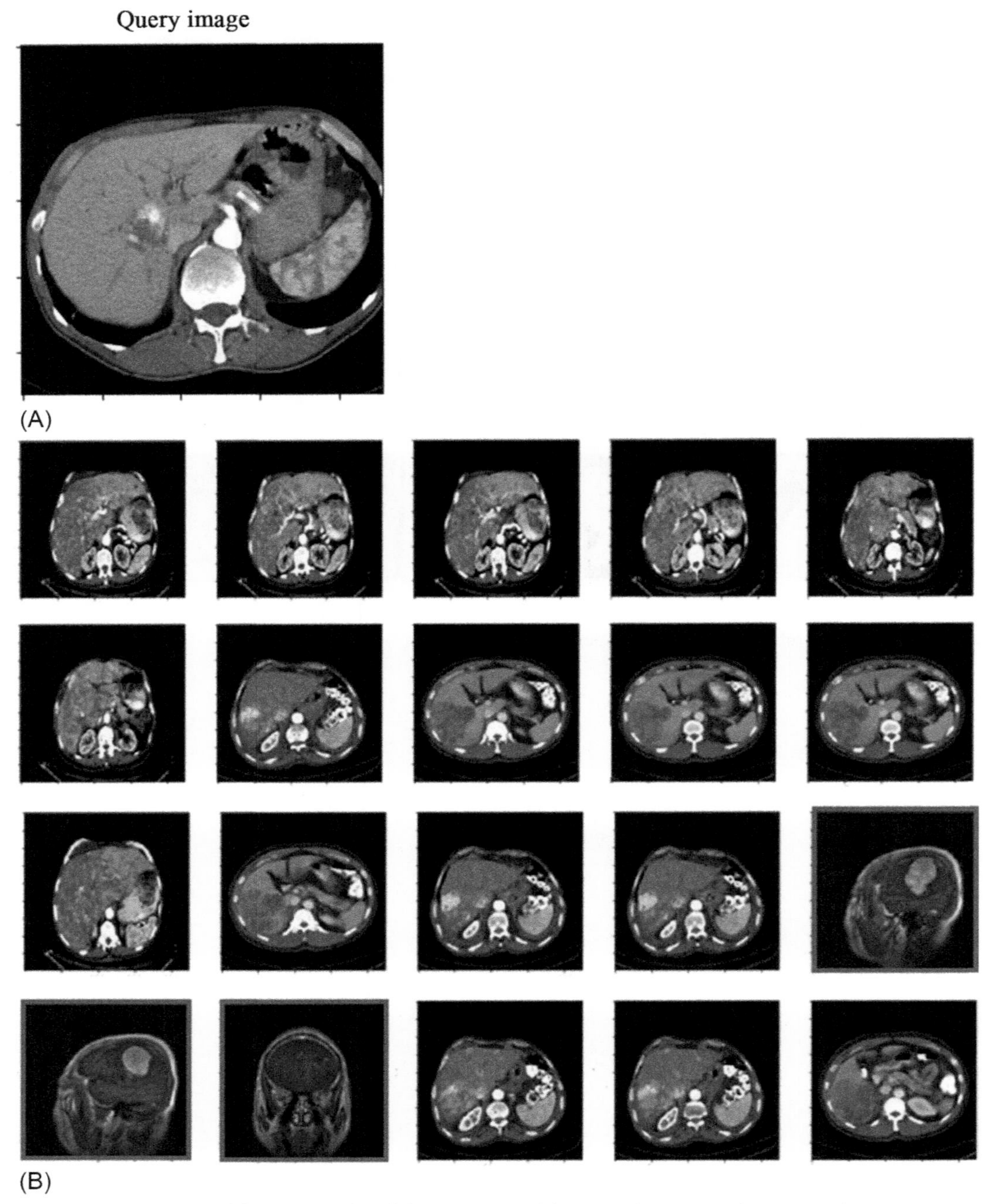

FIG. 3.21 Images retrieved for Liver CT class: (A) query image, (B) retrieved images with LeNet, and

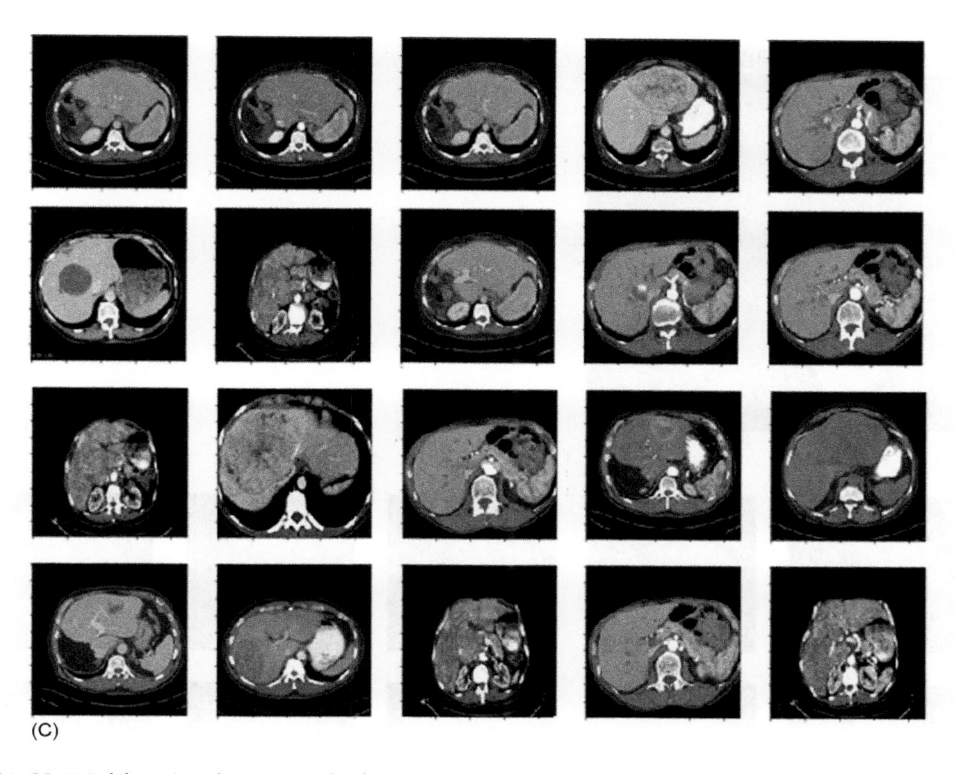

FIG.3.21, CONT'D (C) retrieved images with AlexNet.

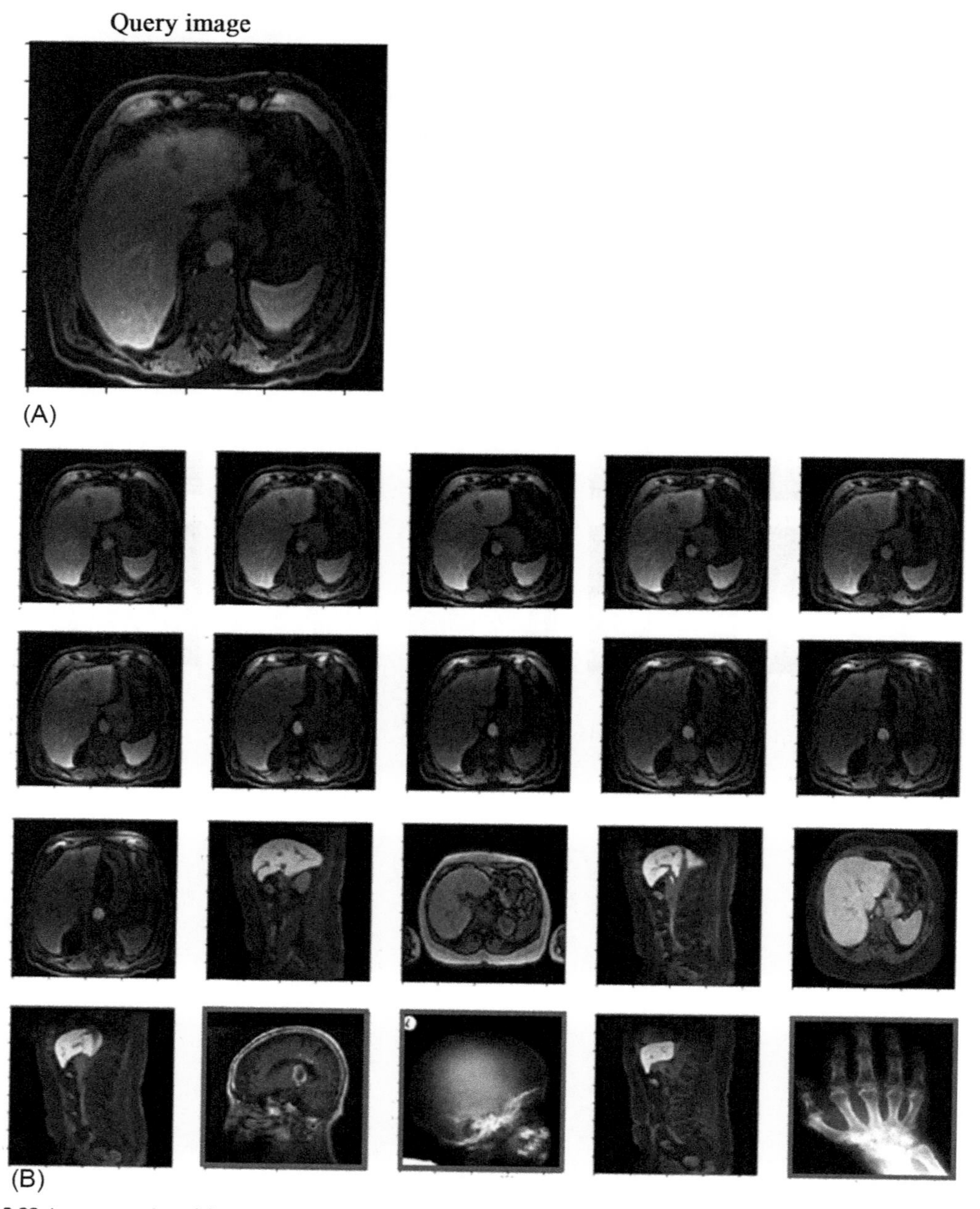

FIG. 3.22 Images retrieved for Liver MRI: (A) query image, (B) retrieved images with LeNet, and

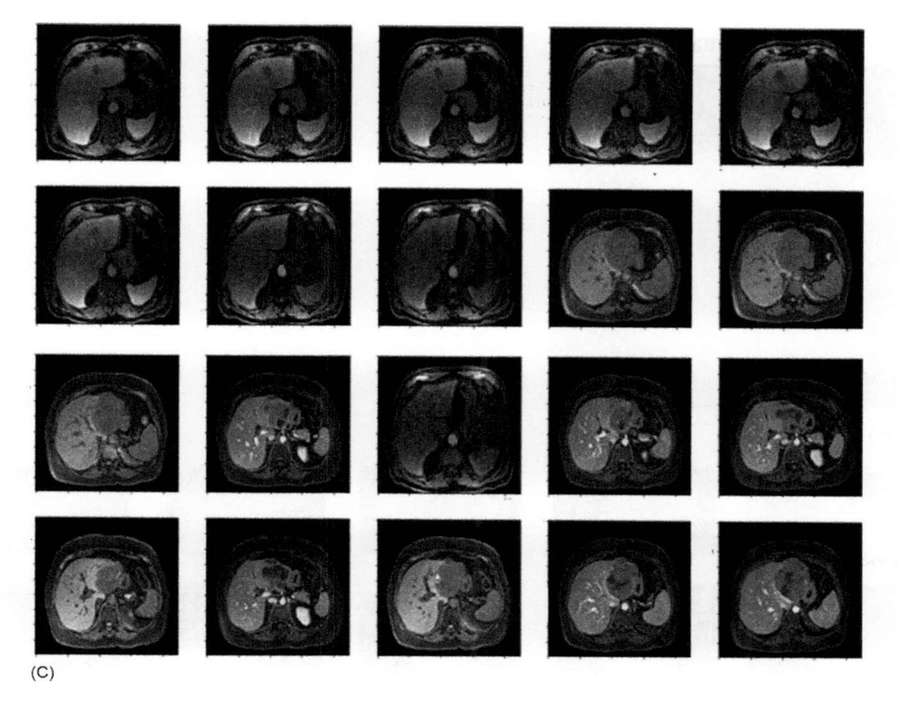

FIG.3.22, CONT'D (C) retrieved images with AlexNet.

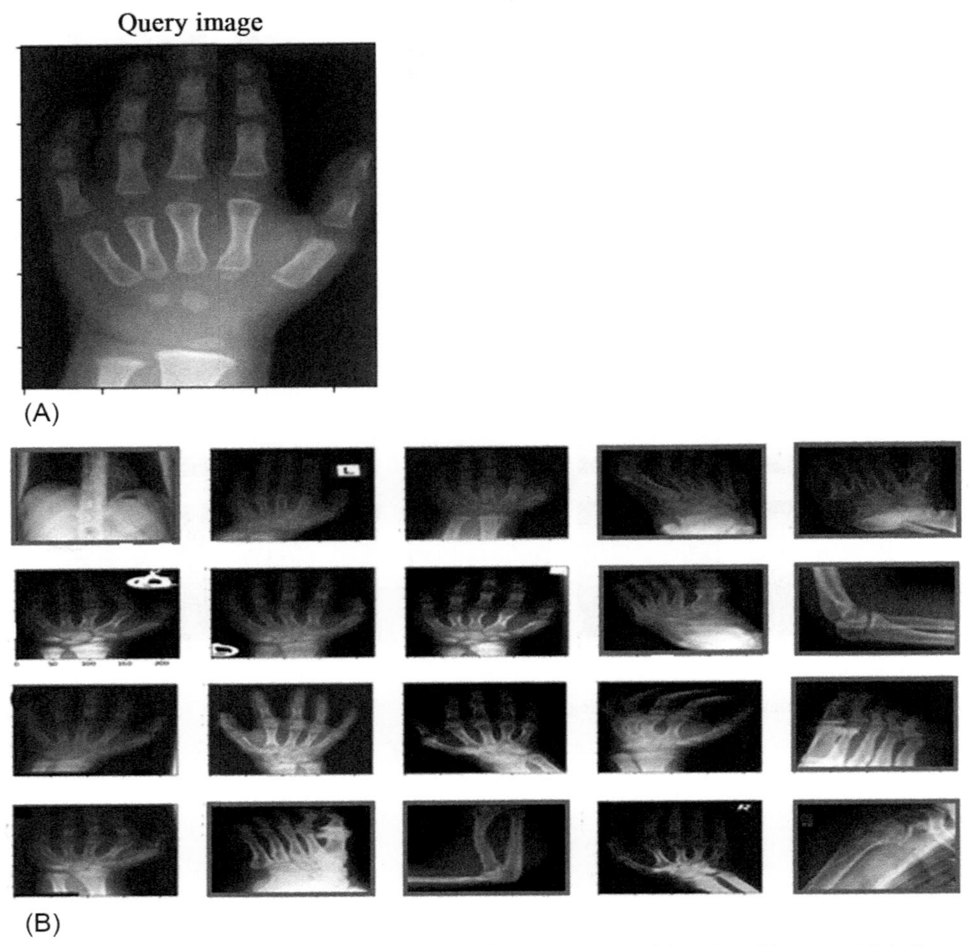

FIG. 3.23 Images retrieved for Multianatomy X-ray: Hand (A) query image, (B) retrieved images with LeNet, and

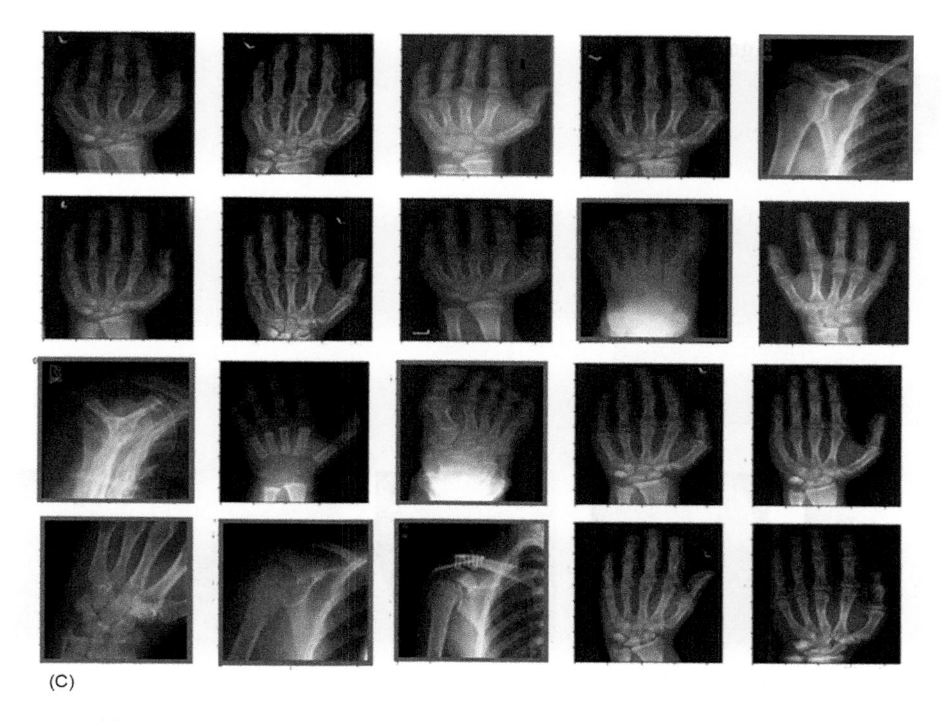

FIG.3.23, CONT'D (C) retrieved images with AlexNet.

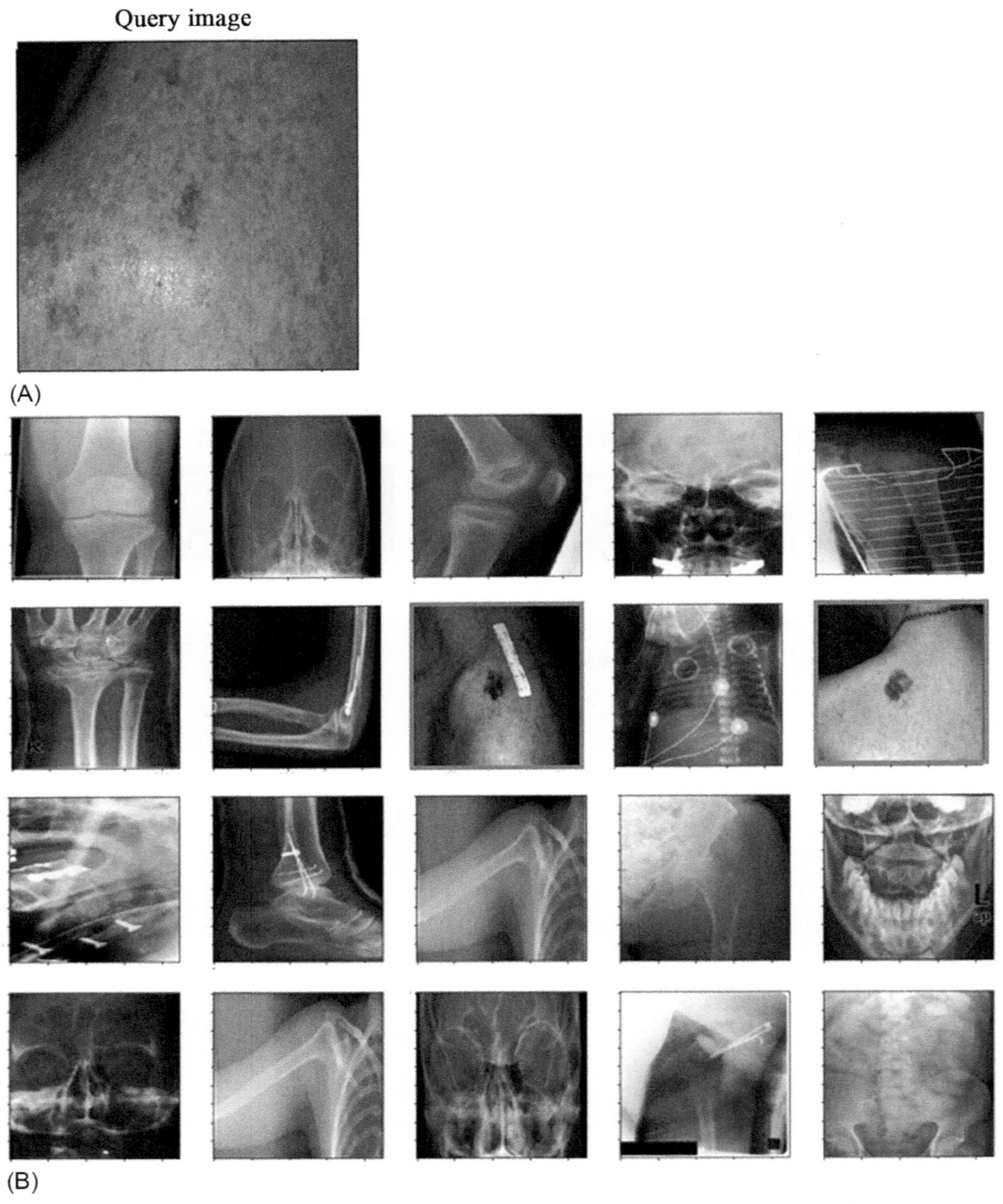

FIG. 3.24 Images retrieved for Skin Clinical: (A) query image, (B) retrieved images with LeNet, and

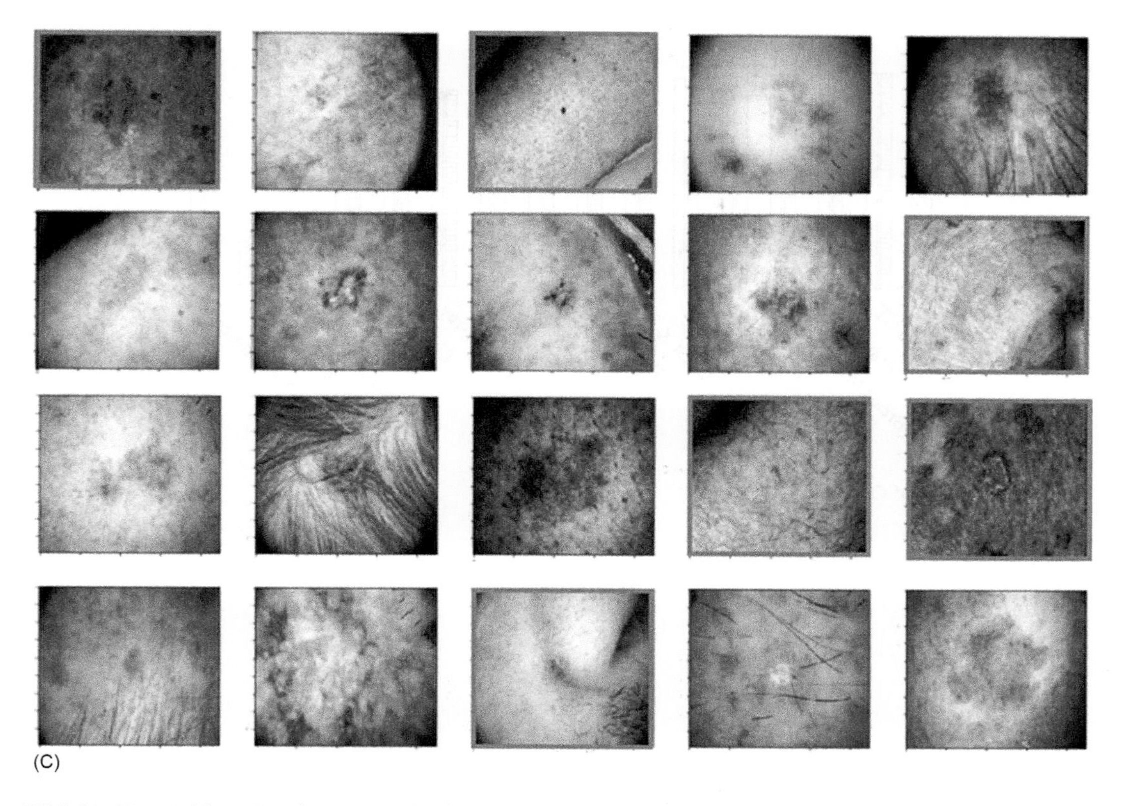

FIG.3.24, CONT'D (C) retrieved images with AlexNet.

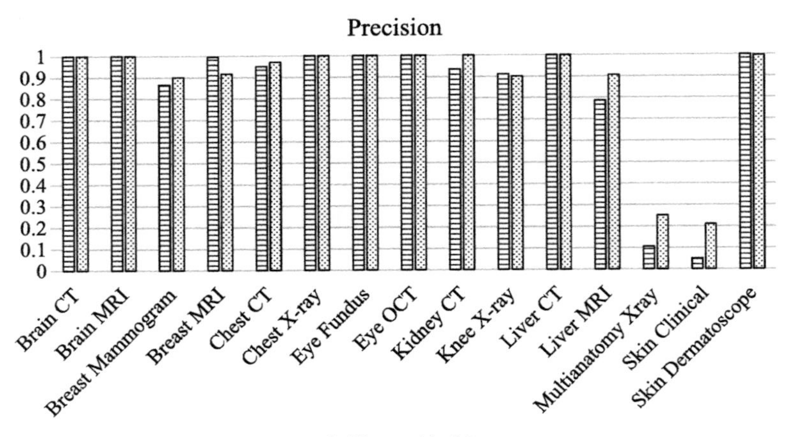

FIG. 3.25 Precision of CBMIR for 15 classes.

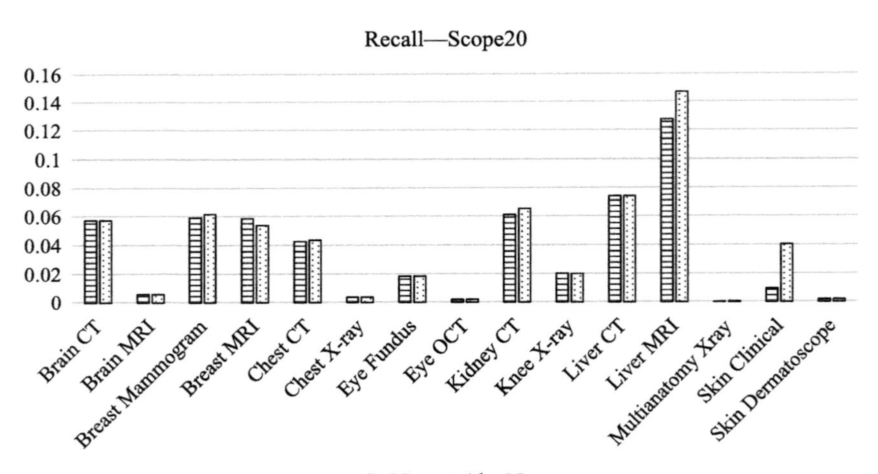

FIG. 3.26 Recall of CBMIR for 15 classes.

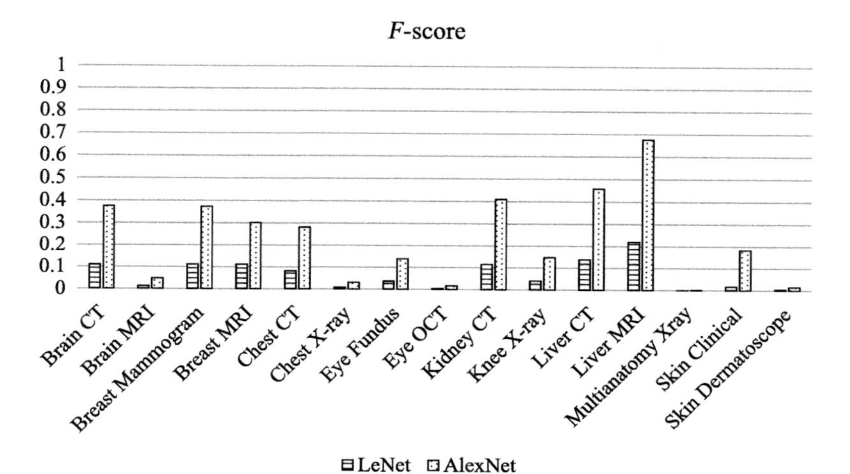

FIG. 3.27 F score of CBMIR for 15 classes.

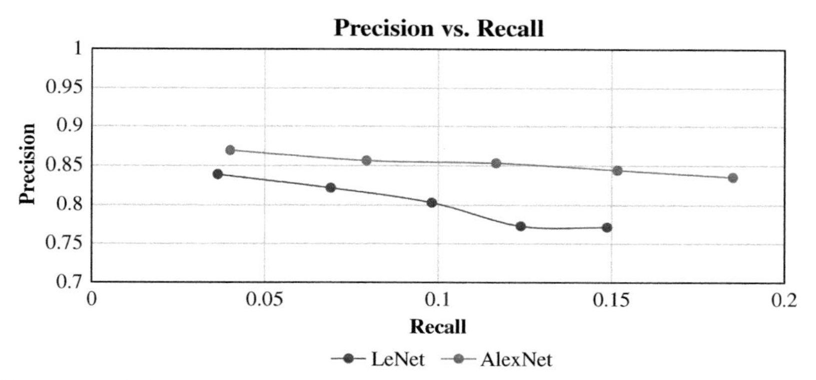

FIG. 3.28 Precision vs. Recall for image retrieval.

Table 3.8 Precision table for different scope sizes for LeNet and AlexNet.

	LeNet					AlexNet				
Types of images	**Scope 20**	**Scope 40**	**Scope 60**	**Scope 80**	**Scope 100**	**Scope 20**	**Scope 40**	**Scope 60**	**Scope 80**	**Scope 100**
Brain CT	1.00	1.00	1.00	0.97	0.97	1.00	1.00	1.00	1.00	1.00
Brain MRI	1.00	1.00	1.00	1.00	1.00	1.00	1.00	1.00	1.00	1.00
Breast Mammogram	0.87	0.84	0.78	0.72	0.72	0.90	0.88	0.88	0.87	0.85
Breast MRI	1.00	1.00	0.99	0.97	0.97	0.92	0.81	0.81	0.79	0.78
Chest CT	0.95	0.91	0.87	0.79	0.79	0.97	0.93	0.93	0.92	0.90
Chest X-ray	1.00	1.00	1.00	1.00	1.00	1.00	1.00	1.00	1.00	1.00
Eye Fundus	1.00	1.00	1.00	1.00	1.00	1.00	1.00	1.00	1.00	1.00
Eye OCT	1.00	1.00	1.00	1.00	1.00	1.00	1.00	1.00	1.00	1.00
Kidney CT	0.94	0.92	0.88	0.85	0.85	1.00	1.00	1.00	0.99	0.99
Knee X-ray	0.91	0.89	0.88	0.85	0.85	0.90	0.90	0.90	0.89	0.89
Liver CT	1.00	0.98	0.97	0.96	0.96	1.00	1.00	1.00	1.00	1.00
Liver MRI	0.79	0.68	0.57	0.40	0.40	0.91	0.90	0.90	0.86	0.81
Multianatomy X-ray	0.11	0.10	0.08	0.07	0.05	0.25	0.24	0.23	0.22	0.20
Skin Clinical	0.05	0.03	0.03	0.03	0.03	0.21	0.22	0.22	0.21	0.18
Skin Dermatoscope	1.00	1.00	1.00	1.00	1.00	1.00	1.00	1.00	1.00	1.00
MAP	**0.84**	**0.82**	**0.80**	**0.77**	**0.77**	**0.87**	**0.86**	**0.86**	**0.85**	**0.84**

Table 3.9 Comparison of AlexNet and LeNet.

	AlexNet	LeNet
Training time	6931 s	407.2 s
Classification accuracy	99.29%	98.82%
Features extracted	1000	84
Retrieval time	~0.27 s	~0.21 s
Mean average precision	86.9%	83.9%

3.5 Summary and conclusion

State-of-the-art CNN architectures LeNet and AlexNet were implemented for a large medical dataset of 49,473 images from seven different imaging modalities, anatomical information, and orientation for the CBIR task. The number of epochs for training was selected to be 100 as a higher number of epochs make the model to overfit the data and unable to generalize for new image data. Different optimizers SGD, AdaGrad, and Adam for back-propagation algorithm to minimize the error function were experimented, and Adam was found to give high validation accuracy. LeNet architecture was time-saving in terms of training but lacked the ability to extract valuable features as the resolution of images were shrunken to 32×32 size. AlexNet consumed more time in training as the number of layers were higher, increasing the computational cost and time complexity. The number of features extracted was higher for AlexNet, which accounted for the increased retrieval time as shown in Table 3.9.

AlexNet can handle a complex image dataset of higher size and resolution, and provide greater classification accuracy. This work was a direct adaptation of well-established CNN architectures for content-based medical image retrieval. Further experiments by tuning of hyperparameters and the number of convolutional layers could improve the MAP for a large multimodal medical image database.

References

[1] The digital universe of opportunities: rich data and the increasing value of the internet of things, in: EMC Digital Universe with Research and Analysis by IDC, 2014. Available at: https://www.emc.com/analyst-report/digital-universe-healthcare-vertical-report-ar.pdf. (Accessed 12 November 2018).

[2] M. Alkhawlani, M. Elmogy, H. El Bakry, Text-based, content-based, and semantic-based image retrievals: a survey, Int. J. Comput. Inform. Technol. 4 (2015) 58–66.

[3] A. Kumar, J. Kim, W. Cai, M. Fulham, D. Feng, Content-based medical image retrieval: a survey of applications to multidimensional and multimodality data, J. Digit. Imaging 26 (6) (2013) 1025–1039.

[4] A. Qayyum, S.M. Anwar, M. Awais, M. Majid, Medical image retrieval using deep convolutional neural network, Neurocomputing 266 (2017) 8–20.

[5] V. Jeyakumar, B. Kanagaraj, A medical image retrieval system in PACS environment for clinical decision making, in: Intelligent Data Analysis for Biomedical Applications Challenges and Solutions Intelligent Data-Centric Systems, 2019, pp. 121–146.

[6] A. Farooq, S.M. Anwar, M. Awais, S. Rehman, A deep CNN based multi-class classification of Alzheimer's disease using MRI, in: 2017 IEEE International Conference on Imaging Systems and Techniques (IST), 2017, pp. 1–6.

[7] S. Khan, S.-P. Yong, A deep learning architecture for classifying medical images of anatomy object, in: Proceedings of APSIPA Annual Summit and Conference, 2017, pp. 1661–1668.

[8] G. Patry, G. Gauthier, B. Lay, J. Roger, D. Elie, Software products and custom development in the fields of image processing and analysis, in: ADCIS, 2019. Available at: http://www.adcis.net/en/Download-Third-Party/Messidor.html. (Accessed 13 November 2018).

[9] J.E.S. Sklan, A.J. Plassard, D. Fabbri, B.A. Landman, Toward content-based image retrieval with deep convolutional neural networks, in: Medical Imaging 2015: Biomedical Applications in Molecular, Structural, and Functional Imaging, vol. 9417, International Society for Optics and Photonics, 2015, p. 94172C.

[10] L. Deng, D. Yu, Deep learning, Signal Process. 7 (2014) 3–4.

[11] CS231n Convolutional Neural Networks for Visual Recognition, Available at: http://cs231n.github.io/convolutional-networks/#conv. (Accessed 12 November 2018).

[12] N.K. Manaswi, Deep Learning With Applications Using Python, Springer Nature, 2018.

[13] Radiopaedia.org, Radiopaedia's Mission is to Create the Best Radiology Reference the World has Ever Seen and to Make it Available For Free, For Ever, For All, Available at: https://radiopaedia.org/. (Accessed 12 November 2018).

[14] C.R. Jack, The Alzheimer's disease neuroimaging initiative (ADNI): MRI methods, J. Magn. Reson. Imaging 27 (4) (2008) 685–691.

[15] J. Cheng, Brain Tumor Dataset. FigShare, 2017. Available at: https://figshare.com/articles/brain_tumor_dataset/1512427/5. https://doi.org/10.6084/m9.figshare.1512427.v5.

[16] J. Suckling, J. Parker, D. Dance, S. Astley, I. Hutt, C. Boggis, I. Ricketts, et al., Mammographic Image Analysis Society (MIAS) Database v1.21 [Dataset], https://www.repository.cam.ac.uk/handle/1810/250394, 2015.

[17] The Cancer Imaging Archive, http://cancerimagingarchive.net/, 2011.

[18] D. Kermany, K. Zhang, M. Goldbaum, Labeled optical coherence tomography (OCT) and chest X-Ray images for classification, Mendeley Data 2 (2018), https://doi.org/10.17632/rscbjbr9sj.2.

[19] D. Vargus, T. Szirányi, Fast content-based image retrieval using convolutional neural network and hash function, in: Systems, Man, and Cybernetics (SMC), IEEE, 2016, pp. 002636–002640.

[20] K. Nagaraj, V. Jeyakumar, A study on comparative analysis of automated and semiautomated segmentation techniques on knee osteoarthritis X-ray radiographs, in: 2018 International Conference on ISMAC in Computational Vision and Bio-Engineering (ISMAC-IoT, Social, Mobile, Analytics and cloud) (ISMAC—CVB), 2018, pp. 1606–1616.

[21] P. Tschandl, C. Rosendahl, H. Kittler, The HAM10000 dataset, a large collection of multi-source dermatoscopic images of common pigmented skin lesions, Sci. Data 5 (2018) 180161, https://doi.org/10.1038/sdata.2018.161.

[22] ImageCLEF/LifeCLEF, ImageCLEF 2009 medical retrieval task, in: ImageCLEF/LifeCLEF—Multimedia Retrieval in CLEF, 2009. Available at: https://www.imageclef.org/2009/medical. (Accessed 13 November 2018).

[23] International Skin Imaging Collaboration: Melanoma Project Website, https://isic-archive.coml, 2019. (Accessed 12 November 2018).

[24] J. Wan, Deep learning for content-based image retrieval: a comprehensive study, in: Proceedings of the 22nd ACM international conference on Multimedia, 2014, pp. 157–166.

[25] A. Krizhevsky, I. Sutskever, G.E. Hinton, ImageNet classification with deep convolutional neural networks, in: Proceeding NIPS'12 Proceedings of the 25th International Conference on Neural Information Processing Systems, vol. 1, 2012, pp. 1097–1105.

[26] B. Stephen, L. Vandenberghe, Convex Optimization, Cambridge University Press, 2004, pp. 463–466.

[27] D. John, E. Hazan, Y. Singer, Adaptive subgradient methods for online learning and stochastic optimization, J. Mach. Learn. Res. (12) (Jul 2011) 2121–2159.

[28] Kingma, Diederik, Ba, Jimmy. (2014). Adam: a method for stochastic optimization. International Conference on Learning Representations. Ar Xiv:1412.6980. (2014).

[29] M. Chowdhury, S.R. Bulo, R. Moreno, M.K. Kundu, O. Smedby, An efficient radiographic image retrieval system using convolutional neural network, in: 23rd International Conference on Pattern Recognition (ICPR), 2016, pp. 3134–3139.

Further reading

Y. LeCun, L. Bottou, Y. Bengio, Reading checks with graph transformer networks, in: International Conference on Acoustics, Speech, and Signal Processing, Munich, vol. 1, 1997, pp. 151–154.

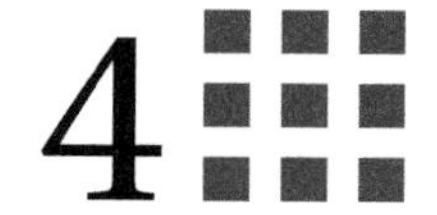

A systematic approach for identification of tumor regions in the human brain through HARIS algorithm

P. Naga Srinivasu[a], T. Srinivasa Rao[a], Valentina Emilia Balas[b]

[a]DEPARTMENT OF CSE, GIT, GITAM DEEMED TO BE UNIVERSITY, VISAKHAPATNAM, INDIA
[b]DEPARTMENT OF AUTOMATION AND APPLIED INFORMATICS, AUREL VLAICU UNIVERSITY OF ARAD, ARAD, ROMANIA

4.1 Introduction

In recent years, there has been significant improvement in the field of medical imaging, and the procedures of medical treatment and diagnosis have been revamped. In this context, image analyzis has become one of the remarkable stages in diagnosis for identification of abnormalities through computed tomography (CT), magnetic resonance imaging (MRI), and positron emission tomography (PET) scans. An image is analyzed and scrutinized by recognizing various objects in the image through segmentation that helps in the planning of surgery or nursing.

Medical image segmentation is done with numerous approaches, and all output we could obtain through a majority of approaches were almost identical. A majority of the approaches differed in the computational efforts, addressing the partial volume effects and anatomic variability and complexity. MRI segmentation through genetic algorithm proposed by Saha and Bandyopadhyay [1], Naga Srinivasu et al. [2], and Dubey [3] had a limitation that the segmentation is done based on distance measures, which may not be an appropriate assignment of pixels, as it never guarantees to produce the best optimal solution. Sulaiman et al. [4] and Sudha et al. [5] have suggested an approach based on fuzzy C means algorithm for the segmentation of images, but it has limitations such as the inception number of segments and the membership estimation for each element for all the segments is a tedious task that needs an exceptionally high computational exercise.

Hiralal and Menon [6] and Deng [7] in their respective articles have elaborated on a semiautomated approach based on region growing for the segmentation of an MRI image,

Deep Learning Techniques for Biomedical and Health Informatics. https://doi.org/10.1016/B978-0-12-819061-6.00004-5

which has limitations over the consideration of the initial seed point selection. If the points are incorrectly chosen, their algorithms end up with a misinterpreted conclusion. Song [8] has proposed an approach through a graph cuts-based method that segments the image based on the Eigenvectors with respect to the populated similarity matrix. In some cases, the afore mentioned approach ends up in the wrong assignment of pixels.

Chandra et al. [9] proposed a particle swam optimization-based MR image segmentation in their article, which is comparatively simple and robust to implement when compared to traditional genetic algorithms. But the main limitation of the approach is that sometimes the pixels may not be assigned to the appropriate cluster due to a local maxima issue that would result in an inappropriate segmentation. Liu et al. [10] suggested an approach based on a support vector machine capable of handling high dimensional data and unstructured data as well. However, it is limited due to the need of training the machine, which engrosses additional time and space. Support vector machine-based approach of segmentation would be difficult to interpret and analyze the outcome because it works effectively for two class problems, but it would be exigent for a multiclass problem. Moreover, the resultant image is susceptible to the underlying training set. Nandi [11] stated that by using morphological operations, the image is segmented, but it suffers from an initial cluster count and it rarely ends up with an optimal number of segments.

Wang et al. [12], Ramakuri et al. [13], and Işın et al. [14, 15] have proposed convolutional neural networks (CNNs) and deep learning-based MRI image segmentation that work with the limitations of training the algorithm; it looks for the elements present in the close vicinity of the network, and it doesn't care about the pixels in far proximity within the image. In some cases, when CNNs would need a tremendous computational effort for adjusting the hyperparameters that assist in feature selection, the afore mentioned approach for the segmentation needs to rigorously train the algorithm for a better outcome. Yu et al. [16] proposed an active contour model that had a limitation of choosing and defining the parameters, which was a demanding task to customize it to a problem-specific approach; however, it didn't apply for an image with a large search space. Apart from the parameter selection, the afore mentioned approach needs more execution time, and there are chances of striking the local minima for optimal use with segments that hold the closest objects.

This chapter is arranged as follows: The first parts are the introduction to the problem and existing approaches to address the problem. Next is the intent of the chapter, and the third section is about image preprocessing to eliminate Gaussian noise from the original image. In the fourth section of the chapter, the HARIS algorithm is elaborated with the objective functions, and finally in the fifth section, experimental results and performance analysis is presented followed by a conclusion and the references.

4.2 The intent of this chapter

The prime objective of this chapter is to bring an efficient approach for the segmentation of MR images by which the practice of identification of damaged tissues could be easily recognized. In earlier mechanisms, the image is segmented through semiautomated approaches like k means, fuzzy C means, morphological operations, and distance-based approaches, or else through automated approaches like genetic algorithm, deep learning, and neural networks followed by an optimization technique for better results. In spite of highly accurate results, by applying two mechanisms simultaneously, the computational effort and the execution time would be very high. To address the issue of high computational effort, we have proposed an approach named HARIS, which we observe would be computationally efficient and technically feasible as implemented through multiobjective functions for segmenting an image with the best optimal number of pixels that would elaborate every minute region in the human brain. It is also observed that the outcomes of the HARIS approach are relatively accurate when examined over its counterparts with minimal computational efforts.

4.3 Image enhancement and preprocessing

Basically, at the time of rendering the MR image, there is a chance the image will be susceptible to noise due to unforeseen latent factors like inappropriate calibration of the equipment or unsuitable shimmer or glitter of the light while capturing. When such images are directly fed to the segmentation algorithm, the resultant outcome may not be optimal due to the presence of noise. Hence the acquired MR image must be preprocessed to address the issue of noise. Generally, the MR images are prone to noises like Gaussian noise and speckle noise, which manifest a notable bump in the quality of the image by affecting pixel intensities. In our work, we proposed an adaptive contourlet transform technique for handling all such noises.

4.3.1 Adaptive contourlet transform

The proposed approach, adaptive contourlet, is a positioning mixed resolution representation strategy that effectively presents the image in very smooth contours in various directions, by which it attains both anisotropy and magnitude as stated by Sivakumar et al. [17] in his article for image normalization. The proposed adaptive approach utilizes dual filter, that is, Laplacian pyramid filter followed by a directional filter. Both filters are superimposed over the original image.

The filters are going to perform selective enhancement of the pixels, or else they will vanquish the undesired wavelengths of the pixels. Laplacian pyramid is employed to attain the extensive decomposition that creates a normalized spectrum of a low-pass model of the original MR image and a model with a high-pass frequency of break points in the image, which gives a glimpse of the details in the image. The Laplacian pyramid is then used to evaluate the differences between the original MR image and the low-pass filtered image. In simple terms, the image is blurred to remove the considerably low frequency noises in the image, and the directional filter is then used to rebuild the image through wedge-shaped frequency partitioning. A direction filter is used to preserve

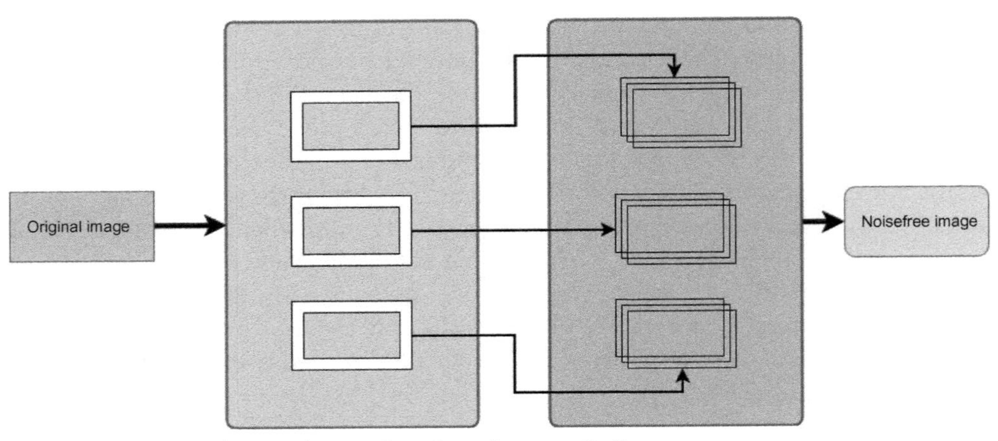

FIG. 4.1 Diagram of layered contourlet transform for noise normalization.

the edge-related information in the image by estimating the first order division of the pixels that lie within the kernel. However, the direction filter could be used in any direction (Fig. 4.1).

By employing the previously discussed filter for noise removal, the error is predicted through the following approach defined by the variable $LP(x,y)$:

$$LP(x,y) = GLP_I(x,y) - \left\{ 4 \sum_{m=-2}^{2} \sum_{n=-2}^{2} \omega(m,n) GLP_i\left(\frac{x-m}{2}, \frac{y-n}{2}\right) \right\} \tag{4.1}$$

From the this equation, the variable $GLP_I(x,y)$ is the Gaussian kernel associated with the coordinates (x,y). $\omega(m,n)$ represents the standard deviation between two coordinates m and n (Figs. 4.2 and 4.3).

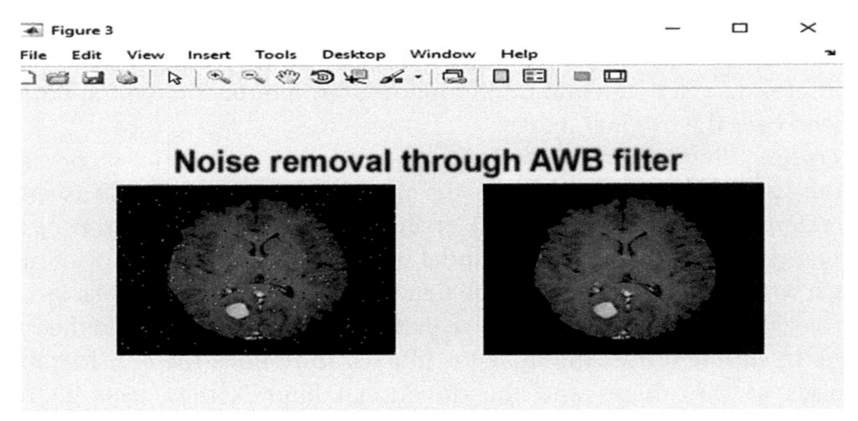

FIG. 4.2 Image on the left represents the original image, and the image on the right represents the noise.

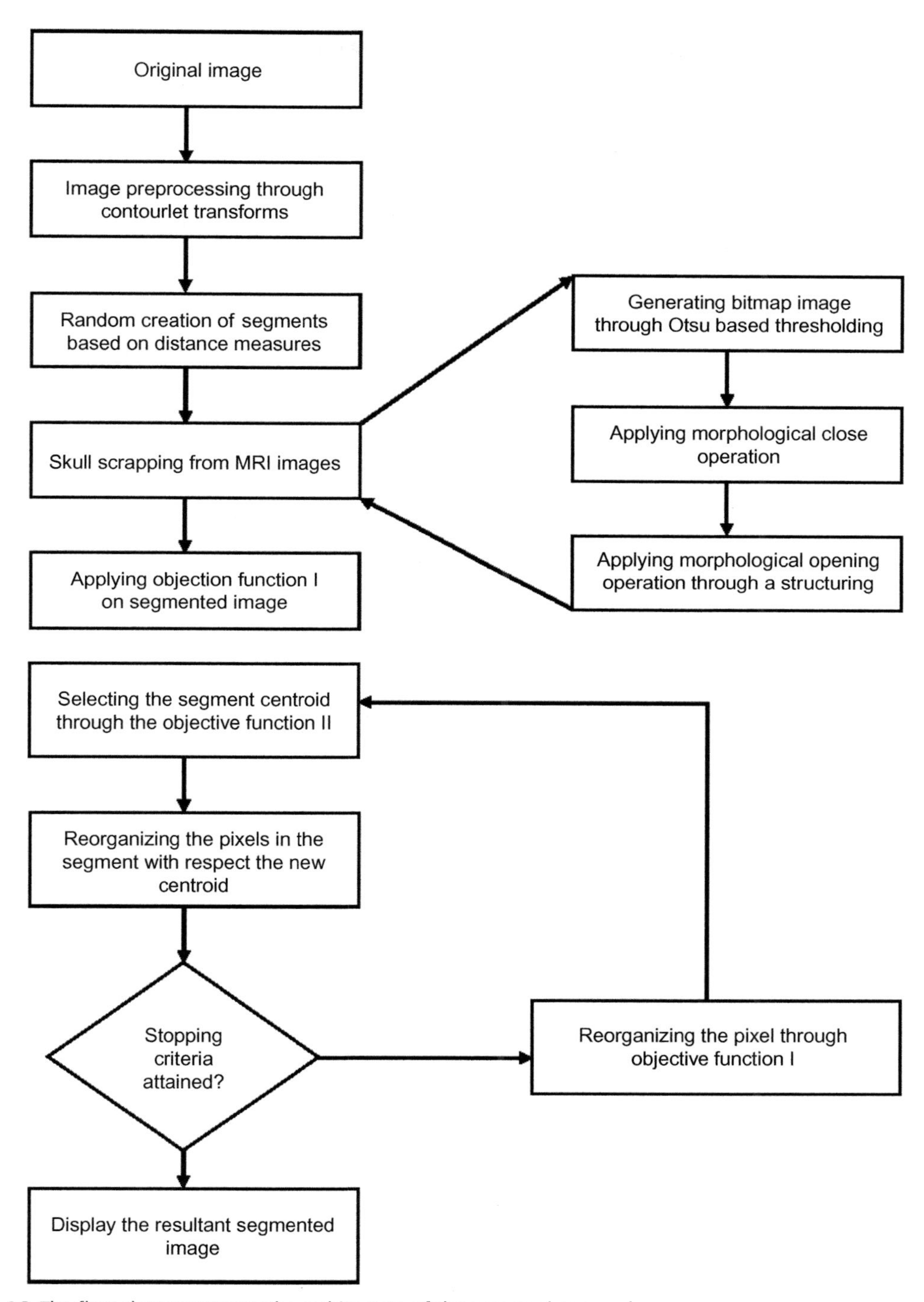

FIG. 4.3 The flow chart represents the architecture of the proposed approach.

4.4 Image preprocessing for skull removal through structural augmentation

In a brain MR image, there would be considerable regions of nonbrain structures like the skull, cerebrospinal fluid, and medulla oblongata that is to be ignored from the MR image for better analysis of the image. The image has to be preprocessed to remove the undesired regions well before it is fed to the segmentation algorithm. Structural augmentation is used for the eviction of the skull region from the T1-weighted MR images as stated by Naga Srinivasu et al. [18] in their article on efficient skull scraping approach for MRI. Because brain tissues appear darker and the bones appear brighter in the MR images, the thresholding of the image pixels is determined to evaluate the bitmap image, and then the morphological operations are performed to remove the skull region in the brain MR image.

Now let the greyscale image be I of size Z whose dimensions are assumed to be $m \times n$ having G_l levels of intensities concerning a neighborhood kernel of size $u \times v$. The size of the image Z is approximated from the total number of pixels in the image, which is assessed as $Z = m \times n$. The threshold of the greyscale is assessed to construct the bitmap image for the resultant greyscale image; the threshold is computed as follows

$$I_t = \frac{f\left(p_{greylevel}, mean_{greylevel}\right)}{Z} \tag{4.2}$$

From this equation, the variable I_t denotes the approximated threshold value, the variable $P_{greylevel}$ denotes the grey-level intensity of the particular pixel in the image, and the variable $mean_{greylevel}$ represents the mean of all the grey-level intensities of the pixels that are part of the kernel assumed as neighborhood (Fig. 4.4).

Based on the threshold value estimated earlier, the binary image I_{Bin} is populated from the greyscale image. The pixel intensities whose values are above the threshold are replaced by 1 in the binary image and 0 in place of the pixel whose intensity values are below the threshold value. Now on the resultant image, the mathematical morphological operations are performed, and the morphological close that performs erosion

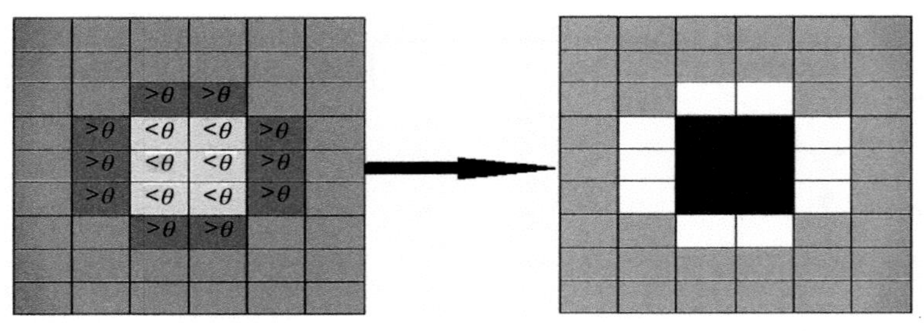

FIG. 4.4 The matrix on the left represents the original image, and the matrix on the right represents the resultant bitmap image. The symbol θ denotes the approximated threshold.

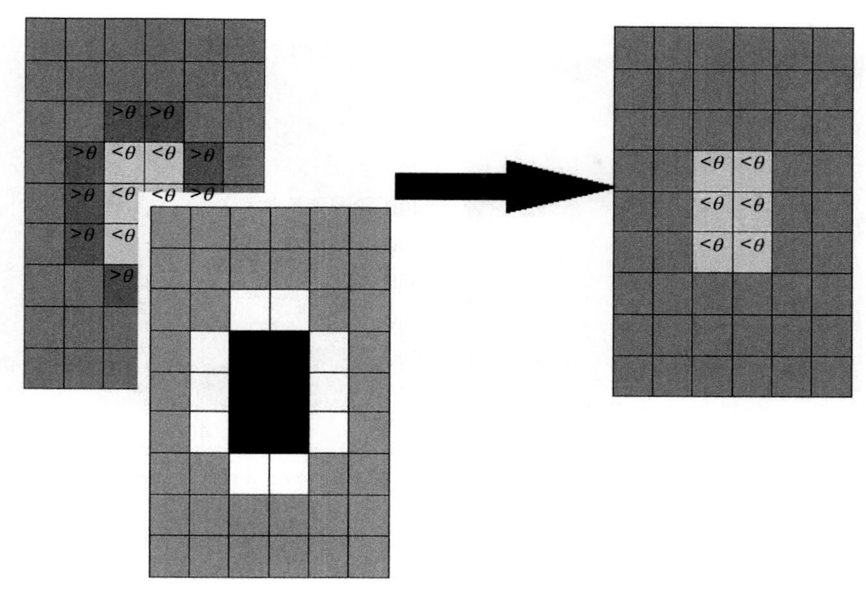

FIG. 4.5 The image represents the morphological closing operation concerning the bitmap image generated, and the matrix on the extreme right represents the resultant skull-free image.

and dilation simultaneously is executed. Dilation operation will fill out the holes in the region of interest, whereas the erosion is going to smoothen the square structuring elements like STREL that consider all eight neighborhood pixels having eight voxels or less (Fig. 4.5).

The morphological opening is performed on the original image concerning the binary image generated earlier as augmentation reference. By performing the opening operation, the skull region in the image is stripped in the resultant MR image, which would make it convenient for the physician to recognize the abnormality in the MR image (Fig. 4.6).

4.5 HARIS algorithm

The proposed approach for image segmentation is computationally efficient compared to its counterparts. HARIS algorithm constitutes a multiobjective function that assists in the identification of initial segment centers, evaluation of the fitness of the pixel in the segment, and refining the achieved results. The HARIS algorithm starts with the identification of the initial segments cluster by using the formula defined here through the elbow methods:

$$N_s = \sum_{s_{n=1}}^{m} \sum_{p_i \in t_n} \| p_i - C_n \| \tag{4.3}$$

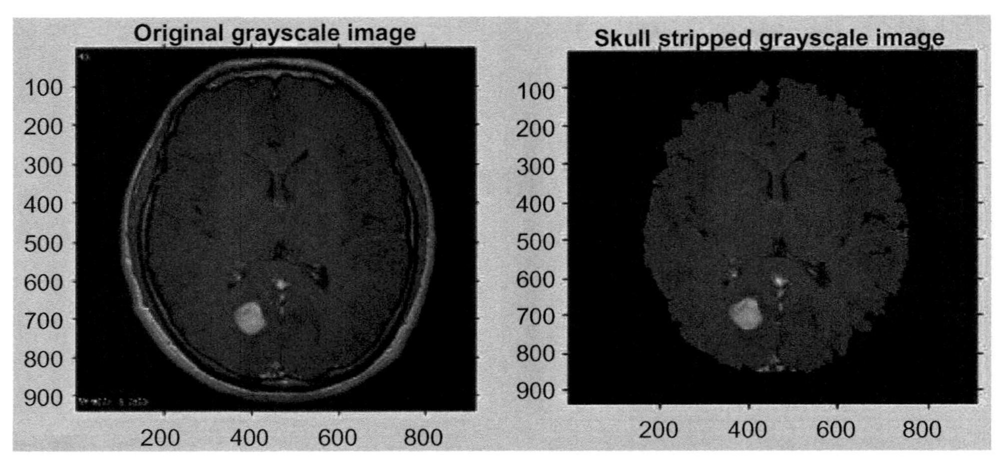

FIG. 4.6 In the image, the left side denotes the original image, and the right side represents the skull-stripped image.

The previously discussed method is applied recursively until there is no further improvement in the value of N_s. In the initial stages of the implementation, the number of segments would be assumed to be 23 from the previous studies. However, the value 23, which was assumed earlier, may not be the final number of clusters, but the optimal number of segments would be evaluated through Eq. (4.3). From the equation, m denotes the initial number of segments assumed or evaluated in the previous iteration; s_n denotes the nth segment in the image whose range lies between 1 to m; p_i represents the pixels involved in the ith segment; and C_n represents the centroid of the nth segment in the image.

The value of C_n in the previous equation is determined by Mac Queen as stated in the paper by Arai and Ridho Barakbah [19] stating that the algorithm can be customized in such a way that it fits with the underlying problem, and the equation is stated thusly:

$$S_{centroid} = \sum_{pixels=1}^{n} \sum_{\mathrm{seg}=1}^{k} M_{pixels,\,\mathrm{seg}} d\left(x_{pixels}, c_{seg}\right) \tag{4.4}$$

From this equation, $S_{centroid}$ represents the estimated centroid of segment S; $M_{pixels,\ seg}$ represents the membership of the pixel in the segment; x_{pixels} represents the xth pixel; C_{seg} represents the cth center in the segment; and $d(x_{pixels}, c_{seg})$ represents the distance measures for evaluating the membership. The formulate of evaluating the distance measure is defined through Mahalanobis distance measure, which could evaluate the nearness of the data concerning the centroid or a particular local maximum. The distance measure could also be understood as the standard deviation from a particular point. In case it is too high, then the pixel will be allotted to some other center whose distance is minimal. The Mahalanobis distance is evaluated through the following distance measure.

The distance is a measure concerning the intensity of the centroid pixel rather than the distance between the coordinates that we generally use to perform in the case of Euclidean distance. Now in Mahalanobis distance measure, the average value of the feature vector is the average of the intensities of the initial population in the segment evaluated as follows:

$$\mu^n = \left\{\mu_1^{(n)}\mu_2^{(n)}\mu_3^{(n)}\ldots\mu_x^{(n)}\right\} = \left\{\sum\nolimits_{x=1}^{i_n} I_1 \sum\nolimits_{x=2}^{i_n} I_2 \ldots \sum\nolimits_{x=n}^{i_n} I_n\right\} \tag{4.5}$$

Where n represents the classes of intensities of the pixels in the initial population, the members in each segment with respect to the segment centroid could be represented as follows:

$$S_n = \sum_{x=1}^{n} \left(I_x^n - \mu_n\right)\left(I_x^n - \mu_n\right)^T \tag{4.6}$$

where

$$I_x^n = \left(I_{x=1}^{(n)}, I_{x=2}^{(n)}, I_{x=3}^{(n)} \ldots I_{x=n}^{(n)}\right)^T \tag{4.7}$$

The Mahalanobis distance for the unexploited pixel I assigned to the segment could be estimated through the following equation:

$$m^2(I, p_n) = \left(I - \mu^{(n)}\right)^T IM_x^{-1}\left(I - \mu^{(n)}\right) \tag{4.8}$$

From this equation, IM_x^{-1} is an inverse matrix evaluated from the unbiased assessed covariance matrix:

$$IM_x^{-1} = (x_i - 1)S_n^{-1} \tag{4.9}$$

By estimating the distance and the degree of acceptance or the degree of the belongingness through the previously discussed Mahalanobis approach, the pixels would be assigned to the segment. However, the estimated membership value would also be used with the objective function.

4.5.1 HARIS algorithm objective function-I

In the multiobjective function-based approach for the segmentation of the medical MR image, the first objective function is used to update the number of regions, that is, the number of segments automatically, and then assigns the pixels based on the membership value evaluated in each iteration. The pixel belongingness is evaluated through the membership of the pixel to the particular seed point of the segment. Seed points in each region are decided through the equation stated here through a variable S_i where i represents the number of regions that lies between 1 and the N.

$$S_i = \frac{\frac{1}{p_{si}}\left(p_1^2 + p_2^2 + \ldots + p_n^2\right)}{\frac{1}{p_{si}}\left(p_1 + p_2 + \ldots + p_n\right)} \tag{4.10}$$

From this equation, the S_i evaluated the contraharmonic mean of all the pixels in the region, and the value is updated every time; p_{si} represents the total number of pixel in the segment i. The contraharmonic mean stated by Srinivas et al. [20] in his paper regards the intensities of all pixels in the group is estimated, and the pixels with the least difference are chosen to be the seed point of the current iteration.

The standard deviation σ is estimated concerning S_i and, based on the value, the pixels are assigned to the segment. The assignment of pixels is carried as follows, and G_T is the global threshold whose value lies between .3 and .6, which is determined by experimental conclusion from previous studies.

$$\left(\sigma = \left|S_i - Intensity\ of\ pixel_{pi}\right|\right) \le G_T \tag{4.11}$$

The objective function for the assessment of pixels and allotment is defined in Eq. (4.12):

$$Obj_{fun} = \left(\alpha \times \frac{T_p}{p_{si}}\right) + \left(\beta \times \frac{T_s}{N_r}\right) \tag{4.12}$$

where α and β are the deciding factors that determine the accuracy and efficiency of the objective function. The value of α is determined by Eq. (4.13), which determines the interclass variance, and β determines the intraclass correlation as determined in Eq. (4.14).T_p in Eq. (4.12) denotes the total number of pixels, and p_{si} in the equation denotes the number of pixels that reside in the segment i whose values ranges from 1 to n. T_s in the equation represents the number of the total number of seeding points, and finally N_r from the equation represents the total number of pixels in the particular region r.

$$\alpha = \frac{\sigma_p^2}{\left(\sigma_p^2 + \frac{\sigma_e^2}{2}\right)} \tag{4.13}$$

$$\beta = \sum_{s_i=0}^{n} \omega_i(t)\sigma_i^2(t) \tag{4.14}$$

In Eq. (7), variable α, which was used Eq. (4.6), is an intraclass coefficient that tells the closeness of pixels that lie within the same segment that has a considerable amount of weightage in deciding the accuracy of the perceptible measure made during the formation of the segments. Variable β discussed in Eq. (4.8), which is a part of Eq. (4.6), has a significant role in the estimation that determines the interclass variance from the centroid pixel through thresholding. The main motivation behind evaluating α and β in the proposed algorithm is to attain a minimum intraclass variance and maximum interclass variance.

The fitness is evaluated for the previous equation to decide whether the numbers of segments are to be further increased or reduced. The fitness is evaluated as follows:

$$fit(p_i) = 1 + abs(Obj(fun))\ \text{if}\ fit(p_i) < 0,\ \text{segments needed to be added} \tag{4.15}$$

$$fit(p_i) = \frac{1}{1 + obj(fun)}\ \text{if}\ fit(p_i) > 0,\ \text{segments needed to be reduced} \tag{4.16}$$

In the proposed algorithm, the number of segments in every iteration would dynamically change based on the fitness of pixels with respect to the segments and based on the values of α and β, which defines the intraclass correlation and interclass variance.

4.5.2 HARIS algorithm objective function-II

The second objective function is used to refine the outcome of the first objective function, and the second objective function decided the optimal seed point of the segment concerning the global best fitness among all the seed points. In each iteration, the seed points of the segments are updated through the objective function stated later. The pixels are assigned to them based on the membership value evaluated through Eq. (4.8), and in each iteration, the value of α and β are evaluated to further decide whether the segments are to be increased or reduced. The second objective function of the HARIS algorithm is stated as follows:

$$seed_i = \text{rand}(0,1) \times fitness\_seed_{i-1} + rand(0,1) \times \left(\text{Global Best}_{seed_{fitness}} - fitness\_seed_{i-1}\right) \tag{4.17}$$

From this equation, *rand*(...) is used to make the values lie between the 0 and 1, and $fitness\_seed_{i-1}$ represents the fitness of the centroid in the previous iteration. The *Global* $Best_{seed\_fitness}$ is the fitness of the centroid in that it holds the maximum fitness value among all the seed points in the image. And the latest seed point is estimated concerning the global best for better accuracy (Fig. 4.7).

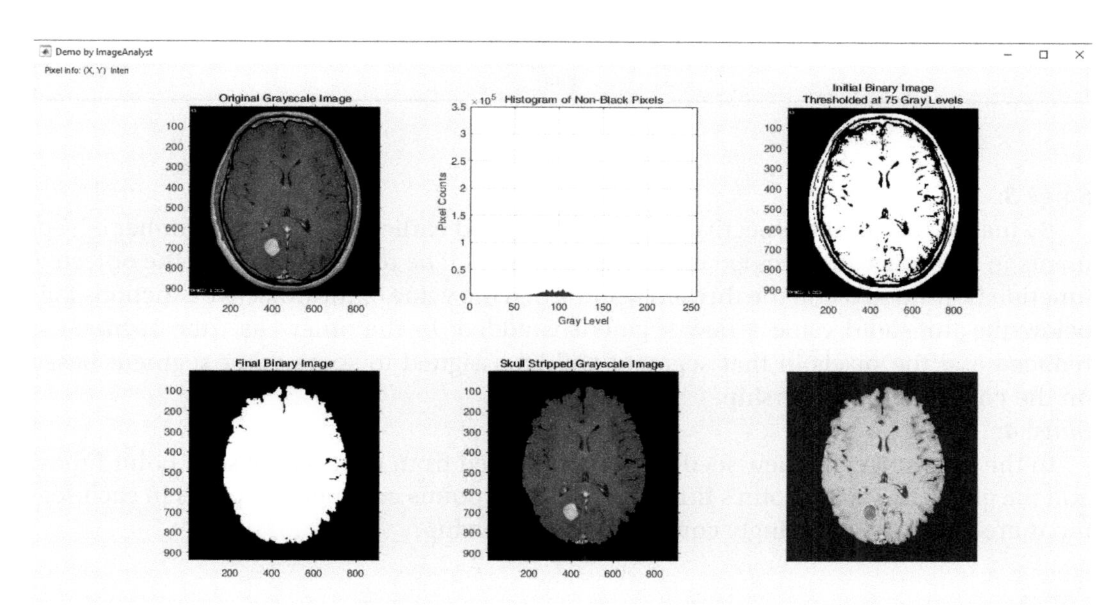

FIG. 4.7 The image illustrates the final image produced by MAT lab upon executing the source code.

4.5.3 Workflow of HARIS algorithm

Stage 1:

In the preliminary stage, the MR image is preprocessed through contourlet transform to remove the noise of the Gaussian noise acquired during image rendering.

Stage 2:

In the next stage, a group of 20 random seed points is identified, and pixels are assigned to them based on the intensity feature through Mahalanobis distance measure.

```
Begin
Set 'n' value
//Calculate the seed point in each region
```
$$S_i = ((2 \times n) + 1)/3$$
```
Set pi and Gt values //Gt value lies between 0.3 and 0.6
```
$$\text{if } |S_i - p_i| \le G_T \text{ then}$$
$$\sigma = |S_i - p_i|$$
```
end if

//Calculate the Interclass Variance 'α'
```
$$\alpha = \sigma_p^2 / \left(\sigma_p^2 + \sigma_e^2/2\right)$$
```
//calculate the Intraclass Correlation 'β'
Assume β = 0
Begin for i from 0 to n do the following
```
$$\beta = \beta + (w_i(t) \times \sigma_i^2(t))$$
```
end for

set Tp, Psi, Ts and Nr values
```
$$obj - fun = \left(\alpha \times \frac{T_p}{p_{si}}\right) + \left(\beta \times \frac{T_s}{N_r}\right)$$
```
Return Obj-fun
End
```

Stage 3:

By making use of the objective function discussed earlier, the optimal number of segments in the image are decided on an iterative basis. The resultant value of the objective function is evaluated and the threshold and when the value of the objective functions falls below the threshold value a new segment is added; in the other case, the segment is reduced and the pixels in that segment will be assigned to some other segment based on the computed membership.

Stage 4:

In the final stage, the new seed point is estimated from the existing seed point fitness and the global best seed point's fitness. As the seed points are updated, pixels in each segment are updated accordingly concerning membership.

```
Get Fitness_Seed, Global_Best_Seed_Fitness
Begin
n=Total_Seg          //Initial Values
segments = 0

Begin for i from 1 to n do the following          // Repeat
```

$$seed_i = [rand(0,1) \times fitness\_seed_{i-1} + rand(0,1) \times (Global\_Best_{Seed\_Fitness} - Fitness\_Seed_{i-1})]$$

```
Segments = evaluate(α,β)
End for
Until segment=0
End
```

The process of stage 3 and stage 4 are repeated indefinitely until no further changes are in the iterations. The two stages work in coherence with each other to attain the best possible solution.

4.6 Experimental analysis and results

The brain MR images of divergent patients with different complications are collected through different sources like open fMRI and SICAS repository for medical MR images, and the proposed HARIS algorithm experiments over the image of size 512×512. The performances of the proposed approach are examined and compared with the existing counterparts, and it is observed that the proposed algorithm exhibits better performance with minimal computational effort. The proposed algorithm has been evaluated through sensitivity, specificity, and Jaccard similarity index (JSI); accuracy and Matthews correlation coefficient (MCC) are used in evaluating the performances of the proposed algorithm (Fig. 4.8).

In the table, the first image from the left is the original image, the second image from left represents the bitmap image generated from the approximated threshold, the third image represents the skull-free image, the fourth image represents the color-mapped image, which is a resultant of the segmented image, and the fifth image represents the actual damaged region that could be fed as an input for the volumetric estimation to assess the damaged region in the human brain.

In the tabulated set of images, the right-most image denotes the actual region of tumor that has been recognized through the segmentation and the texture-based thresholding of the image. The final segmented image is fed as the input for the Ostrogradsky theorem for the volumetric estimation over a 2D MR image. The accuracy and precision that is being computed from the TP, TN, FP, FN are assessed concerning the final segmented MR image. However, the accuracy of the proposed HARIS algorithm completely relies on the quality

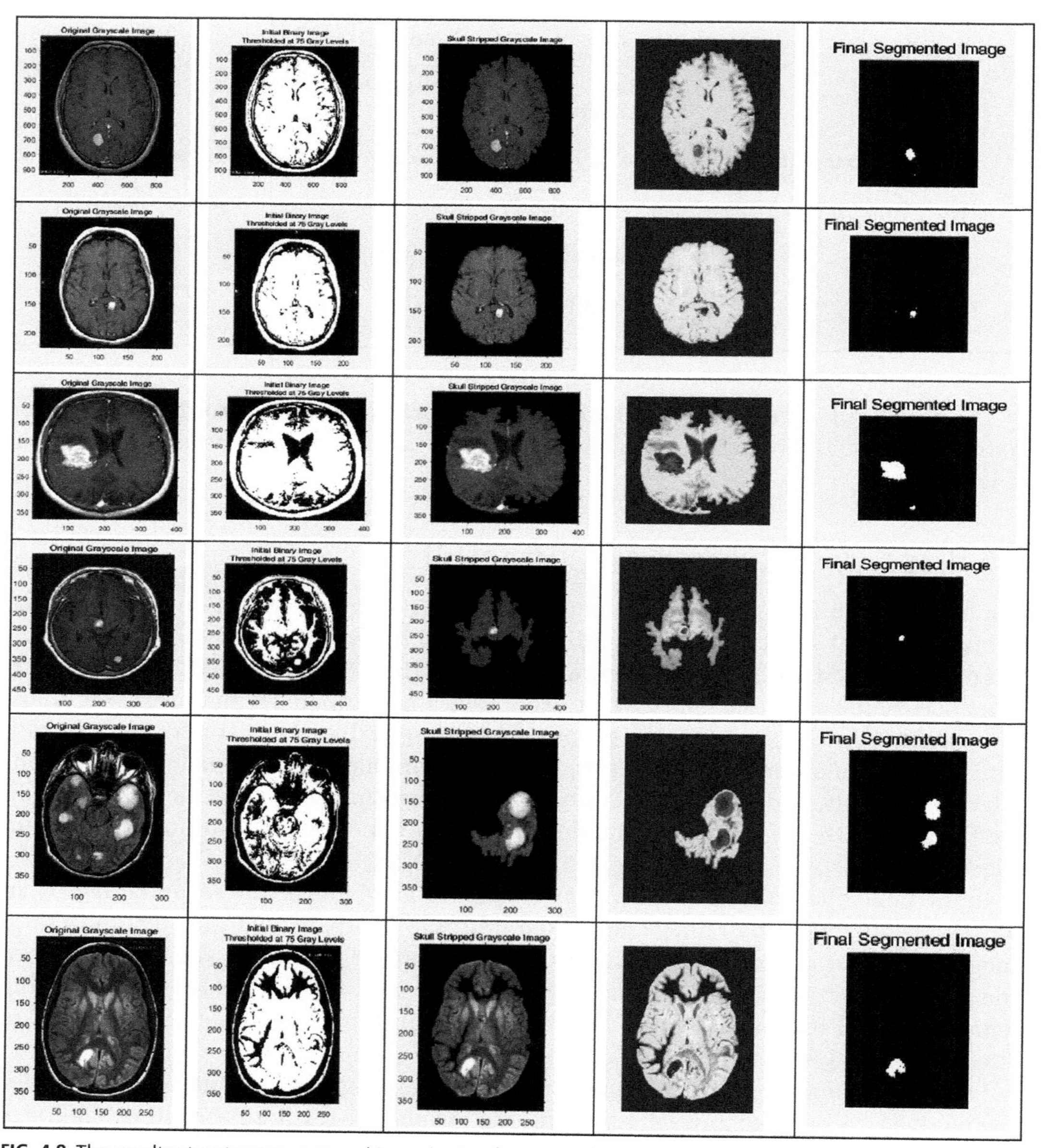

FIG. 4.8 The resultant outcomes captured in each phase on executing source code.

of the segmentation algorithm. The region could be exactly localized only when the segmentation is perfectly performed over the image. Irrespective of the segmentation algorithms and the approach used in the identification of the tumor regions, there would be a significant downfall of the proposed algorithm when experimented over larger-sized images as well as the computational latency.

The performance of the contourlet transform is analyzed through performance analysis metrics like PSNR (peak signal to noise ratio) value, which is compared against the normalized image, which has to be good enough for better results, and the MSE (mean square error) value is the one that designates the standard deviation of the actual value and the estimated value, which has to be low enough for better outcome. Similarly, RMSE is a square root of MSE value, and the IQI is the image quality index that always lies between 0 and 1, which must be big enough for a better image. MCC is the other parameter that we consider for estimating the accuracy and preciseness by evaluating the correlation coefficient of the approximated region against the actual region in the human brain MR image (Fig. 4.9).

The performance of the proposed approach could be analyzed from Table 4.1; the proposed HARIS algorithm is proven to exhibit a better performance irrespective of the preprocessing stage. The proposed algorithm has performed well in low-noise variance levels than higher. From the table, the IQI value of the T1-weighted is exceptionally good for lower noise variance by employing contourlet transforms. The following graph represents the performance of HARIS algorithm with and without contourlet transforms over varied noise levels (Table 4.2).

The four metrics used in evaluating the performances are True Positive, True Negative, False Positive, and False Negative. True positive value depicts the number of times the defected area is correctly chosen, and the True Negative depicts wrongly choosing the area assuming it as a defective area. False Positive in our experimental analysis depicts choosing the nondefective area correctly, and the False Negative depicts wrongly choosing the nondefective area. These metrics are used in analyzing the sensitivity, specificity, JSI, and MCC. The formulas of the metrics are stated as follows:

$$Senstivity = \frac{T_P}{T_P + F_N} \tag{4.18}$$

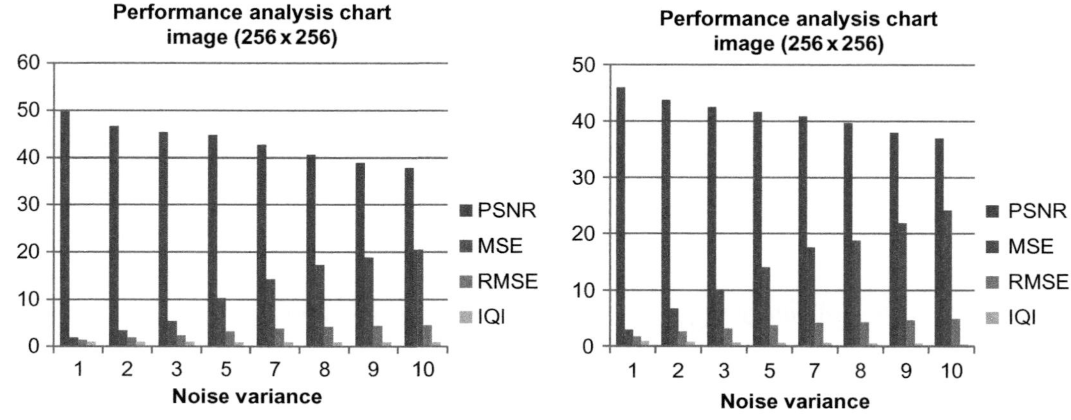

FIG. 4.9 The figure represents the plotted values of PSNR, MSE, RMSE, and IQI from Table 4.1 obtained by experimenting over images of size 256 × 256 and 512 × 512.

Table 4.1 The tabulated values represents the comparative analysis of the contour transforms that evaluate the performance of the algorithm. The left half of the table represents the HARIS algorithm over a preprocessed image, and the right subhalf represents the HARIS algorithm alone.

256 × 256 size image										
Image segmentation through contour transform and HARIS algorithm						**Image segmentation without contour transform and directly with HARIS algorithm**				
S. No.	Noise variance	PSNR value	MSE value	RMSE value	IQI value	Noise variance	PSNR value	MSE value	RMSE value	IQI value
1	10	37.92	20.49	4.52	0.873	10	37.01	24.15	4.91	0.354
2	9	38.92	18.83	4.33	0.875	9	37.97	21.90	4.67	0.493
3	8	40.65	17.28	4.15	0.889	8	39.71	18.78	4.33	0.502
4	7	42.75	14.23	3.77	0.897	7	40.87	17.56	4.19	0.597
5	5	44.78	10.33	3.21	0.912	5	41.66	14.02	3.74	0.621
6	3	45.36	5.36	2.31	0.929	3	42.51	9.97	3.15	0.689
7	2	46.61	3.42	1.84	0.935	2	43.73	6.74	2.59	0.775
8	1	49.72	1.89	1.37	0.947	1	45.96	2.95	1.71	0.886

Table 4.2 The tabulated values represents the comparative analysis of the contour transforms that evaluate the performance of the algorithm. The left half of the table represents the HARIS algorithm over a preprocessed image, and the right subhalf represents the HARIS algorithm alone.

512 × 512 size MR image										
Image segmentation through contour transform and HARIS algorithm						**Image segmentation without contour transform and directly with HARIS algorithm**				
S. No.	Noise variance	PSNR value	MSE value	RMSE value	IQI value	Noise variance	PSNR value	MSE value	RMSE value	IQI value
1	10	36.54	22.12	4.70	0.852	10	36.01	25.47	5.04	0.221
2	9	38.25	20.34	4.50	0.864	9	36.97	22.22	4.71	0.385
3	8	39.97	17.79	4.21	0.878	8	38.71	19.14	4.37	0.442
4	7	41.88	14.65	3.82	0.886	7	39.87	16.47	4.05	0.506
5	5	43.01	10.08	3.17	0.901	5	40.66	13.34	3.65	0.632
6	3	44.68	6.87	2.62	0.918	3	41.51	9.96	3.15	0.698
7	2	45.92	4.88	2.20	0.924	2	42.73	6.91	2.62	0.765
8	1	48.02	2.33	1.52	0.936	1	43.96	2.96	2.64	0.836

$$Specificity = \frac{T_N}{T_N + F_P} \tag{4.19}$$

$$Accuracy = \frac{T_P + T_N}{SUM(T+P)}, sum(T+F) = T_P + F_P + T_N + F_N \tag{4.20}$$

$$\text{JSI} = \frac{T_P}{K}, K = T_P + F_P + F_N \tag{4.21}$$

$$\text{MCC} = \frac{T_P \times T_N - F_P \times F_N}{\sqrt{(T_P + F_P)(T_P + F_N)(T_N + F_P)(T_N + F_N)}} \tag{4.22}$$

From Eq. (4.18), the sensitivity of the proposed HARIS algorithm is evaluated, which determines the accuracy of identifying the abnormality in the human brain correctly, and the specificity evaluates with Eq. 4.19 the actual region of nondamaged region correctly. The JSI value evaluated through Eq. (4.21) represents the similarity of the pixel that lies within the same segment, and it also represents how distant the pixels are from the other segments. The value of JSI is directly proportional to the closeness among the pixel with respect to a particular segment. MCC is the acronym of Matthews.

These results are evaluated against the images of different sizes, and it experiments over the wide range of images. The metrics are assessed against both the images that are preprocessed through contourlet transforms and noncontourlet transforms. And the computed values are tabulated in the tables, and it is observed that HARIS algorithm exhibits a better performance over many existing algorithms (Tables 4.3–4.6 and Fig. 4.10).

Table 4.3 Denote the various performance assessment metrics like sensitivity, specificity, accuracy, JSI, MCC, and among all the approaches, HARIS algorithms exhibit better results in spite of passing original image without preprocessing.

	256 × 256 MRI image (without contourlet transform)				
	Sensitivity	**Specificity**	**Accuracy**	**JSI**	**MCC**
HARIS algorithm	0.8312432	0.855487	0.770034	0.89435	0.821019
Twin centric GA with SGO	0.8302324	0.854843	0.755261	0.88916	0.821329
Using CNN	0.8123232	0.848542	0.762983	0.87343	0.819854
GA with TLBO	0.802327	0.843378	0.692165	0.85371	0.804221

SGO, social group optimization; *TLBO*, teacher learner-based optimization.

Table 4.4 Denote the various performance assessment metrics like sensitivity, specificity, accuracy, JSI, MCC, and among all the approaches, HARIS algorithms exhibit better results over a 256 × 256 sized image.

	256 × 256 MRI image (through contourlet transform)				
	Sensitivity	**Specificity**	**Accuracy**	**JSI**	**MCC**
HARIS algorithm	0.824215	0.875291	0.796457	0.86871	0.811035
Twin centric GA with SGO	0.819252	0.865752	0.755261	0.85456	0.809681
Using CNN	0.819386	0.859255	0.762983	0.85396	0.819854
GA with TLBO	0.799848	0.836598	0.692165	0.84915	0.786399

Table 4.5 Denote the various performance assessment metrics like sensitivity, specificity, accuracy, JSI, MCC, and among all the approaches, HARIS algorithms exhibit better results in spite of passing the original image without preprocessing.

	512 × 512 MRI image (without contourlet transform)				
	Sensitivity	**Specificity**	**Accuracy**	**JSI**	**MCC**
HARIS algorithm	0.796532	0.852183	0.756886	0.85112	0.79883
Twin centric GA with SGO	0.791002	0.851141	0.745311	0.84987	0.79141
Using CNN	0.781211	0.848221	0.749281	0.84119	0.78547
GA with TLBO	0.771321	0.838974	0.724323	0.83090	0.77565

Table 4.6 Denote the various performance assessment metrics like sensitivity, specificity, accuracy, JSI, MCC. Among all the approaches, HARIS algorithms exhibit a better result.

	512 × 512 MRI image (through contourlet transform)				
	Sensitivity	**Specificity**	**Accuracy**	**JSI**	**MCC**
HARIS Algorithm	0.819685	0.875291	0.789487	0.86171	0.81161
Twin centric GA with SGO	0.808356	0.865436	0.765251	0.85691	0.79025
Using CNN	0.796536	0.859657	0.744287	0.84898	0.79214
GA with TLBO	0.785961	0.853522	0.682569	.0.83712	0.77595

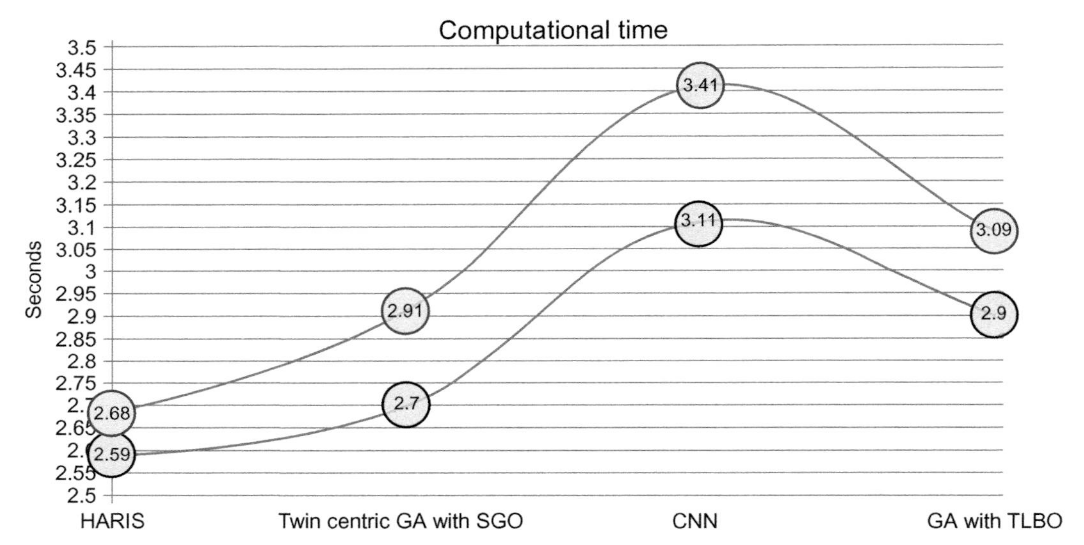

FIG. 4.10 The graph represents the computational time consumed by the proposed algorithm when experimented over 256 × 256 and 512 × 512 images. The blue color (gray color in print version) rings correspond to 256 × 256 and purple rings denotes the computational time of 512 × 512.

Through the proposed method, the severity of the damaged region is being assessed through the Hausdorff distance measure that gives the clear picture about the largest of all the possible distances to a specific point in one segment to the closest point on some other segment. Hausdorff distance is one of the best and predominant ways to decide the efficiency and accuracy of the segmentation algorithm. In the proposed approach, the Hausdorff distance measure is used in approximating the correctness of the estimated value and the actual value. The region of the tumor has been further categorized into Tumor Core (CT), Whole Tumor (WT) and Enhanced Tumor (ET).

Hausdorff distance has been employed in assessing how accurately the proposed method could locate the region of the tumor in the human brain. Hausdorff distance would give the statistics of actual deviations in the approximated region and the actual region, and Hausdorff distance is mathematically represented through the formula stated here:

$$H_d\left(\text{seg}_x, \text{seg}_y\right) = \max_{px\in \text{seg}_x}\left\{\min_{py\in \text{seg}_y}\{e_d(p_x, p_y)\}\right\} \tag{4.23}$$

In Eq. (4.23), the p_x and p_y are the pixels in the centroid of the segments seg_x and seg_y. $e_d(p_x,p_y)$ is the Euclidean distance between the two points p_x and p_y that belongs to two different segments of the image.

	Tumor core (TC)	Whole tumor (WT)	Enhanced tumor (ET)
T1-weighted MR image of size 128 × 128 (without contourlet transform)	9.453 ± 0.714	3.246 ± 0.781	8.443 ± 0.872
T1-weighted MR image of size 128 × 128 (with contourlet transform)	8.254 ± 0.872	4.564 ± 0.948	7.872 ± 0.984
T1-weighted MR image of size 256 × 256 (without contourlet transform)	8.901 ± 0.912	4.293 ± 0.659	7.128 ± 0.753
T1-weighted MR image of size 256 × 256 (with contourlet transform)	8.072 ± 0.512	5.191 ± 0.729	6.891 ± 0.165
T1-weighted MR image of size 512 × 512 (without contourlet transform)	7.982 ± 0.795	6.017 ± 0.318	6.087 ± 0.720
T1-weighted MR Image of size 512 × 512 (with contourlet transform)	7.541 ± 0.872	7.192 ± 0.665	5.918 ± 0.968

4.7 Conclusion

It is observed experimentally that the proposed HARIS algorithm has shown a superior outcome compared to the existing approach with minimum computational efforts. The proposed approach has progressed through 20 initial random seed points, and the seed points are identified through Mac Queen's approach so that the algorithm could converge faster rather doing it right from scratch as the genetic algorithm-based approaches, as many morphological-based approaches had performed. Moreover, the seed points are chosen strategically through contraharmonic means and objective function stated earlier rather than picking them based on the membership and distances. The algorithm has given notable weightage to intraclass correlation and interclass variance in the first objective function, which grants convergence of the algorithm, unlike other approaches. And the resultant segmented image is better optimized through the second objective function that motivates the algorithm to function better than its equivalents with lesser computational effort.

4.8 Future scope

The proposed Heuristic approach for real-time image segmentation algorithm is proven to be exceptionally good in segmenting the image, and the proposed approach has experimented over T1-weighted and T2-weighted images. The results of the proposed algorithm are to be further optimized when it experimented over the T2 FLAIR MRI and PD MRI. The tissues in the FLAIR and PD MRIs are not elaborated like in T1- and T2-weighted. The contrast of the T1-weighted and T2-weighted images is reduced to attain the proton density weight for PD weighted and FLAIR MRI, which has become a challenging task for the proposed approach for segmenting the MRI image with higher accuracy. This issue could be efficiently addressed by incorporating bilateral filters with canny filter for better preprocessing the pixels, and the objective function-II has to be redefined in such way that every minor object in the MRI is elevated perfectly with a fewer number of iterations.

References

[1] S. Saha, S. Bandyopadhyay, MRI brain image segmentation by fuzzy symmetry based genetic clustering technique, in: 2007 IEEE congress on Evolutionary Computation, Singapore, 2007.

[2] P. Naga Srinivasu, G. Srinivas, T. Srinivas Rao, An automated brain MRI image segmentation using generic algorithm and TLBO, Int. J. Control Theory Appl. 9 (32) (2016).

[3] R. Dubey, A review on MRI image segmentation techniques, Int. J. Adv. Res. Electron. Commun. Eng. 4 (5) (2015).

[4] S.N. Sulaiman, N.A. Non, I.S. Isa, N. Hamzah, Segmentation of brain MRI image based on the clustering algorithm, in: IEEE Symposium on Industrial Electronics and Applications, Kota Kinabalu, 2014, pp. 60–65.

[5] V.K. Sudha, R. Sudhakar, V.E. Balas, Fuzzy rule-based segmentation of CT brain images of hemorrhage for compression, Int. J. Adv. Intell. Parad. 4 (3/4) (2012) 256–267.

[6] R. Hiralal, H.P. Menon, A survey of brain MRI image segmentation methods and the issues involved, in: J. Corchado Rodriguez, S. Mitra, S. Thampi, E.S. El-Alfy (Eds.), Intelligent Systems Technologies and Applications 2016. Advances in Intelligent Systems and Computing, vol. 530, Springer, Cham, 2016.

[7] W. Deng, W. Xiao, H. Deng, J. Liu, MRI brain tumor segmentation with region growing method based on the gradients and variances along and inside of the boundary curve, in: 3rd International Conference on Biomedical Engineering and Informatics, Yantai, 2010, pp. 393–396.

[8] Z. Song, N. Tustison, B. Avants, J.C. Gee, Integrated graph cuts for brain MRI segmentation, in: R. Larsen, M. Nielsen, J. Sporring (Eds.), Medical Image Computing and Computer-Assisted Intervention—MICCAI, vol. 4191, Springer, Berlin, Heidelberg, 2006.

[9] S. Chandra, R. Bhat, H. Singh, A PSO based method for detection of brain tumors from MRI, in: World Congress on Nature & Biologically Inspired Computing', (NaBIC), Coimbatore, 2009, pp. 666–671.

[10] Y.-t. Liu, H.-x. Zhang, P.-h. Li, Research on SVM-based MRI image segmentation, J. China Univ. Posts Telecommun. 18 (2) (2011) 129–132.

[11] A. Nandi, Detection of human brain tumor using MRI image segmentation and morphological operators, in: IEEE International Conference on Computer Graphics, Vision and Information Security, Bhubaneswar, 2015, pp. 55–60.

[12] Y. Wang, Z. Sun, C. Liu, W.P.J. Zhang, MRI image segmentation by fully convolutional networks, in: IEEE International Conference on Mechatronics and Automation, Harbin, 2016, pp. 1697–1702.

[13] S.K. Ramakuri, C. Chakrabothy, S. Ghosh, B. Gupta, Performance analysis of eye-state characterization through single electrode EEG device for medical application, in: 2017 Global Wireless Summit (GWS), Cape Town, 2017, pp. 1–6.

[14] A. Işın, C. Direkoglu, M. Sah, Review of MRI-based brain tumor image segmentation using deep learning methods, Procedia Comput. Sci. 102 (2016) 317–324.

[15] A. Işın, C. Direkoğlu, M. Şah, Review of MRI-based brain tumor image segmentation using deep learning methods, Procedia Comput. Sci. 102 (2016) 317–324.

[16] W. Yu, F. Franchetti, Y. Chang, T. Chen, Fast and robust active contour for image segmentation, in: 2010 IEEE International conference on Image Processing, Hong Kong, 2010, pp. 641–644.

[17] R. Sivakumar, G. Balaji, R.S.J. Ravikiran, R. Karikalan, S. Saraswatijanaki, Image denoising using contourlet transform, in: Second International Conference on Computer and Electrical Engineering, Dubai, 2009, pp. 22–25.

[18] P. Naga Srinivasu, G. Srinivas, T. Srinivasa Rao, Valentina Emalia Balas, 'A novel approach for assessing the damaged region in MRI through improvised GA and SGO.' Int. J. Adv. Intell. Parad. Indersci. in press.

[19] K. Arai, A. Ridho Barakbah, Hierarchical K-Means: an algorithm for centroid initialisation for k-means, in: Reports of the Faculty of Science and Engineering, vol. 36(1), Saga University, 2007.

[20] G. Srinivas, P.N. Srinivasu, T.S. Rao, C. Ramesh, Harmonic and contra-harmonic mean-centric JPEG compression for an objective image quality enhancement of noisy images, in: Smart Computing and Informatics. Smart Innovation, Systems, and Technologies, vol. 78, Springer, Singapore, 2018.

Further reading

E. Abdel-Maksoud, M. Elmogy, R. Al-Awadi, Tumor segmentation based on a hybrid clustering technique, Egypt. Inform. J. 16 (1) (2015) 71–81.

Z. Akkus, A. Galimzianova, A. Hoogi, D.L. Rubin, B.J. Erickson, Deep learning for brain MRI segmentation: state of the art and, future directions, J. Digit. Imag. 30 (4) (2017) 449–459.

M. Arulraj, A. Nakib, Y. Cooren, P. Siarry, Multicriteria image thresholding based on multiobjective particle swarm optimization, Appl. Math. Sci. 8 (3) (2014) 131–137.

N.B. Bahadure, A.K. Ray, H.P. Thethi, Image analysis for MRI based brain tumor detection and feature extraction using biologically inspired BWT and SVM, Int. J. Biomed. Imag. 2017 (2017), 9749108. 12 pp.

N.S. Datta, H.S. Dutta, K. Majumder, S. Chatterjee, N.A. Wasim, A Survey on the Application of Multi-Objective Optimization Methods in Image Segmentation: Evolutionary to Hybrid Framework. 2018, https://doi.org/10.1007/978-981-13-1471-1_12.

K. Deb, Multi-objective optimization using evolutionary algorithms: an introduction, in: L. Wang, A. Ng, K. Deb (Eds.), Multi-Objective Evolutionary Optimization for Product design and Manufacturing, Springer, London, 2011.

Y. Fang, Z. Zhen, Z. Huang, C. Zhang, Multi-objective fuzzy clustering method for image segmentation based on variable-length intelligent optimization algorithm, in: Cai Z., Hu C., Kang Z., Liu Y. (Eds.), Advances in Computation and Intelligence. ISICA 2010. Lecture Notes in Computer Science, vol. 6382, Springer, Berlin, Heidelberg, 2010.

R. Lavanyadevi, M. Machakowsalya, J. Niveditha, A.N. Kumar, Brain tumor classification and segmentation in MRI images using PNN, in: 2017 IEEE International Conference on Electrical, Instrumentation and Communication Engineering, Karur, 2017, pp. 1–6.

N. Madesh, H.P. Menon, Automated segmentation of brain parts from MRI image slices, Int. J. Pure Appl. Math. 114 (11) (2017).

P. Moeskops, M.A. Viergever, A.M. Mendrik, L.S. de Vries, M.J.N.L. Benders, I. Isgum, Automatic segmentation of MR brain images with a convolutional neural network, IEEE Trans. Med. Imag. 35 (5) (2017) 1252–1261.

A. Nakib, H. Oulhadj, P. Siarry, Image segmentation based on multi-objective optimization, Pattern Recogn. Lett. 29 (2) (2008) 161–172.

S. Pereira, A. Pinto, V. Alves, C.A. Silva, Brain tumor segmentation using convolutional neural networks in MRI images, IEEE Trans. Med. Imag. 35 (5) (2016) 1240–1251.

C. Tsai, B.S. Manjunath, R. Jagadeesan, Automated segmentation of brain MR images, Pattern Recogn. 28 (12) (1995).

F. Wei, T. Ming, J. Hong-bing, Adaptive watermark scheme based on contourlet transform, in: International Symposium on Computer Science and Computational Technology, Shanghai, vol. 2, 2008.

W. Zhao, L. Wang, Y. Shi, X. Xi, Y. Yin, Y. Tang, A multi-objective framework for brain MRI threshold segmentation, in: 2016 8th International Conference on Information Technology in Medicine and Education (ITME), Fuzhou, 2016, pp. 20–24.

H. Zhao, X. Zhao, T. Zhang, Y. Liu, A new contourlet transform with adaptive directional partitioning, IEEE Signal Process. Lett. 24 (6) (2017) 843–847.

5

Development of a fuzzy decision support system to deal with uncertainties in working posture analysis using rapid upper limb assessment

Bappaditya Ghosh[a], Subhashis Sahu[b], Animesh Biswas[a]
[a]DEPARTMENT OF MATHEMATICS, UNIVERSITY OF KALYANI, KALYANI, INDIA
[b]DEPARTMENT OF PHYSIOLOGY, UNIVERSITY OF KALYANI, KALYANI, INDIA

5.1 Introduction

Work-related risks and injuries among the workers are increasing significantly at worksites. In most worksites, workers have to go through several hazardous working postures. As a consequence, a large number of workers suffer from work-related musculoskeletal disorders (WMSDs). Globally, musculoskeletal disorder (MSD) is one of the most important causes of temporary work disability. According to the World Health Organization, a significant inducement of working activities by workers [1] is one of the most primary reasons for WMSDs. In 2015, the United States Department of Labor stated that almost 31% of all occupational injuries and illness are caused due to WMSDs [2]. Several occupational risk factors, such as pain in nerves, bones, ligaments, joints, tendons, etc., also cause WMSDs [3, 4]. In India, the epidemiological studies on several jobsites show that the occupation-based prevalence of WMSDs among the workers is about 90% [5]. Unscientific working postures [6] are also a reason for increasing the number of work-related injuries, particularly muscle injuries, at several jobsites. It was already established that there is a close relationship between working postures and the prevalence of musculoskeletal problems. The proper justification and analysis of these working postures can significantly reduce these musculoskeletal problems [7].

There are mainly three approaches for analyzing working postures, viz., self-report, direct measurement, and observational methods [8]. Among these approaches, observational methods are widely used for analyzing working postures, particularly in the industrial context [9]. There are also several observational methods. Among them, one of the

Deep Learning Techniques for Biomedical and Health Informatics. https://doi.org/10.1016/B978-0-12-819061-6.00005-7

earliest observational methods is the Ovako Working Posture Assessment System (OWAS) [10]. Other methods include rapid upper limb assessment (RULA) [11], rapid entire body assessment (REBA) [12], loading on the upper body assessment (LUBA) [13], etc. Among the previously mentioned methods, RULA established its credential due to its simplicity and user-friendly accessibility in the rapid analysis process of mainly static working postures where the upper limbs of the body are involved. But there is a need to revalidate the analysis process, especially in an Indian context [14]. The main drawback of this method is that the method demands exact values of the input parameters (e.g., body joint angles, force/load, etc.) rather than a range of values. These input parameter values are measured manually and, during this process, some sort of inexactness and uncertainty intrinsically occurred. Also, the boundaries of body joint angles, force/load, etc., are discrete. As a consequence, a step jump in computing the score value against each of the inputs occurs in the border region, which will be described in detail in Section 5.3.

A decision support system (DSS) is developed in this chapter, by introducing the concept of a fuzzy expert system [15] to deal with the uncertainties associated with the posture analysis process through RULA. A fuzzy system [16] is a part of soft computing that emulates a systematic reasoning process considering human perception in an expert manner. This new approach will be capable to capture the uncertainties associated with the traditional method of posture analysis. Few research works had been done regarding the issues of posture analysis. For the purpose of performance assessment of an integrated health, safety, environment, and ergonomics system, a fuzzy system was successfully introduced by Azadeh et al. [17]. An integration process between fuzzy logic and RULA was performed by Rivero et al. [18] to assess the risk of workers engaged at several workplaces. The concept of fuzzy logic was introduced by Golabchi et al. [19] in the posture-based ergonomic analysis for field observation and assessment of construction manual operations. A new and easy way of posture analysis through visual management was invented by Savino et al. [20]. Debnath and Biswas [21] introduced an interval type-2 fuzzy analytic hierarchy process for assessment of occupational risks in construction sites in the recent past.

Fuzzy DSS is developed using a fuzzy inference system (FIS) to enrich the posture analysis process through RULA by capturing all kinds of uncertainties associated with the input parameters. In developing FIS, the multigranular linguistic term set [22] is used to represent the membership functions (MFs) of the input parameters of the proposed system. To find out the most suitable operations between antecedents and consequents, Pearson's correlation coefficient [23] is taken into account. Using the developed DSS, a case study is performed concerning the health-related problems of the workers who are engaged in Sal leaf plate-making units and who suffer from several MSDs. To establish the reliability and efficiency of the proposed methodology, the body part discomfort (BPD) scale is used to compare the results of fuzzy DSS with the existing methods of posture analysis.

5.2 RULA method

RULA [11] is an assessment tool to analyze working postures of the workers in which mostly the upper section of the body is in action. In this process, at first, the awkward postures of the body parts are identified from still photographs at the time of action and are analyzed through a RULA worksheet [11]. The whole procedure is summarized as follows:

Step 1: The position of body parts, viz., upper arm, lower arm, wrist, neck, trunk, and legs, are traced and their corresponding body joint angles (except the legs) are measured. Then, whether the legs and feet are supported or not is identified.
Step 2: The scores corresponding to upper arm, lower arm, wrist, wrist twist, neck, trunk, and legs are calculated as per the RULA worksheet.
Step 3: The scores achieved in Step 2 corresponding to upper arm, lower arm, wrist, and wrist twist are put into Table A [11], and the scores corresponding to neck, trunk, and legs are put into Table B [11].
Step 4: In this step, the muscle use score and force/load score are added to each of the scores obtained from Tables A and B to get posture scores A and B, respectively.
Step 5: Both scores found in Step 4 are put into Table C [11] to obtain the RULA score, which falls in one of the four action levels [11].

5.3 Uncertainties occur in analyzing the working posture using RULA

In spite of being one of the easiest and most useful ergonomic techniques for posture analysis, RULA [11] involves some sort of uncertainties that occur in the handling of real-life situations. Few of them are mentioned as follows:

1. In RULA, the boundaries of body joint angles, force/load, etc., are discrete. As a result, a step jump in the border region occurs when computing the score value against each of the inputs. Therefore, some sort of uncertainty occurs between two ranges of body joint angles, force/load, etc. Due to this reason, sometimes two sets of very close values of inputs give substantially different output values, and two sets of considerably different values of inputs give the same output value. This fact is described in Table 5.1 by considering three sets of inputs.

Table 5.1 RULA scores corresponding to three sets of inputs.

Input	Upper arm	Lower arm	Wrist	Wrist twist	Neck	Trunk	Legs	Muscle use score	Force/load score (lb)	RULA score
Input 1	19°	99°	14°	1	9°	0°	1	1	21	4
Input 2	21°	101°	16°	1	11°	1°	1	1	23	7
Input 3	170°	180°	85°	1	80°	90°	1	1	150	7

It is seen from Table 5.1 that a significant change in RULA score is found for Input 1 and Inpu 2, although the differences of body joint angles and force/load are very small. On the other hand, in spite of huge differences between the input parameters, the same RULA score is found for Input 2 and Input 3.

2. During the scoring process of body parts through the RULA worksheet, the user has to put exact values of input parameters, viz., body joint angles, force/load, etc. Therefore, some sort of uncertainties inevitably occurs due to human errors in evaluating these values.
3. The boundaries of action levels are also discrete, which neglects the intermediate action between two successive action levels and, as a result, the exact percentage of severity of a body part cannot be concluded.

Considering these discrepancies, the main focus of this study is to develop a DSS using RULA that can analyze the working postures and postural risks of the workers by minimizing the uncertainties associated with the evaluation process.

The entire research methodology is discussed in the next section.

5.4 Research methodology

This study focuses on capturing all kinds of uncertainties associated with the scoring process through RULA and developing a DSS using fuzzy logic. Fuzzy logic deals with the uncertainties and imprecision associated with ambiguous information by bringing out a significant transition among different continuous variables with unsharp boundaries [15]. The fuzzy DSS is developed by converting all the ranges of input parameters into continuous MFs having sharp boundaries. This new approach provides a significant transition from the traditional approach to a more reliable modified approach, as it considers all the possible uncertainties associated with the posture analysis process.

In developing the fuzzy DSS, a Mamdani fuzzy inference system (MFIS) [24, 25] is generated to deal with uncertainties associated with the traditional approach. An MFIS is a special type of expert system based on the concept of fuzzy sets, fuzzy if-then rules, and fuzzy reasoning.

The design of an MFIS is discussed as follows.

5.4.1 MFIS

An MFIS consists of a knowledge base, rule base, and an inference engine that executes the whole inference procedure. The step-by-step formation of an MFIS is as follows:

Step 1: Selection of input and output parameters

Let A_i $(i=1,2,\ldots,n)$ be n input parameters defined on the respective universe of discourses X_i $(i=1,2,\ldots,n)$ and C be the desired output parameter defined on the universe of discourse Z.

Step 2: Formation of rule base

The rule base [26] in an inference system defines the relation between the input and output parameters. A typical rule base [27, 28] of an MFIS consisting of m number of rules is of the following form:

$$Rule^{j}: \text{If } x_1 \text{ is } F^{j}_{A_1}, x_2 \text{ is } F^{j}_{A_2}, \ldots, \text{and } x_n \text{ is } F^{j}_{A_n}, \text{then } z = F^{j}_{C}$$

where $F^{j}_{A_i}$ is the qualitative descriptor of A_i $(i=1,2,\ldots,n)$ in the form of a fuzzy number and F^{j}_{C} is the qualitative descriptor of C, which is also a fuzzy number for the jth $(j=1, 2, \ldots, m)$ rule of the inference system. Here, x_i and z are the crisp input values taken from sets X_i $(i=1,2,\ldots,n)$ and Z, respectively.

Step 3: Computation of firing strength of each rule

Firing strength of a rule represents the strength of the rule to which the rule matches with the inputs. Here, the firing strength, γ^{j}, for the jth rule is evaluated using conjunction method as follows:

$$\gamma^{j} = conjuction\left[\mu_{F^{j}_{A_1}}(x_1), \mu_{F^{j}_{A_2}}(x_2), \ldots, \mu_{F^{j}_{An}}(x_n)\right]$$

where $\mu_{F_{Ai}}{}^{j}(x_i)$ is the degree of membership in the fuzzy set $F^{j}_{A_i}$ corresponding to the crisp inputs $x_i(i=1,2,\ldots,n; j=1,2,\ldots,m)$.

Step 4: Determination of fuzzy output of each rule

Now, the fuzzy output of each rule is determined by using a suitable T-conorm. The fuzzy output for the jth $(j=1,2,\ldots,m)$ rule is determined as follows:

$$\mu^{j}_{C}(z) = T - conorm\left(\gamma_j, \mu_{F^{j}_{C}}(z)\right)$$

Step 5: Evaluation of aggregated output

The aggregated output of the MFIS is evaluated using a suitable aggregation operator as follows:

$$\mu_{AGG_C}(z) = Aggregation^{m}_{j=1}\mu^{j}_{C}(z)$$

Step 6: Defuzzification of the aggregated output

In the last step, the aggregated output is defuzzified using a defuzzification operator to get the final output as follows:

$$Final\ output = Defuzzification\ operator\ (AGG_C)$$

To enrich the methodology, a sensitivity analysis is carried out by changing the operators used in the proposed MFIS.

In the next subsection, the fuzzy DSS is developed.

5.4.2 Development of fuzzy DSS

In the proposed methodology, the MFIS consists of an expert system with 15 input parameters, 9 output parameters, 95 MFs, 299 if-then rules, and 9 rule blocks. Here, two types of fuzzy numbers, viz., trapezoidal fuzzy numbers (TrFNs) and triangular fuzzy numbers (TFNs) are used for input variables. The crossover point of two adjacent MFs is considered as either the boundary point of two adjacent ranges of inputs (e.g., body joint angles, force/load, etc.) or the average of two adjacent numbers, which is used for labeling Table A, Table B, etc. and their entries in the existing method.

In this chapter, for partitioning the input parameters, the concept of a multigranular linguistic term set [22] is used. A typical multigranular linguistic term set with n number of linguistic terms is defined as:

$$L^n = \{ l_i^n \mid i \in \{1, 2, \ldots, n\} \}$$

where l_i^n represents the ith linguistic term of the set.

The whole development procedure of the fuzzy DSS is described below. To understand the whole methodology easily for the user, the computation procedure of Total Upper arm score is discussed in detail, whereas the remaining scoring procedures are discussed very briefly.

5.4.2.1 Arm and Wrist analysis

Step 1: Computation of Total Upper arm score

Step 1a: Calculation of Upper arm score In this step, an MFIS is generated with one input parameter, viz., *Upper arm angle* and one output parameter, viz., *Upper arm score* to compute the *Upper arm score.* The universe of discourse of *Upper arm angle* is considered as $[-90°, 180°]$, as the least value of *Upper arm angle* can be −90 degrees, and Upper arm can be rotated through an angle of maximum 180 degrees. The linguistic term set for *Upper arm angle* is considered as $\{l_i^5 \mid i = 1, 2, 3, 4, 5\}$. The MFs, as shown in Fig. 5.1, corresponding to these linguistic terms are represented by the respective TrFNs as $<-90, -90, -30, -10>$, $<-30, -10, 10, 30>$, $<10, 30, 35, 55>$, $<35, 55, 80, 100>$, and $<80, 100, 180, 180>$. On the other hand, the linguistic term set for *Upper arm score* is considered as $\{l_i^4 \mid i = 1, 2, 3, 4\}$. The MFs, as shown in Fig. 5.2, corresponding to these linguistic terms are represented by the respective TFNs as $<0, 1, 2>$, $<1, 2, 3>$, $<2, 3, 4>$, and $<3, 4, 4>$.

The rule base consisting of five inference rules is used to execute the whole inference process, which is summarized as follows:

Rule 1: If *Upper arm angle* is l_1^5, then *Upper arm score* is l_2^4.
Rule 2: If *Upper arm angle* is l_2^5, then *Upper arm score* is l_1^4.
Rule 3: If *Upper arm angle* is l_3^5, then *Upper arm score* is l_2^4.
Rule 4: If *Upper arm angle* is l_4^5, then *Upper arm score* is l_3^4.
Rule 5: If *Upper arm angle* is l_5^5, then *Upper arm score* is l_4^4.

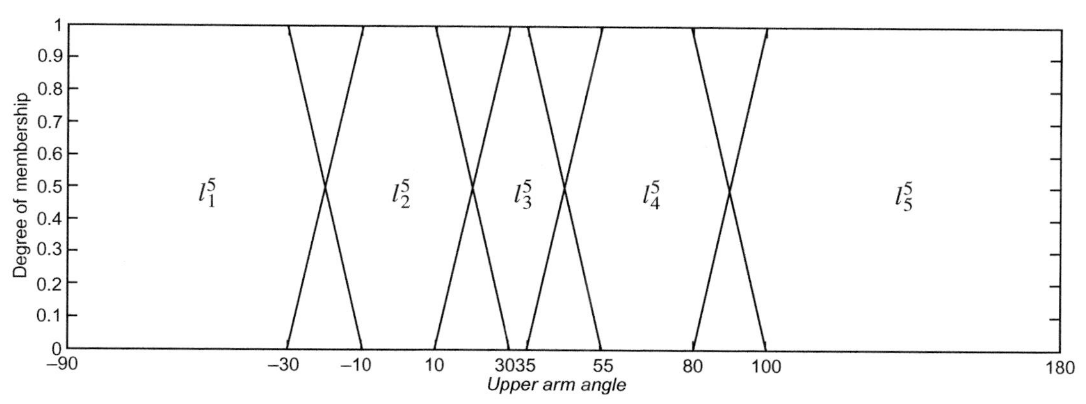

FIG. 5.1 MFs corresponding to the linguistic terms representing *Upper arm angle*.

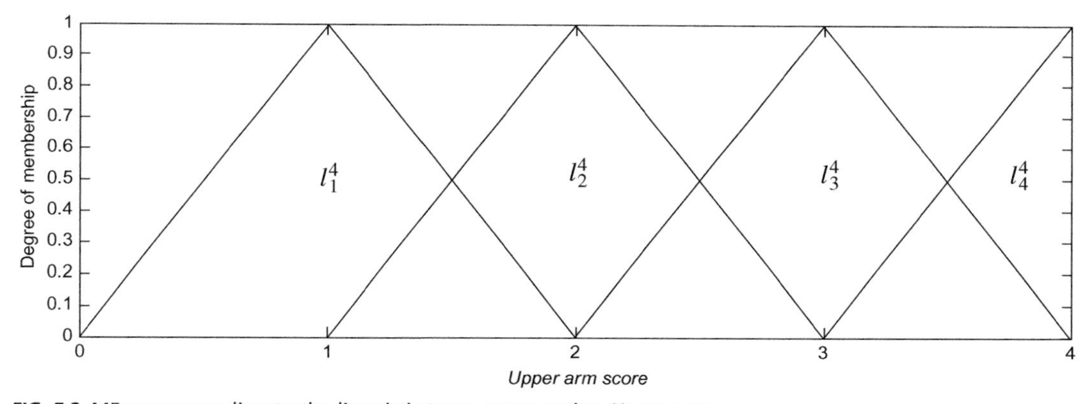

FIG. 5.2 MFs corresponding to the linguistic terms representing *Upper arm score*.

Step 1b: Adjustment score If the shoulder is raised, add 1.
If upper arm is abducted, add 1.
If arm is supported or person is leaning, subtract 1.
If no such situation occurs, add 0.

Finally, Total Upper arm score is evaluated as follows:

$$\text{Total Upper arm score} = \textit{Upper arm score} + \text{Adjustment score}$$

Step 2: Computation of Total Lower arm score

Step 2a: Calculation of Lower arm score The linguistic term sets for input, *Lower arm angle*, and output, *Lower arm score*, are, respectively, $\{l_i^3 \mid i=1, 2, 3\}$ and $\{l_i^2 \mid i=1, 2\}$. Three TrFNs, viz., $<0, 0, 50, 70>$, $<50, 70, 90, 110>$, and $<90, 110, 180, 180>$, and two TFNs, viz., $<0, 1, 2>$ and $<1, 2, 2>$, are considered to represent the MFs corresponding to the linguistic hedges of *Lower arm angle* and *Lower arm score*, respectively.

The rules of the MFIS for computation of *Lower arm score* are given as follows:

Rule 1: If *Lower arm angle* is l_1^3, then *Lower arm score* is l_2^2.
Rule 2: If *Lower arm angle* is l_2^3, then *Lower arm score* is l_1^2.
Rule 3: If *Lower arm angle* is l_3^3, then *Lower arm score* is l_2^2.

Step 2b: Adjustment score If arm is working across midline or out to side of body, add 1. If no such situation occurs, add 0.

Finally, Total Lower arm score is evaluated as follows:

$$\text{Total Lower arm score} = \textit{Lower arm score} + \text{Adjustment score}$$

Step 3: Computation of Total Wrist score

Step 3a: Calculation of Wrist score The linguistic term sets for input, *Wrist angle*, and output, *Wrist score*, are considered as $\{l_i^5 \mid i=1, 2, 3, 4, 5\}$ and $\{l_i^3 \mid i=1, 2, 3\}$, respectively. The TrFNs, viz., $<-90, -90, -25, -5>$, $<-25, -5, -5, 0>$, $<-5, 0, 0, 5>$, $<0, 5, 10, 25>$, and $<10, 25, 90, 90>$, and the TFNs, viz., $<0, 1, 2>$, $<1, 2, 3>$, and $<2, 3, 3>$, are considered to represent the MFs corresponding the linguistic hedges of *Wrist angle* and *Wrist score*, respectively.

The rules of the MFIS for computation of *Wrist score* are given as follows:

Rule 1: If *Wrist angle* is l_1^5, then *Wrist score* is l_3^3.
Rule 2: If *Wrist angle* is l_2^5, then *Wrist score* is l_2^3.
Rule 3: If *Wrist angle* is l_3^5, then *Wrist score* is l_1^3.
Rule 4: If *Wrist angle* is l_4^5, then *Wrist score* is l_2^3.
Rule 5: If *Wrist angle* is l_5^5, then *Wrist score* is l_3^3.

Step 3b: Adjustment score If wrist is bent from midline, add 1. If no such situation occurs, add 0.

Finally, Total Wrist score is evaluated as follows:

$$\text{Total Wrist score} = \textit{Wrist score} + \text{Adjustment score}$$

Step 4: Wrist twist score

If wrist is twisted in midrange, then Wrist twist score is 1.
If wrist is at or near end of range, then Wrist twist score is 2.

Step 5: Evaluation of Posture score A

In this step, *Posture score A* is evaluated by generating an MFIS with four input parameters, viz., *Total Upper arm score, Total Lower arm score, Total Wrist score*, and *Wrist twist score*; and one output parameter, viz., *Posture score A*. The linguistic term sets for four input parameters are $\{l_i^6 \mid i=1, 2, 3, 4, 5, 6\}$, $\{l_i^3 \mid i=1, 2, 3\}$, $\{l_i^4 \mid i=1, 2, 3, 4\}$, and $\{l_i^2 \mid i=1, 2\}$, respectively. The MFs corresponding to these linguistic term sets are represented by the TFNs as $\{<0,1,2>, <1,2,3>, <2,3,4>, <3,4,5>, <4,5,6>, <5,6,6>\}$, $\{<0,1,2>, <1,2,3>, <2,3,3>\}$, $\{<0,1,2>, <1,2,3>, <2,3,4>, <3,4,4>\}$ and $\{<1,1,1>, <2,2,2>\}$, respectively. The linguistic term set for the output parameter, *Posture score*

A, is considered as $\{l_i^9 \mid i=1, 2, 3, 4, 5, 6, 7, 8, 9\}$, and the MFs corresponding to these linguistic terms are <0, 1, 2>, <1, 2, 3>, <2, 3, 4>, <3, 4, 5>, <4, 5, 6>, <5, 6, 7>, <6, 7, 8>, <7, 8, 9>, and <8, 9, 9>.

The rule base for the evaluation of *Posture score A* is represented in Table 5.2 as follows:

Step 6: Computation of Muscle use score

If posture is mainly static (i.e., held more than 10 min) or if the action is repeated as 4X per minute, then add 1.

Step 7: Evaluation of *Force/load score*

In evaluating Force/load score, two ways of load handling by the workers are mainly considered, viz., intermittent way and static or repeated way. Thus, considering these two load handling conditions, two MFIS are generated for evaluation of *Force/load score.*

For intermittent cases, the linguistic term set is considered as $\{l_i^3 \mid i=1, 2, 3\}$ for both the input parameter, *Force/load,* and output parameter, *Force/load score.* The MFs corresponding to the linguistic terms of input parameters are <0, 0, 2.2, 6.6>, <2.2, 6.6, 19.8, 24.2>, and <19.8, 24.2, 220, 220>, respectively. On the other hand, the MFs corresponding to the linguistic terms of output parameters are <0, 0, 0, 0.5>, <0, 0.5, 1.5, 3>, and <2, 3, 3, 3>, respectively.

Table 5.2 Rule base for the evaluation of *Posture score A*.

Posture score A		Total Wrist score							
		l_1^4 Wrist twist score		l_2^4 Wrist twist score		l_3^4 Wrist twist score		l_4^4 Wrist twist score	
Total Upper arm score	Total Lower arm score	l_1^2	l_2^2	l_1^2	l_2^2	l_1^2	l_2^2	l_1^2	l_2^2
l_1^6	l_1^3	l_1^9	l_2^9	l_2^9	l_2^9	l_2^9	l_3^9	l_3^9	l_3^9
	l_2^3	l_2^9	l_2^9	l_2^9	l_2^9	l_3^9	l_3^9	l_3^9	l_3^9
	l_3^3	l_2^9	l_3^9	l_3^9	l_3^9	l_3^9	l_3^9	l_4^9	l_4^9
l_2^6	l_1^3	l_2^9	l_3^9	l_3^9	l_3^9	l_3^9	l_4^9	l_4^9	l_4^9
	l_2^3	l_3^9	l_3^9	l_3^9	l_3^9	l_3^9	l_4^9	l_4^9	l_4^9
	l_3^3	l_3^9	l_4^9	l_4^9	l_4^9	l_4^9	l_4^9	l_5^9	l_5^9
l_3^6	l_1^3	l_3^9	l_3^9	l_4^9	l_4^9	l_4^9	l_4^9	l_5^9	l_5^9
	l_2^3	l_3^9	l_4^9	l_4^9	l_4^9	l_4^9	l_4^9	l_5^9	l_5^9
	l_3^3	l_4^9	l_4^9	l_4^9	l_4^9	l_4^9	l_5^9	l_5^9	l_5^9
l_4^6	l_1^3	l_4^9	l_4^9	l_4^9	l_4^9	l_4^9	l_5^9	l_5^9	l_5^9
	l_2^3	l_4^9	l_4^9	l_4^9	l_4^9	l_4^9	l_5^9	l_5^9	l_5^9
	l_3^3	l_4^9	l_4^9	l_4^9	l_5^9	l_5^9	l_5^9	l_6^9	l_6^9
l_5^6	l_1^3	l_5^9	l_5^9	l_5^9	l_5^9	l_5^9	l_6^9	l_6^9	l_7^9
	l_2^3	l_5^9	l_6^9	l_6^9	l_6^9	l_6^9	l_7^9	l_7^9	l_7^9
	l_3^3	l_6^9	l_6^9	l_6^9	l_7^9	l_7^9	l_7^9	l_7^9	l_8^9
l_6^6	l_1^3	l_7^9	l_7^9	l_7^9	l_7^9	l_7^9	l_8^9	l_8^9	l_9^9
	l_2^3	l_8^9	l_8^9	l_8^9	l_8^9	l_8^9	l_9^9	l_9^9	l_9^9
	l_3^3	l_9^9	l_9^9	l_9^9	l_9^9	l_9^9	l_9^9	l_9^9	l_9^9

For static or repeated cases, all the information of input and output parameters are the same except the MFs corresponding to the linguistic terms of output parameters, which are changed to $<0, 0, 0, 1.5>$, $<0, 1.5, 2.5, 3>$, and $<2.5, 3, 3, 3>$.

The rules of the MFIS for evaluation of *Force/load score* (for both load handling conditions) are given as follows:

Rule 1: If *Force/load* is l_1^3, then *Force/load score* is l_1^3.
Rule 2: If *Force/load* is l_2^3, then *Force/load score* is l_2^3.
Rule 3: If *Force/load* is l_3^3, then *Force/load score* is l_3^3.

Step 8: Calculation of Wrist and Arm score
Finally, Wrist and Arm score is calculated as

$$\text{Wrist and Arm score} = \textit{Posture score A} + \text{Muscle use score} + \textit{Force/load score}$$

5.4.2.2 Neck, trunk, and leg analysis

Step 9: Computation of Total Neck score
***Step 9a: Calculation of* Neck score** Here, an MFIS is generated to compute *Neck score* with one input parameter, viz., *Neck angle,* and one output parameter, viz., *Neck score.* The linguistic term set corresponding to both *Neck angle* and *Neck score* is considered as $\{l_i^4 \mid i=1, 2, 3, 4\}$. The MFs corresponding to the linguistic terms for both the input and output parameters are represented by the respective sets as $\{<-45, -45, -5, 5>, <-5, 5, 5, 15>, <5, 15, 15, 25>, <15, 25, 90, 90>\}$ and $\{<0, 1, 2>, <1, 2, 3>, <2, 3, 4>, <3, 4, 4>\}$, respectively.

The rules of the MFIS for computation of *Neck score* are given as follows:

Rule 1: If *Neck angle* is l_1^4, then *Neck score* is l_4^4.
Rule 2: If *Neck angle* is l_2^4, then *Neck score* is l_1^4.
Rule 3: If *Neck angle* is l_3^4, then *Neck score* is l_2^4.
Rule 4: If *Neck angle* is l_4^4, then *Neck score* is l_3^4.

Step 9b: Adjustment score If neck is twisted or side bending, add 1.
If no such situation occurs, add 0.

Finally, Total Neck score is computed as follows:

$$\text{Total Neck score} = \textit{Neck score} + \text{Adjustment score}$$

Step 10: Computation of Total Trunk score
***Step 10a: Calculation of* Trunk score** The linguistic term sets for both the input, *Trunk angle,* and output, *Trunk score,* is considered as $\{l_i^4 \mid i=1, 2, 3, 4\}$. The MFs corresponding to the linguistic terms for both the input and output parameters are represented by the respective sets as $\{<0, 0, 0, 5>, <0, 5, 15, 25>, <15, 25, 55, 65>, <55, 65, 120, 120>\}$ and $\{<0, 1, 2>, <1, 2, 3>, <2, 3, 4>, <3, 4, 4>\}$, respectively.

The rule base of MFIS for computation of *Trunk score* consists of four inference rules, which are given as follows:

Rule 1: If *Trunk angle* is l_1^4, then *Trunk score* is l_1^4.
Rule 2: If *Trunk angle* is l_2^4, then *Trunk score* is l_2^4.
Rule 3: If *Trunk angle* is l_3^4, then *Trunk score* is l_3^4.
Rule 4: If *Trunk angle* is l_4^4, then *Trunk score* is l_4^4.

Step 10b: Adjustment score If Trunk is twisted or side bending, add 1.
If no such situation occurs, add 0.

Finally, Total Trunk score is computed as follows:

$$\text{Total Trunk score} = \textit{Trunk score} + \text{Adjustment score}$$

Step 11: Computation of Leg score

If legs and feet are supported, then Leg score is 1.
If not, then Leg score is 2.

Step 12: Evaluation of *Posture score B*

In this step, *Posture score B* is evaluated by generating an MFIS with three input parameters, viz., *Total Neck score, Total Trunk score,* and *Leg score*; and one output parameter, viz., *Posture score B.* The linguistic term sets for the first two input parameters is $\{l_i^6 \mid i=1, 2, 3, 4, 5, 6\}$ and for the third is $\{l_i^2 \mid i=1, 2\}$. The MFs corresponding to these linguistic term sets are represented by the TFNs as $\{<0, 1, 2>, <1, 2, 3>, <2, 3, 4>, <3, 4, 5>, <4, 5, 6>, <5, 6, 6>\}$ and $\{<1, 1, 1>, <2, 2, 2>\}$, respectively. The linguistic term set for the output parameter, *Posture score B,* is considered as $\{l_i^9 \mid i=1, 2, 3, 4, 5, 6, 7, 8, 9\}$, and the MFs corresponding to these terms are $<0, 1, 2>, <1, 2, 3>, <2, 3, 4>, <3, 4, 5>, <4, 5, 6>, <5, 6, 7>, <6, 7, 8>, <7, 8, 9>$, and $<8, 9, 9>$.

The rule base for the evaluation of *Posture score B* is represented in Table 5.3 as follows:

Step 13: Evaluation of Neck, Trunk, and Leg score

Finally, Neck, Trunk, and Leg score is calculated as:

$$\text{Neck, Trunk and Leg score} = \textit{Posture score B} + \text{Muscle use score} + \textit{Force/load score}$$

Table 5.3 Rule base for the evaluation of *Posture score B*.

Posture score B	**Total Trunk score**											
	l_1^6		l_2^6		l_3^6		l_4^6		l_5^6		l_6^6	
Total Neck score	**Leg score**		**Leg score**		**Leg score**		**Leg score**		**Leg score**		**Leg score**	
	l_1^2	l_2^2	l_1^2	l_2^2	l_1^2	l_2^2	l_1^2	l_2^2	l_1^2	l_2^2	l_1^2	l_2^2
l_1^6	l_1^9	l_3^9	l_2^9	l_3^9	l_3^9	l_4^9	l_5^9	l_5^9	l_6^9	l_6^9	l_7^9	l_7^9
l_2^6	l_2^9	l_3^9	l_2^9	l_3^9	l_4^9	l_5^9	l_5^9	l_5^9	l_6^9	l_7^9	l_7^9	l_7^9
l_3^6	l_3^9	l_3^9	l_3^9	l_4^9	l_4^9	l_5^9	l_5^9	l_6^9	l_6^9	l_7^9	l_7^9	l_7^9
l_4^6	l_5^9	l_5^9	l_5^9	l_6^9	l_6^9	l_7^9	l_7^9	l_7^9	l_7^9	l_7^9	l_8^9	l_8^9
l_5^6	l_7^9	l_7^9	l_7^9	l_7^9	l_7^9	l_8^9	l_8^9	l_8^9	l_8^9	l_8^9	l_8^9	l_8^9
l_6^6	l_8^9	l_8^9	l_8^9	l_8^9	l_8^9	l_8^9	l_8^9	l_9^9	l_9^9	l_9^9	l_9^9	l_9^9

5.4.2.3 Posture analysis

Step 14: Evaluation of *RULA score*

In this step, *RULA score* is evaluated by generating an MFIS with two input parameters, viz., *Wrist and Arm score* and *Neck, Trunk, and Leg score*; and one output parameter, viz., *RULA score*. The linguistic term sets for the two input parameters are considered as $\{l_i^8 \mid i=1, 2, 3, 4, 5, 6, 7, 8\}$ and $\{l_i^7 \mid i=1, 2, 3, 4, 5, 6, 7\}$. The MFs corresponding to these linguistic term sets are represented by the TrFNs as {<0, 1, 1, 2>, <1, 2, 2, 3>, <2, 3, 3, 4>, <3, 4, 4, 5>, <4, 5, 5, 6>, <5, 6, 6, 7>, <6, 7, 7, 8>, <7, 8, 13, 13>} and {<0, 1, 1, 2>, <1, 2, 2, 3>, <2, 3, 3, 4>, <3, 4, 4, 5>, <4, 5, 5, 6>, <5, 6, 6, 7>, <6, 7, 13, 13>}, respectively. The linguistic term set for the output parameter, *RULA score*, is considered as $\{l_i^7 \mid i=1, 2, 3, 4, 5, 6, 7\}$ and the MFs corresponding to these terms are <0, 1, 2>, <1, 2, 3>, <2, 3, 4>, <3, 4, 5>, <4, 5, 6>, <5, 6, 7>, and <6, 7, 7>.

The rule base for the evaluation of *RULA score* is represented in Table 5.4 as follows:

Step 15: Identifying the Action level

According to the RULA score, four action levels, viz., Action level 1, Action level 2, Action level 3, and Action level 4, as presented in Table 5.5, are considered corresponding to four respective actions, viz., acceptable posture, change of posture may be needed, posture needed to be changed soon, and implement change of posture for measurement of discomfort level of whole body working in a specific working posture. The linguistic hedges representing these action levels are shown in Fig. 5.3.

The whole methodology of the proposed fuzzy DSS is presented through a flowchart as shown in Fig. 5.4.

Table 5.4 Rule base for the evaluation of *RULA score*.

RULA score		Neck, Trunk, and Leg score						
		l_1^7	l_2^7	l_3^7	l_4^7	l_5^7	l_6^7	l_7^7
Wrist and Arm score	l_1^8	l_1^7	l_2^7	l_3^7	l_3^7	l_4^7	l_5^7	l_5^7
	l_2^8	l_2^7	l_2^7	l_3^7	l_4^7	l_4^7	l_5^7	l_5^7
	l_3^8	l_3^7	l_3^7	l_3^7	l_4^7	l_4^7	l_5^7	l_6^7
	l_4^8	l_3^7	l_3^7	l_3^7	l_4^7	l_5^7	l_6^7	l_6^7
	l_5^8	l_4^7	l_4^7	l_4^7	l_5^7	l_6^7	l_7^7	l_7^7
	l_6^8	l_4^7	l_4^7	l_5^7	l_6^7	l_6^7	l_7^7	l_7^7
	l_7^8	l_5^7	l_5^7	l_6^7	l_6^7	l_7^7	l_7^7	l_7^7
	l_8^8	l_5^7	l_5^7	l_6^7	l_7^7	l_7^7	l_7^7	l_7^7

Table 5.5 Action levels with their corresponding TrFNs.

Action level	TrFN
Action level 1	<0, 0, 2, 3>
Action level 2	<2, 3, 4, 5>
Action level 3	<4, 5, 6, 7>
Action level 4	<6, 7, 7, 7>

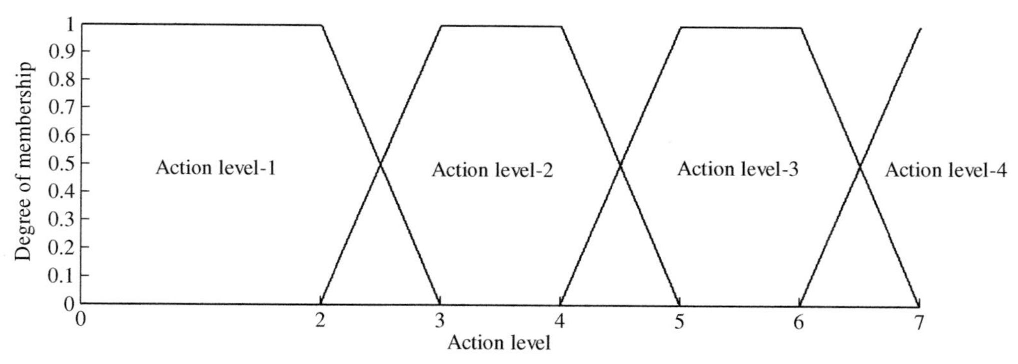

FIG. 5.3 Linguistic hedges representing the action levels.

5.4.3 Selection of most suitable operations associated with the proposed fuzzy DSS using MFIS

To find out the most suitable operations for MFIS, a sensitivity analysis is carried out by varying all the operations associated with the model. This analysis is performed by selecting 10 arbitrary working postures (as shown in Fig. 5.5) having respective RULA scores 7, 7, 6, 7, 7, 7, 6, 7, 7, and 7. The most suitable operations for the proposed fuzzy DSS are selected according to the highest value of Pearson's correlation coefficient [23], which is evaluated between the previously mentioned scores and the scores obtained from the proposed fuzzy DSS. The values of correlation coefficient are shown in Table 5.6.

This sensitivity analysis shows that the most suitable operation for the proposed methodology is found in Case 13, which proposes Minimum operation for conjunction, Maximum operation for disjunction, Product operation for Implication, Sum operation for aggregation, and Centroid operation for the defuzzification method. This high value, 0.9759, of Pearson's correlation coefficient implies that the proposed methodology is consistent and, as a consequence, the developed method appears as a reliable transition from the existing method to the newer one in the context of posture analysis.

To establish the application potentiality, validity, and reliability of the developed fuzzy DSS, a case study is performed concerning the health-related problems of the female workers who are engaged in Sal leaf plate-making units and who suffer from several MSDs due to awkward working postures.

5.5 Analysis of postures of the female workers engaged in Sal leaf plate-making units: A case study

Sal leaf plate making is a common occupation of tribal people in West Bengal, India [29]. The workers engaged with this profession collect Sal leaves from the nearby forest, stitch and dry it, and the final products are made with a semiautomatic machine [29]. Here, a case study is carried out among the female workers engaged at three different Sal leaf

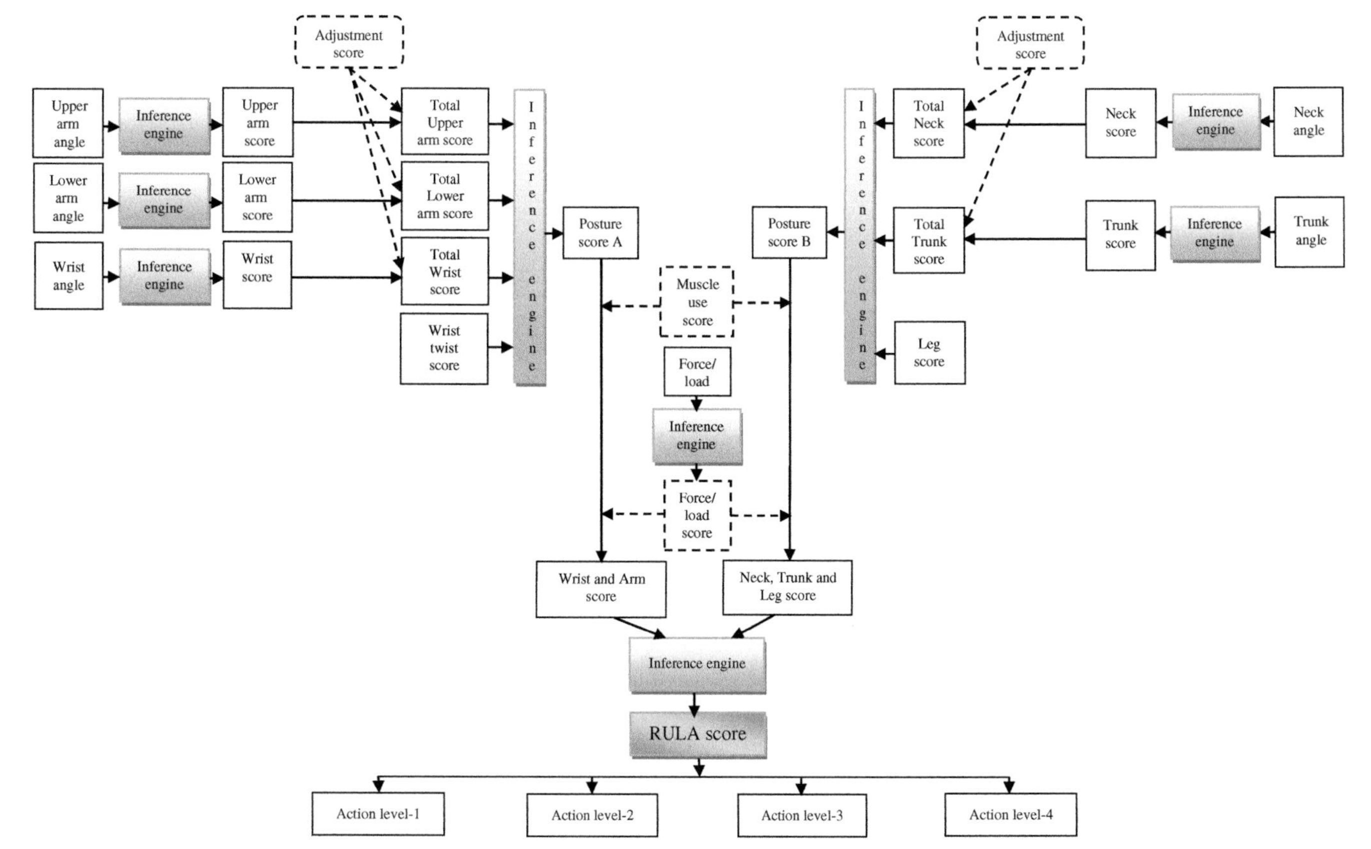

FIG. 5.4 Flowchart of the proposed fuzzy DSS.

FIG. 5.5 Working postures for selection of most suitable operations for fuzzy DSS.

plate-making units in West Bengal, India. For data collection, a team of eight members spent five days in those units. After careful observation of all kinds of works done by the female workers, a modified Nordic questionnaire (MNQ) was administrated to them and asked them to give all the relative information about their works. Through this questionnaire, the most discomfort-prone body part was found to be the lower back. The discomfort ratings of the lower back were collected using BPD scale. Finally, the consistent data are considered by comparing the values through the BPD scale with the information given by the workers through MNQ.

For validation of the developed methodology, 10 awkward working postures are identified where the lower back is mostly affected and are shown in Fig. 5.6.

Now, the posture analysis through the existing method corresponding to Posture I will be described in detail. The analyses corresponding to other postures can be performed in a similar manner.

Upper arm score, Lower arm score, Wrist score, and Wrist twist score corresponding to Posture I are calculated as 2, 1, 3, 1, respectively. Using these scores, Posture score A is evaluated as 3. Afterward, assigning Muscle use score and Force/load score as 1, 0, respectively, Wrist and Arm score is calculated as 4. Again, from the RULA worksheet, Neck score, Trunk score, and Leg score are found as 2, 3, 1, respectively, and thus Posture score B is evaluated as 4. Adding Muscle use score and Force/load score with Posture score B, Neck, Trunk, and Leg score is found as 5. Finally, RULA score for Posture I is calculated as 5, which belongs to Action Level 3 of the existing RULA method.

In the next section, the entire results of posture analyses corresponding to these postures are evaluated through the proposed fuzzy DSS.

Table 5.6 Sensitivity analyses for selection of most suitable operations.

Case	Conjunction	Disjunction	Implication	Aggregation	Defuzzification	Pearson's correlation coefficient
1	Minimum	Maximum	Minimum	Maximum	Centroid	0.8747
2	Minimum	Maximum	Minimum	Maximum	Bisector	0.8807
3	Minimum	Maximum	Minimum	Maximum	Middle of maximum	−0.1667
4	Minimum	Maximum	Minimum	Maximum	Smallest of maximum	0.5422
5	Minimum	Maximum	Minimum	Sum	Centroid	0.6721
6	Minimum	Maximum	Minimum	Sum	Bisector	0.8385
7	Minimum	Maximum	Minimum	Sum	Middle of maximum	−0.1667
8	Minimum	Maximum	Minimum	Sum	Smallest of maximum	0.3485
9	Minimum	Maximum	Product	Maximum	Centroid	0.9617
10	Minimum	Maximum	Product	Maximum	Bisector	0.8729
11	Minimum	Maximum	Product	Maximum	Middle of maximum	−0.1667
12	Minimum	Maximum	Product	Maximum	Smallest of maximum	0.6667
13	Minimum	Maximum	Product	Sum	Centroid	0.9759
14	Minimum	Maximum	Product	Sum	Bisector	0.7663
15	Minimum	Maximum	Product	Sum	Middle of maximum	−0.1667
16	Minimum	Maximum	Product	Sum	Smallest of maximum	0.6667
17	Product	Maximum	Minimum	Maximum	Centroid	0.7115
18	Product	Maximum	Minimum	Maximum	Bisector	0.5833
19	Product	Maximum	Minimum	Maximum	Middle of maximum	−0.1667
20	Product	Maximum	Minimum	Maximum	Smallest of maximum	0.3941
21	Product	Maximum	Minimum	Sum	Centroid	0.6939
22	Product	Maximum	Minimum	Sum	Bisector	0.7485
23	Product	Maximum	Minimum	Sum	Middle of maximum	−0.1667
24	Product	Maximum	Minimum	Sum	Smallest of maximum	0.6522
25	Product	Maximum	Product	Maximum	Centroid	0.6667
26	Product	Maximum	Product	Maximum	Bisector	0.6667
27	Product	Maximum	Product	Maximum	Middle of maximum	−0.1667
28	Product	Maximum	Product	Maximum	Smallest of maximum	0.6667
29	Product	Maximum	Product	Sum	Centroid	0.7331
30	Product	Maximum	Product	Sum	Bisector	0.6667
31	Product	Maximum	Product	Sum	Middle of maximum	−0.1667
32	Product	Maximum	Product	Sum	Smallest of maximum	0.6667

5.6 Results and discussion

The primary inputs of fuzzy DSS, viz., body joint angles, Force/load, etc. are the same as the existing methods of posture analysis. The steps for evaluation of final results of fuzzy DSS corresponding to Posture I is summarized as follows:

5.6.1 Arm and Wrist analysis

Step 1: Upper arm score is computed as 0.473, whereas adjustment score is found as 1. Thus, Total Upper arm score is evaluated as:

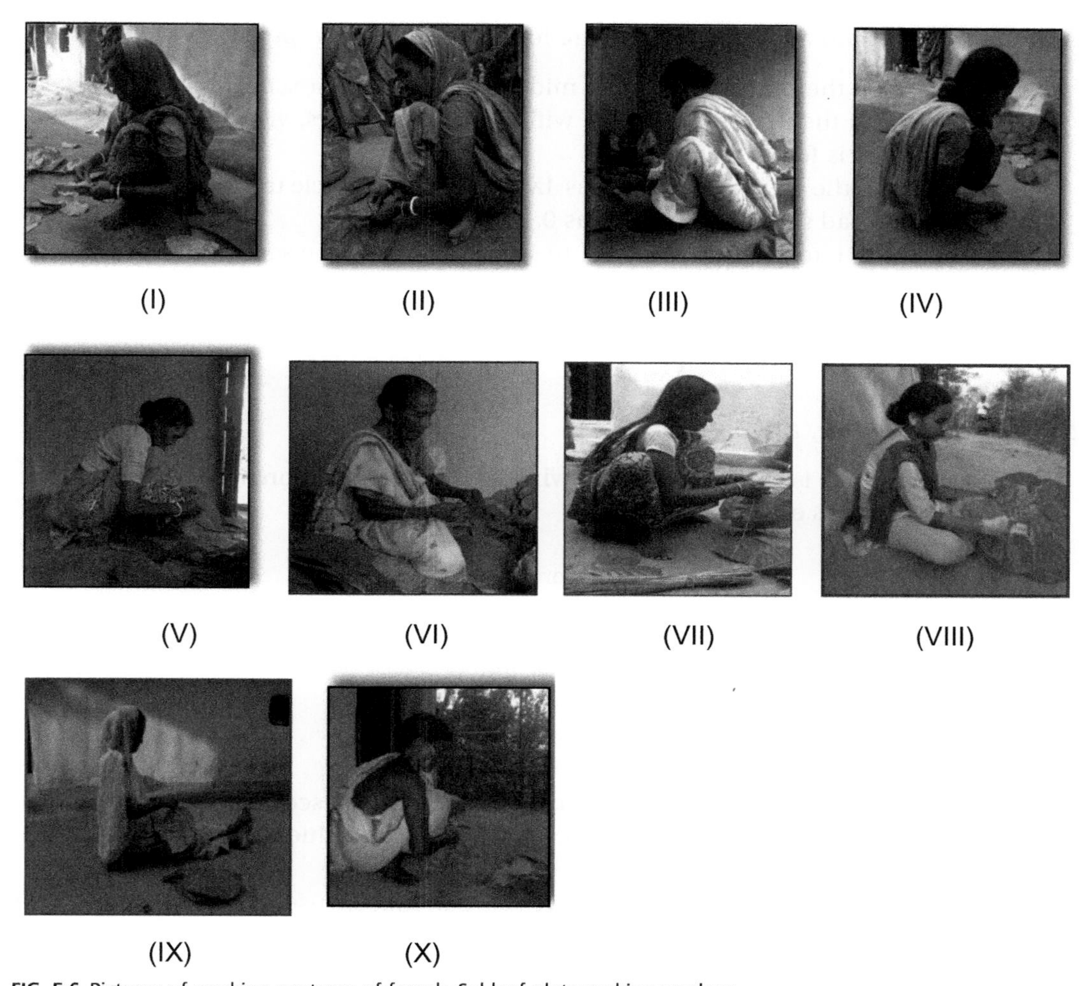

FIG. 5.6 Pictures of working postures of female Sal leaf plate-making workers.

$$\text{Total Upper arm score} = 0.473 + 1 = 1.473$$

Step 2: Lower arm score is computed as 1, whereas adjustment score is found 0. Thus, Total Lower arm score is evaluated as:

$$\text{Total Lower arm score} = 1 + 0 = 1$$

Step 3: Total Wrist score is computed as:

$$\text{Total Wrist score} = 2 + 1 = 3$$

as Wrist score and adjustment score are found as 2 and 1, respectively.

Step 4: Because the wrist is twisted in mid-range, Wrist twist score is found as 1.
Step 5: Through the inference process with four input values, viz., 1.473, 1, 3, and 1, Posture score A is found as 2.47.
Step 6: Because the action is repeated as 4X per minute, Muscle use score is found as 1.
Step 7: Force/load score is evaluated as 0.157.
Step 8: According to the value of Posture score A, Muscle use score, and Force/load score, Wrist and Arm score is evaluated as:

$$\text{Wrist and Arm score} = 2.47 + 1 + 0.157 = 3.627$$

5.6.2 Neck, trunk, and leg analysis

Step 9: Neck score is computed as 1.9, whereas adjustment score is found as 0. Thus, Total Neck score is evaluated as:

$$\text{Total Neck score} = 1.9 + 0 = 1.9$$

Step 10: Trunk score is computed as 3, whereas adjustment score is found as 0. Thus, Total Trunk score is evaluated as:

$$\text{Total Trunk score} = 3 + 0 = 3$$

Step 11: Because legs and feet are properly supported, Leg score is found as 1.
Step 12: Through the inference process with three input values, viz., 1.9, 3, and 1, Posture score B is found as 3.9.
Step 13: According to the value of Posture score B, Muscle use score, and Force/load score, Neck, Trunk, and Leg score is evaluated as:

$$\text{Neck, Trunk and Leg score} = 3.9 + 1 + 0.157 = 5.057$$

5.6.3 Posture analysis

Step 14: Through the MFIS with two input values obtained from Step 8 and Step 13 as 3.627 and 5.057, respectively, RULA score is evaluated as 4.72.
Step 15: The RULA score obtained through the proposed methodology belongs to two adjacent action levels, viz., Action level 2 and Action level 3 with degrees of membership 0.28 and 0.72, respectively.

The analyses corresponding to the remaining postures are performed through the proposed methodology in a similar manner and are listed in Table 5.7. The BPD scale is used to measure the discomfort of the lower back corresponding to these postures.

Table 5.7 Final results of posture analyses.

Posture	Existing RULA score	Action level	Final result of fuzzy DSS	Action level: degree of membership	Score through BPD scale
I	5	Action level 3	4.72	Action level 2: 0.28 Action level 3: 0.72	7
II	7	Action level 4	6.17	Action level 3: 0.83 Action level 4: 0.17	7
III	6	Action level 3	6.03	Action level 3: 0.97 Action level 4: 0.03	8
IV	7	Action level 4	6.64	Action level 3: 0.36 Action level 4: 0.64	9
V	7	Action level 4	6.64	Action level 3: 0.36 Action level 4: 0.64	9
VI	7	Action level 4	6.64	Action level 3: 0.36 Action level 4: 0.64	8
VII	7	Action level 4	6.64	Action level 3: 0.36 Action level 4: 0.64	9
VIII	6	Action level 3	5.68	Action level 3: 1	7
IX	4	Action level 2	3.62	Action level 2: 1	6
X	7	Action level 4	6.64	Action level 3: 0.36 Action level 4: 0.64	9

To establish the consistency and steadiness of the developed fuzzy DSS, the Pearson's correlation coefficient between the existing RULA scores and the final results of fuzzy DSS is evaluated as 0.9810, which reflects that the proposed model is consistent enough with the existing process.

The Pearson's correlation coefficient between the results of fuzzy DSS and the scores through BPD scale is found as 0.8565, whereas the Pearson's correlation coefficient between the existing RULA scores and the scores through BPD scale is found as 0.7911. This fact proves the validity and reliability of the developed methodology for analyzing working postures of the workers.

It is also seen from Table 5.7 that the final scores obtained through fuzzy DSS for all the postures, except Posture VIII and Posture IX, belong to two different adjacent action levels rather than one action level determined by the existing one. This fact establishes the strength of the proposed methodology in dealing with the uncertainties associated in the border region between two adjacent action levels. The degree of membership of the final score in an action level corresponding to a posture indicates the percentage of action need to treat with the discomfort associated with that posture.

5.7 Conclusions

The high value of Pearson's correlation coefficient between the existing RULA scores and the final results of fuzzy DSS also establishes the strength of the transition process from the traditional method to a more reliable technique of posture analysis. The higher value of Pearson's correlation coefficient of the scores through the BPD scale with the results of fuzzy DSS compared to the existing method indicates that the proposed model is very much capable in capturing the inexactness and uncertainties associated with the existing method of posture analysis. The value also shows that the proposed methodology successfully resolves the uncertainties occurred in the border region between input parameters.

The inclusion of a fuzzy expert system with the existing RULA method enriches the process of posture analysis in many ways. First, a slight change in the input values does not hamper the final result at all, which was a major problem faced in the traditional method. Second, the overlapping boundaries of the MFs of input variables are capable to capture all kinds of uncertainties associated with the boundary regions. Third, the overlapping boundaries of the four action levels minimizes the inexactness and the uncertainties associated with the final results, as it includes the intermediate action between two successive action levels. Fourth, the higher degree of membership of the final score in an action level corresponding to a posture, signifying that the posture needs that action more than others.

The developed DSS is capable of analyzing working postures of the workers by putting only the inputs in an easy manner, even in the absence of decision makers, so that necessary remedies can be taken in advance, which can reduce WMSDs by maintaining a healthy work environment. Because WMSDs are common globally, the data on an informal sector is very limited. So, proper evaluation and quantification will plan a preventive strategy in all the sectors. Finally, it is hoped that this study will change the approaches of posture analyses by identifying the awkward postures of the workers more accurately, which will be very helpful for enriching the working environment of the jobsites as well as for taking proper precautions of the workers well in advance.

Acknowledgments

The authors are thankful to the workers for filling in the supplied MNQ voluntarily. The authors remain grateful to the concerning managers and other related people of the Sal leaf plate-manufacturing units who shared their opinions. The authors also express their sincere gratitude to the anonymous reviewers for their comments and suggestions in improving the quality of the chapter.

References

[1] R.C. Kessler, P.A. Berglund, W.T. Chiu, A.C. Deitz, J.I. Hudson, V. Shahly, S. Aguilar-Gaxiola, J. Alonso, M.C. Angermeyer, C. Benjet, R. Bruffaerts, G. de Girolamo, R. de Graaf, J.M. Haro, J. Kovess-Villa, C. Sasu, K. Scott, M.C. Viana, M. Xavier, The prevalence and correlates of binge eating disorder in the world health organization world mental health surveys, Biol. Psychiatry 73 (9) (2013) 904–914.

[2] Bureau of Labor Statistics, https://www.bls.gov/news.release/osh2.t04.htm, 2015 Accessed 19 February 2018.

[3] D. Bernal, J. Campos-Serna, A. Tobias, S. Vargas-Prada, F.G. Benavides, C. Serra, Work-related psychosocial risk factors and musculoskeletal disorders in hospital nurses and nursing aides: a systematic review and meta-analysis, Int. J. Nurs. Stud. 52 (2) (2015) 635–648.

[4] D. Kee, S.R. Seo, Musculoskeletal disorders among nursing personnel in Korea, Int. J. Ind. Ergon. 37 (3) (2007) 207–212.

[5] S. Yasobant, S. Mohanty, Musculoskeletal disorders as a public health concern in India: a call for action, Physiotherapy 12 (2018) 46–47.

[6] S. Sahu, Musculoskeletal pain among female labourers engaged in manual material handling task in informal sectors of West Bengal, India, Asian Pac. Newslett. Occup. Health Saf. 17 (3) (2010) 58–60.

[7] S. Sahu, S. Chattopadhyay, K. Basu, G. Paul, The ergonomic evaluation of work related musculoskeletal disorder among construction labourers working in unorganized sector in West Bengal, India, J. Hum. Ergol. 39 (2) (2010) 99–109.

[8] G. Li, P. Buckle, Current techniques for assessing physical exposure to work-related musculoskeletal risks, with emphasis on posture-based methods, Ergonomics 42 (1999) 674–695.

[9] M.È. Chiasson, D. Imbeau, K. Aubry, A. Delisle, Comparing the results of eight methods used to evaluate risk factors associated with musculoskeletal disorders, Int. J. Ind. Ergon. 42 (5) (2012) 478–488.

[10] O. Karhu, P. Kansi, I. Kuornika, Correcting working postures in industry: a practical method for analyses, Appl. Ergon. 8 (4) (1977) 199–201.

[11] L. McAtamney, E.N. Corlett, RULA: a survey method for the investigation of work related upper limb disorders, Appl. Ergon. 24 (2) (1993) 91–99.

[12] S. Hignett, L. McAtamney, Rapid entire body assessment (REBA), Appl. Ergon. 31 (2000) 201–205.

[13] D. Kee, W. Karwowski, LUBA—an assessment technique for postural loading on the upper body based on joint motion discomfort and maximum holding time, J. Appl. Ergon. 32 (2001) 357–366.

[14] M. Sett, S. Sahu, Ergonomics evaluation of the tasks performed by the female workers in the unorganized sectors of the manual brick manufacturing units in India, Ergon. SA 22 (1) (2010) 2–16.

[15] L.A. Zadeh, The concept of a linguistic variable and its application to approximate reasoning-II, Inf. Sci. 8 (4) (1975) 301–357.

[16] J. Debnath, A. Biswas, S. Presobh, K.N. Sen, S. Sahu, Fuzzy inference model for assessing occupational risks in construction sites, Int. J. Ind. Ergon. 55 (2016) 114–128.

[17] A. Azadeh, I.M. Fam, M. Khoshnoud, M. Nikafrouz, Design and implementation of a fuzzy expert system for performance assessment of an integrated health, safety, environment (HSE) and ergonomics system: the case of a gas refinery, Inf. Sci. 178 (2008) 4280–4300.

[18] L.C. Rivero, G.R. Rodriguez, M.D.R. Perez, C. Mar, Z. Juárez, Fuzzy logic and RULA method for assessing the risk of working, Proc. Manuf. 3 (2015) 4816–4822.

[19] A. Golabchi, S.U. Han, A.R. Fayek, A fuzzy logic approach to posture-based ergonomic analysis for field observation and assessment of construction manual operations, Can. J. Civ. Eng. 43 (4) (2016) 294–303.

[20] M. Savino, A. Mazza, D. Battini, New easy to use postural assessment method through visual management, Int. J. Ind. Ergon. 53 (2016) 48–58.

[21] J. Debnath, A. Biswas, Assessment of occupational risks in construction sites using interval type-2 fuzzy analytic hierarchy process, Lect. Notes Networks Systs. 11 (2018) 283–297.

[22] L. Li, F. Liang, Q. Qin, X. Li, Behavioural multi-criteria decision making with multi-granularity 2-dimension fuzzy linguistic variables, in: Proceedings of the 28th Chinese Control and Decision Conference, (2016 CCDC), Yinchuan, China, 2016, pp. 5688–5692.

[23] K. Pearson, Notes on regression and inheritance in the case of two parents, Proc. R. Soc. Lond. 58 (1895) 240–242.

[24] M. An, W. Lin, A. Stirling, Fuzzy-reasoning-based approach to qualitative railway risk assessment, Proc. Inst. Mech. Eng. F J. Rail. Rapid Transit 220 (2) (2006) 153–167.

[25] E.H. Mamdani, S. Assilian, An experiment in linguistic synthesis with a fuzzy logic controller, Int. J. Man Mach. Stud. 7 (1) (1975) 1–13.

[26] A. Biswas, D. Majumder, S. Sahu, Assessing morningness of a group of people by using fuzzy expert system and adaptive neuro fuzzy inference model, Commun. Comput. Inf. Sci. 140 (2011) 47–56.

[27] H.K. Lee, First Course on Fuzzy Theory and Applications, Springer-Verlag, Berlin, Heidelberg, 2005.

[28] D. Majumder, J. Debnath, A. Biswas, Risk analysis in construction sites using fuzzy reasoning and fuzzy analytic hierarchy process, Proc. Tech. 10 (2013) 604–614.

[29] M. Dey, S. Sahu, Ergonomics survey of leaf plate making activity of tribal women, in: S. Gangopadhyay (Ed.), Proceeding of International Ergonomics Conference, HWWE, 2009-Ergonomics for Everyone, 2009, pp. 433–440.

6

Short PCG classification based on deep learning

Sinam Ajitkumar Singh[a], Takhellambam Gautam Meitei[a], Swanirbhar Majumder[b]

[a]DEPARTMENT OF ELECTRONICS AND COMMUNICATION ENGINEERING, NORTH EASTERN REGIONAL INSTITUTE OF SCIENCE AND TECHNOLOGY, NIRJULI, INDIA

[b]DEPARTMENT OF INFORMATION TECHNOLOGY, TRIPURA UNIVERSITY, AGARTALA, INDIA

6.1 Introduction

Health problems associated with the heart is one of the major causes of global mortality. Nearly half of the deaths that occur in the European continent are due to heart-related ailments [1]; other than this, 34.3% is the share of deaths that occur in America [2]. Unluckily, 50% of the world's cardiovascular diseases have occurred in the Asian region as both the top two countries of the world in terms of population are from this region [3]. Prediction of heart diseases at an early stage is necessary for immediate and consistent treatment, which further helps in reducing the risk factor. With the invention of the stethoscope by Hyacinth in 1816 [4], auscultation approach has become the widely used method for prediction of cardiovascular diseases. This is done via monitoring and analyzing the heart sounds based on the stethoscope. Advanced signal processing techniques are used in an analysis of heart auscultation via cheap and efficient techniques. The method of auscultation analysis can be divided into two steps [5]. First, to collect the heart sound as recorded data from the subject and second, prediction of the heart sound as normal or abnormal. But the physician requires extensive practice and skill for the clinical analysis of heart auscultation [6]. Furthermore, as reported in the literature [7–9], the accuracy of prediction by the physicians and medical students is about 20%–40% whereas by expert cardiologists, it is approximately 80% [7].

The recent studies reveal that analysis of heart auscultation is inadequate. This is because it has failed to detect the qualitative and quantitative properties of the heart sound signals [10, 11]. The heart sound visualization based on a graphical waveform is known as a phonocardiogram (PCG). Systole is the time interval between the first heart sound and the beginning of the corresponding cycle of the second heart sound. Diastole is the time interval between the second heart sound to the next S1.

The drawback of the heart auscultation analysis has been overcome by phonocardiography as it helps in providing essential features for identification of cardiovascular disorders. The typical waveform of any standard normal heart sound is conventional to the

Deep Learning Techniques for Biomedical and Health Informatics. https://doi.org/10.1016/B978-0-12-819061-6.00006-9

cardiologist. Thus, any notable abnormality in the waveform implies pathological analysis of the PCG signal will be required. Martinez et al. [12] had explained that the study based on PCG is a challenging task. This was due to the presence of several variables that affect the heart sound. Additionally, time-frequency characteristics of the PCG signal are not enough to identify the pathological condition. Hence, several researchers have analyzed multidomain characteristics to overcome the aforesaid drawbacks. Some shortcomings of PCG signal for clinical analysis are as follows:

1. Failure to analyze the frequency characteristics.
2. Failure to distinguish among different frequencies and shortage of knowledge about energy variations.
3. Failure to identify noise and presence of murmurs.
4. Failure to recognize the specific boundaries of the heart sound.

A normal PCG signal is produced by the closing of the heart valves, whereas abnormal murmurs are produced by turbulence of the flowing of blood across the heart valves. A murmur is one of the most familiar events of cardiovascular activities. A murmur can easily be distinguished from normal heart sound due to its distinguishing characteristics like noise and prolonged duration. A murmur can be of two types, innocent or benign murmur. Innocent murmurs are harmless, which are usually due to external factors. Benign murmurs are harmful and deadly. They are primarily due to a defect of the heart valve. However, a murmur can be a symptom of mental illness as well. A murmur can be inherent or from a person subjected to certain drugs. Valvular stenosis and insufficiency are two abnormal conditions that cause turbulence in blood flow. The valve is said to be stenosed when the valve surface is thickened and has blockages that partially result in obstruction or prevents the path of blood ejection due to deposition of calcium. When the valve cannot close effectively, it causes heave, ejection, or reverse leakage of blood; this condition is called valvular insufficiency. A ventricular septal defect is a type of murmur that occurs when there is a hole between the left and right ventricular wall. An abnormality such as ventricular septal defect, pulmonary stenosis, tricuspid insufficiency, aortic stenosis, and mitral insufficiency can cause systolic murmurs. The physician evaluates a heart murmur by identifying the factors like timing, quality, location, intensity, duration, and pattern [13].

6.1.1 Heart sound analysis

Before modeling a PCG-based classification system, the basic knowledge for generation of cardiac sound is necessary. A cardiac sound is an acoustic signal generated due to blood flow across the valves and transmission toward the chest, which can be recorded with the help of an electronic stethoscope [4]. With the help of a digital stethoscope, the recordings of PCG signals can be analyzed. The process can be broadly divided into four steps. The algorithm for analyzing PCG signal is illustrated in Fig. 6.1. The generalized steps for analyzing the heart sound are as follows:

1. preprocessing of the raw PCG signal to remove unwanted noise and murmur;
2. segmentation of the cardiac cycle;

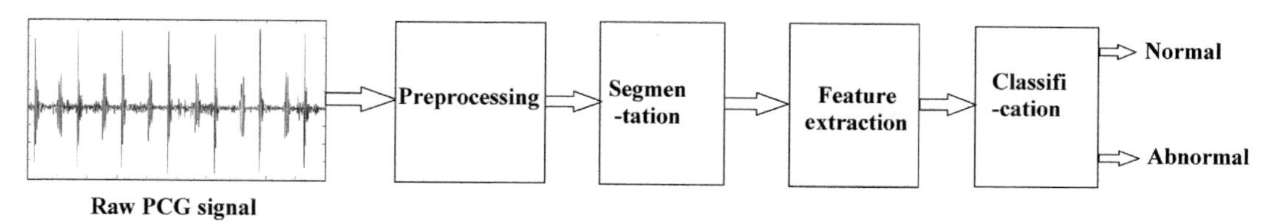

FIG. 6.1 Generalized algorithm for analyzing PCG signal.

3. feature extraction based on the segmented cardiac cycle; and
4. classification using extracted features.

The preprocessing step involves the localization of the heart sound. It includes downsampling of the PCG signal along with filtering to remove noise that corrupts the raw PCG signal. This is followed by normalization of the resultant filtered signal. The localization and segmentation of PCG signal too were important for analyzing the cardiovascular activities [4, 14–16]. Ahlstorm et al. [15] explained that the recorded PCG signal has a transient waveform with unwanted noise and murmur, which may need further localization. Localization of the PCG signal involves the tasks of determining the time of the event of PCG during the cardiac cycle. The preprocessed signal is applied for further segmentation in the cardiac cycle. In the segmentation step, the boundaries of the digitized heart sound are analyzed. The segmentation process has been classified into two types, namely direct segmentation and indirect segmentation. The direct segmentation process involves detecting the cardiac cycle boundaries from the PCG signal, whereas the indirect segmentation involves detecting the heart sound boundaries using electrocardiogram (ECG) signal as a reference. The feature extraction process based on segmented heart sound generates an output that tends to be more informative for further analysis from each cardiac cycle. In the classification step, the anomaly of the heart sound is identified and categorized based on the parameters recognized in the feature extraction step. The generalized algorithm for classifying the PCG signals has been illustrated in Fig. 6.1.

6.2 Materials and methods

This section describes the literature about the analysis of a PCG signal along with the limitation of segmentation steps. This is the process used mainly prior to the extraction of features for heart anomaly classification. This section also discusses the overview of the proposed model that comprises a preprocessing step for eliminating the unwanted noises and scalogram conversion using a continuous wavelet transform (CWT) approach.

6.2.1 Related works

Prediction of heart abnormality based on cardiac sound or PCG has become a popular research goal for numerous researchers. Brief literature about heart sound processing methods and observation of existing methods have been presented by Clifford et al.

[17]. Leng et al. [18] and Noponen et al. [19] displayed a brief survey of the different abnormalities of the PCG signal and its corresponding features. Leng et al. [18] further discussed heart auscultation analysis stressing methods of collection of PCG sound recordings along with an analysis of recorded PCG signals using computer-aided diagnosis (CAD) tools. A detailed explanation of the features selection and classification approaches have been presented by Lui et al. [20] and Marascio et al. [21].

Automated PCG classification based on CAD tools mainly depends on segmentation, extraction of features, and classification based on different models. Many researchers have used an ECG signal as a reference for segmentation of the PCG signal [22–24]. Ahlstrom et al. [23] extracted the features after segmentation of the cardiac cycle based on ECG reference. They followed it up with murmur classification based on an artificial neural network (ANN). Jabbari et al. [24] also extracted the features using matching pursuits after segmentation of a PCG signal based on ECG reference and similarly followed up by murmur classification using three-layer multilayer perceptron (MLP).

Lately, segmentation of cardiac cycles without ECG reference has been implemented in numerous case studies. Segmentation of PCG using envelope analysis is one of the traditional approaches for deriving features and classification of the heart sound. Segmentation of the cardiac cycle based on the envelope analysis is performed using the following three steps. First, the envelope of the heart sound is derived. This is followed up by identification of the peaks of the heart sound. Finally, the cardiac cycle is classified based on the peak conditioning approach. Many researchers have realized different algorithms for segmenting cardiac cycle based on envelope analysis. Shannon energy [25], Hilbert transforms [26], and deep recurrent neural network [27] are the popular different algorithms employed for envelope-based segmentation. Furthermore, another approach for segmentation uses a statistical model. An author in Schmidt et al. [28] adopted the duration-dependent hidden Markov models (HMMs) algorithm for the segmentation of the PCG signal. Gupta et al. [29] segmented the heart sound based on the k-means clustering algorithm.

Analyzing the heart sound signal in a different domain explains different physiological and pathological characteristics that allow dominant feature extraction. Recent studies have reported that the time-frequency domain analysis captures essential characteristics for extracting a feature against time representation or frequency representation [23]. Based on different transform domains, qualitative and quantitative estimation of the heart sound signals were analyzed. Feature extraction based on the time-frequency domain were investigated using short-time Fourier transform initially [30], wavelet transforms [5, 10], S transform [11], and Wigner-Ville distribution [31]. The most popular algorithm among the feature extraction approach was short-time Fourier transform that provides better resolution in a time and frequency domain using a fixed window. In the literature for analyzing the PCG signals, it was reported that the wavelet transform is the best fit for analyzing multiresolution signals [10]. The types of a wavelet transform that have been widely used for extracting the features are namely CWTs [32, 33], discrete wavelet transform [5, 10], and Mel-scale wavelet transform [34].

For classification of the heart sound signals, extraction of dominant features is very important. Most of the literature focuses on extracting features based on the diagnosed heart diseases followed by optimization to decrease the complexity of the model. Besides time-frequency-based features extraction, there are other features, namely entropy, energy, Mel-frequency cepstral coefficient (MFCC), and high-order statistics that helps in obtaining important information of the nonstationary heart sound signals, which further enhances the performance of the classification. The time domain features have been derived from timing characteristics of the heart sound. It includes the timing location of the first heart sound, systolic period, second heart sound, and diastolic period. The feature extraction in the time domain has been implemented after segmentation of the cardiac cycle. By analyzing the frequency domain features, the limitation of various physiological characteristics in the time domain have been solved. Extraction of features based on the frequency domain has been investigated by using zero-crossing analysis and bandpass filter bank [10, 11].

Classification is the final step for predicting the outcomes. This is done by feeding the extracted features to the relevant classifier. Various classification approaches that have been implemented as per the literature are K-nearest neighbors (KNN) [35], support vector machine (SVM) [25, 36–39], MLP [23, 24, 29], convolutional neural network (CNN) [40, 41], ensemble classifier [5, 42], discriminant analysis [43], and HMM [34]. Based on the dataset of the PhysioNet/CinC 2016 Challenge, various methods based on classifiers for predicting the anomaly of the heart sound have been realized. Most of the approaches executed on this dataset have been based on prior segmentation of the cardiac cycle before extracting features [36–38, 40–45]. Only a few papers based on the unsegmented approach have been implemented for classification [35, 46]. The brief study about various methods implemented by different researchers based on PhysioNet 2016 PCG dataset has been provided in Table 6.1.

Kamson et al. [47] had used an extended logistic regression-hidden semi-Markov model (HSMM) for segmentation of heart sound and reported an F1 score of 98.36%. Messner et al. [27] segmented the cardiac cycle of the PCG signal after the extraction of an envelope and spectral features followed by the deep learning algorithm and reported an average F1 score of 96%. Tang et al. [38] predicted the PCG signal using SVM classifier based on multidomain features with an average accuracy of 88%. Bradley et al. [36] also reported the use of an SVM classifier with an improved classification accuracy of 89.26%. Wei et al. [45] classified the heart sound using the CNN with a classification accuracy of 91.50%.

6.2.2 Limitation of segmentation

Any algorithm used for segmentation of the cardiac cycle has some limitations despite enhancing the performance of the system. It adds to the complexity and computational burden of the system. The demerits of realizing segmentation of the cardiac cycle based on the PCG signal are as follows:

Table 6.1 Detailed literature based on dataset provided by PhysioNet 2016 Challenge.

Year	Refs.	Analysis/ transform	Feature extraction method	Feature detailed/characteristics	Classifier
2017	[36]	FFT	Unsupervised sparse coding	Segmented sparse coefficient vector and time domain	SVM
2018	[40]	SSP technique	CNN		Modified AlexNet
2017	[43]	CQA	WPD CWT Shannon energy MFCC	Time, time-frequency, and perceptual domains	FDA-ANN
2017	[42]	HSMM		Time, frequency, wavelet, and statistical domain	Ensemble classifier
2017	[37]	DWT Shannon energy STFT	Tensor decomposition	Scaled spectrograms	SVM
2017	[44]	FFT bandpass filter Hilbert transformation	CS transformation	Time and statistical properties	Probability assessment
2018	[38]	High-pass filter	Discrete Fourier transform	Multidomain features	SVM
		HSMM	MFCC		
2018	[45]	HSMM	MFCC	Cepstral domain	CNN
2017	[41]		MFSC		CNN
2018	[35]	Wavelet analysis	Curve fitting	Time-frequency, spectral, and entropy domain	KNN
		Power spectral density	Fractal dimension		
2018	[46]		FFT	Spectral amplitude and wavelet entropy	Decision tree
			Wavelet analysis		

FFT, fast Fourier transform; *SVM*, support vector machine; *SSP*, spike signal processing; *CNN*, convolutional neural network; *CQA*, cycle quality assessment; *WPD*, wavelet package decomposition; *CWT*, continuous wavelet transform; *MFCC*, Mel-frequency cepstral coefficient; *FDA-ANN*, Fisher's discrimination analysis-artificial neural network; *HSMM*, hidden semi-Markov model; *DWT*, discrete wavelet transform; *STFT*, short-time Fourier transform; *CS*, continuous sound; *MFSC*, Mel-frequency spectral coefficient; *KNN*, K-nearest neighbors.

1. Segmentation based on ECG reference
   - **(a)** The segmentation using ECG signal as a reference requires both PCG and ECG recordings. However, the acquisition of both ECG and PCG is an inconvenient task in the case of a newborn child.
   - **(b)** The price of the simultaneous recording of both ECG and PCG is high.
2. Segmentation based on envelope detection (Hilbert transform, Homomorphic filtering, and Shannon energy)

   (a) The peak of the first heart sound is usually overlooked resulting to the misclassification of extra fault peak due to the influence of undesired noise and murmur.
   (b) For an inaccurate analysis applied during a peak conditioning in which the systolic period is shorter than the diastolic period, this hypothesis is wrong in case of a heart patient and newborn babies.
3. Segmentation based on statistical model (HMM, K mean, and HSMM)
   (a) Based on some characteristics like the energy of time-frequency, cardiac cycle, and correlation, the PCG signal was actually segmented. But the characteristics of the PCG signals vary between infants to aged people as well as persons with healthy hearts and cardiac patients.

6.2.3 Database

The PhysioNet/CinC Challenge of 2016 provides a platform by contributing a heart sound database for several researchers to analyze the anomaly of the heart using PCG recordings. The principal purpose of the challenge is to encourage the researchers to develop novel approaches for classification of the cardiovascular abnormalities from the PCG recordings collected from various environments (clinical or nonclinical). The record names are prefixed starting from "A" through "F" consisting of six different datasets. The PCG recordings were collected from both abnormal as well as healthy subjects in various environments like home care or clinical health care centers. PhysioNet provided a total of 3240 PCG recordings, including both training and hidden test datasets [17]. The PCG recordings lasted from 5 to 120 seconds with a sampling frequency of 2 kHz. To avoid overfitting of the system, the training samples and the testing samples were made mutually exclusive to each other. The heart sounds were collected from 764 subjects and stored as a .wav format. Four peculiar locations (pulmonic area, mitral area, aortic area, and tricuspid area) were selected from nine different locations for collecting the heart sound recordings. The datasets were tagged either as 1 for normal or −1 for abnormal. The abnormal recordings were from subjects with a chronic heart abnormality, whereas normal PCG recordings were from healthy subjects. Out of 3240 PCG recordings, 665 PCG recordings were collected from cardiac patients and remaining 2575 from healthy subjects. The detailed study of the database was analyzed by Liu et al. [20].

6.2.4 Overall system design

A passband filter was used to eliminate the undesired noises. Some of the recorded PCG signals contained an unwanted spike of higher amplitude, so to remove those redundant amplitudes we used a spike removal approach. The resultant signals were converted to scalogram images using the CWT method. These scalogram images are used to train and validate a neural network using CNN-based deep learning algorithm for predicting heart sound abnormality. The model of the proposed classification method is illustrated in Fig. 6.2.

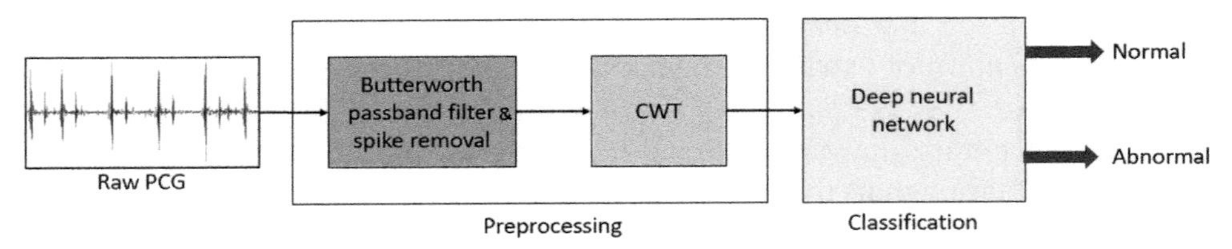

FIG. 6.2 Block diagram of the proposed model.

6.2.5 Preprocessing

To obtain a consistent analysis, the first 5 seconds of PCG recordings from each data had been selected for the study. Unwanted noise and murmur have an impact on raw PCG signals. PCG signals are affected by numerous noises, such as stethoscope movement, talking and breathing, etc. These can be cleaned using a bandpass filter in between. In this study, a Butterworth bandpass filter (fourth order) with a cut-off frequency of 25 and 400 Hz has been used to remove the undesirable noise as used by Schmidt et al. [28]. To exclude the excessive amplitude of PCG (frictional spike), steps used by Naseri et al. [48] have been realized. This is shown in the following table. Fig. 6.3 illustrates the preprocessing of the raw PCG signals.

Spike removal

(a) Based on the 0.5-second window, divide the PCG recording.
(b) Compute the maximum absolute amplitude (MAA) for each sliding window.
(c) If the value of MAA equal three times the median values of MAA then advance to step (d) else continue.
  (i) Determine the window with the highest MAA value.
  (ii) With the previous window as reference MAA, the noise spikes are computed from the respected location.
  (iii) Determine the last zero-crossing point using the starting location of the noise spike, which is just before the MAA point.
  (iv) Determine the first zero-crossing point using the end location of the noise spike which is just after the MAA point.
  (v) The determined noise spike is displaced by zeros.
  (vi) Start again from step (b).
(d) Tasks completed.

6.2.6 Continuous wavelet transform

CWT is an approach that presents an overcomplete description of the signal by dilation and shift of the mother wavelet, which varies continuously. For analyzing the nonstationary signals, CWT is widely used as compared with discrete wavelet transform (DWT). This is because DWT is unable to utilize all possible scales that compresses and extends

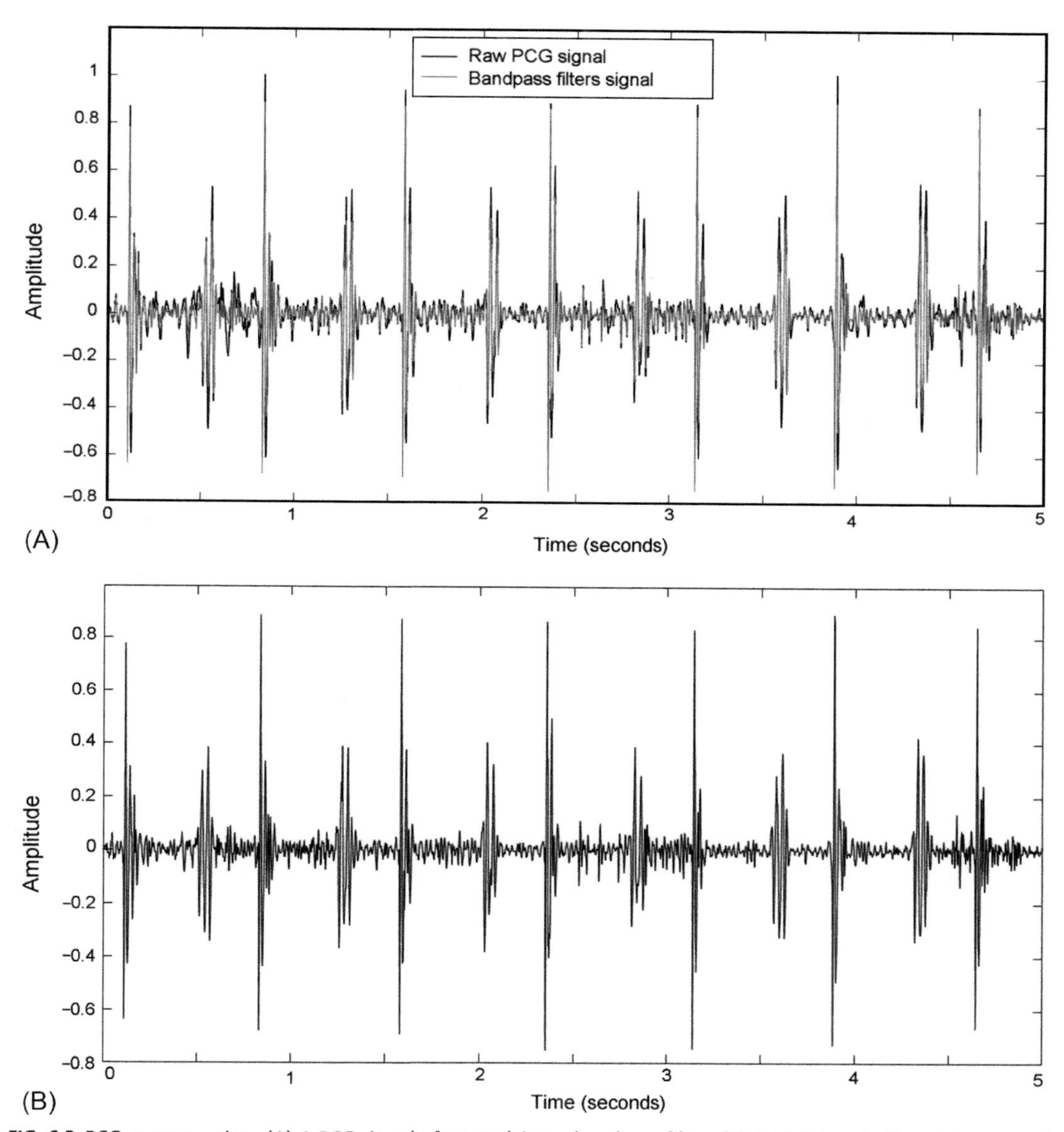

FIG. 6.3 PCG preprocessing. (A) A PCG signal after applying a bandpass filter. (B) A PCG signal after spike removal.

the mother wavelet due to the discretization process [32]. As a result, it tends to lose some useful information that may lead to extraction of false features from the source. Ergen et al. [32] analyzed the comparative study of different wavelets using CWT. They further proved that the Morlet wavelet is most suitable for analyzing the time-frequency representation of the PCG signals.

Based on a mother wavelet function, CWT decomposes the signal by contraction and transformation. By decomposing the PCG signal, CWT renders high resolution in terms of both time and frequency. According to wavelet properties, these can be classified as orthogonal, bi-orthogonal, complex, or noncomplex. In a linear representation of time-frequency representation, a continuous version of CWT can be expressed as

$$CWT_n^x(s,n) = \frac{1}{\sqrt{s}} \int_{-\infty}^{\infty} x(t)\phi\left(\frac{t-n}{s}\right) dt \tag{6.1}$$

where $\phi(t)$ denotes mother wavelet function, and s and n are the scale and shift parameters, respectively. The t and $x(t)$ denote the duration of the time and the original signal, respectively. CWT estimates the scale rather than frequency; hence the plot of CWT provides time-scale representation. The scale parameters represent the inverse of the frequency. In the literature, analysis of time-frequency characteristics of the PCG signal produces optimal results with the use of Morlet wavelet [32, 49]. Hence in this study, CWT is implemented for generating scalogram images using Morlet wavelet function. The Morlet wavelet function can be expressed as

$$\phi(t) = \frac{1}{\sqrt{d\pi}} e^{-\left(\frac{t}{d}\right)^2} e^{j2\pi f_c t} \tag{6.2}$$

where f_c is the center frequency of the Morlet wavelet and $j = \sqrt{-1}$. Hence, one scalogram image was generated for every preprocessed PCG signal based on CWT approach.

Fig. 6.4 represents the comparison of a time-frequency scalogram of abnormal and normal PCG recording after analyzing with the CWT method. Fig. 6.4A and C represents a pathological heart sound and healthy heart sound, respectively. The density of the time-series component has been represented with the color codes in a corresponding scalogram as shown in Fig. 6.4B and D. After analyzing the 2D signal as shown in Fig. 6.4B and D, from these images, the difference between normal PCG and a pathological PCG signal could easily be identified. Across the normal scalogram, the color intensity is smoothly dispersing. But in the case of abnormal scalogram, the color intensity abruptly disperses. CWT can reflect a different type of unique data from the time domain, which helps in a conversion from a 1D time domain to a 2D time-frequency characteristics scalogram representation at a different timing. This can help in prediction of the abnormality with the help of the heart sounds.

6.3 Convolutional neural network

A CNN (also ConvNet) is a neural network with an inherent machine learning approach based on a deep learning algorithm. It is mainly used to predict the classes of the input images. Most of the traditional methods require preprocessing steps, whereas the CNN determines the characteristics with limited preprocessing. It may include a single or multiple number of convolution layers with respective pooling layers. The predictive layers are

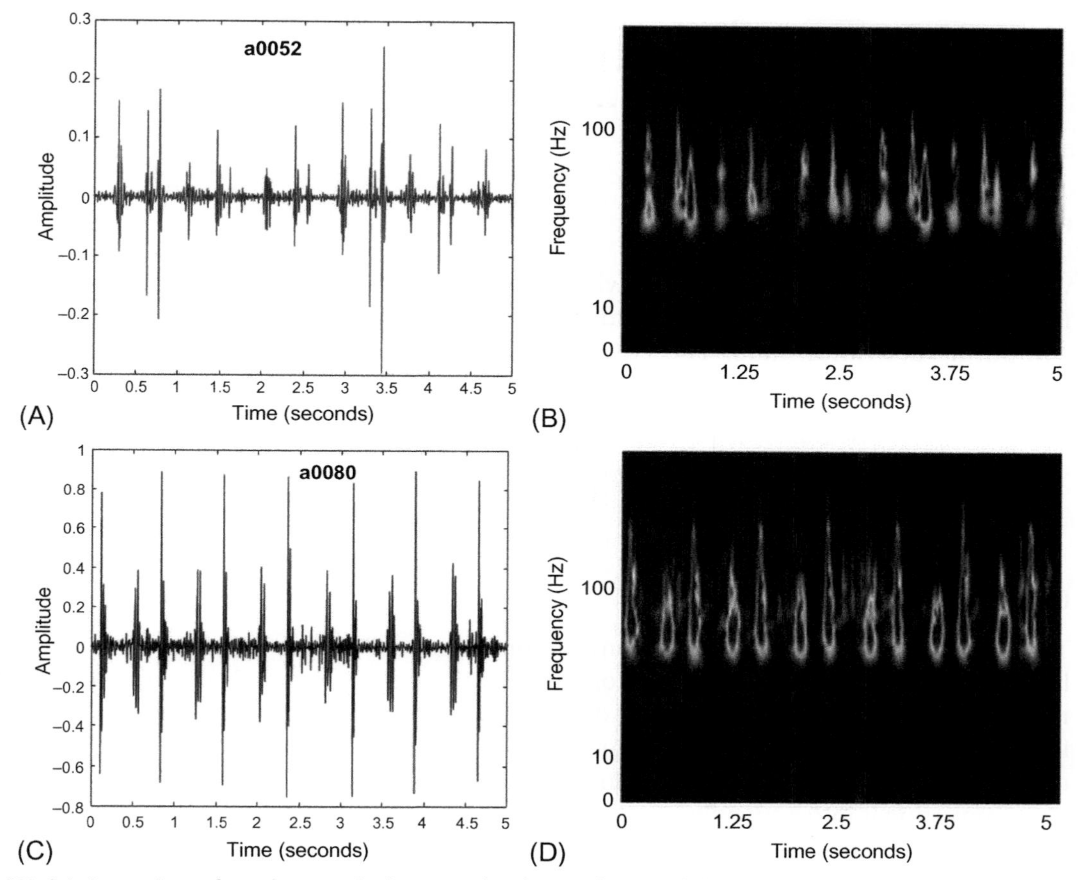

FIG. 6.4 Comparison of a scalogram plot between the abnormal PCG and the normal PCG during 5 seconds. (A) Abnormal PCG signal with annotation a0052 after filtering. (B) Scalogram representation of an abnormal PCG signal with annotation a0052. (C) Normal PCG signal with annotation a0080 after filtering. (D) Scalogram representation of a normal PCG signal with annotation a0080.

also known as fully connected layers due to their dense connections. A typical block diagram of a CNN is shown in Fig. 6.5.

6.3.1 Convolutional layer

The convolutional layer computes the convolutional operation of the input images using kernel filters to extract fundamental features. The kernel filters are of the same dimension but with smaller constant parameters as compared to the input images. As an example, for computing a $35 \times 35 \times 2$ 2D scalogram image, the acceptable filter size is $f \times f \times 2$, where $f = 3, 5, 7$, and so on. But the filter size needs to be smaller compared to that of the input image. The filter mask slides over the entire input image step by step and estimates the dot product between the weights of the kernel filters

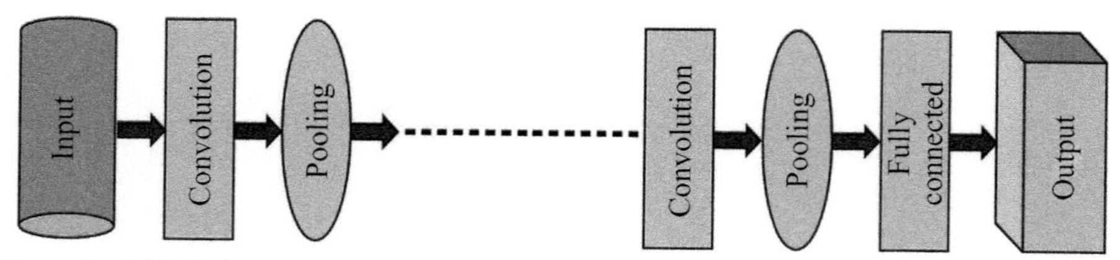

FIG. 6.5 Generalized block diagram of CNN.

with the value of the input image, which results in producing a 2D activation map. Hence, CNN will learn a visual feature. Fig. 6.6 illustrates a simple example of the computing activation map. The equation of the generalized convolutional layer is shown as

$$\begin{aligned}\text{Activation map} &= \text{Input} * \text{Filter} \\ &= \sum_{y=0}^{\text{columns}} \left(\sum_{x=0}^{\text{rows}} \right) \text{Input}\,(x-p, y-q)\ \text{Filter}\,(x,y)\end{aligned} \tag{6.3}$$

6.3.2 Pooling layer

The successive layer just after every convolution layer is called the pooling layer. A pooling layer helps in reducing the dimension of the activation map. It normally is an input for the next convolution layer obtained by the downsampling process. It drives to the loss of some information, which in turn helps the network from reducing both overfitting as well as the computational burden. A chosen window will slide over the entire area of the input step by step converting it into a specific value based on a subsampling method. The subsampling approach works using either an averaging approach or maximum value approach [50]. Due to the excellent performance characteristic, we prefer to choose the maximum value approach. Fig. 6.7 represents a simple operation for reducing the dimension of an activation map using a max pooling approach.

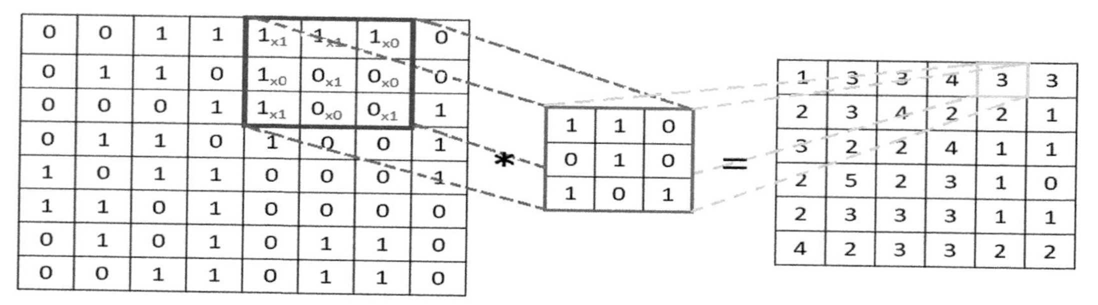

FIG. 6.6 Convolutional layer.

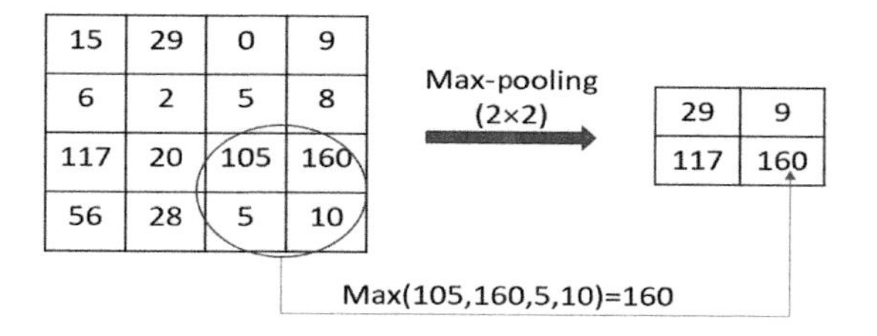

FIG. 6.7 Pooling layer.

6.3.3 Fully connected layer

The output of the preceding layer is connected as an input to the fully connected layer. This layer performs a high-level logical operation by collecting the features from the previous layers. This results in providing the final decision. In this study, the dropout layer has been adopted for eliminating the overfitting of the system. The final changes with the softmax layer provide a binary decision.

6.4 CNN-based automatic prediction

6.4.1 GoogleNet

The architecture of a GoogleNet-based deep learning method was introduced by Szegedy [51] in 2014. It consisted of 22 layers along with multiple inception layers; this helps to enhance the performance further. The inception layer [52] applies parallel filters of sizes $1 \times 1, 3 \times 3$, and 5×5. All the layers use RELU (rectified linear) activation function to provide efficient training performance. The model had been created based on an idea to be of low cost and with high efficiency. Hence, the model can be implemented for any application based on an embedded system or a system for communication. The architecture also consists of Gabor filters of various sizes in series to control multiple scales. The model has been employed as an inception module with dimensionality reduction in spite of the simple traditional version to enhance the classification efficiency of the model. GoogleNet uses 12 times fewer parameters than the popular AlexNet model. GoogleNet provides higher classification accuracy compared with the other models. The fully connected layer has been replaced with pooling layer (average pooling) for classification. The generalized network of GoogleNet is illustrated in Fig. 6.8. Fig. 6.9 represents the inception layer with dimensionality reduction.

6.4.2 Training using GoogleNet

The systems were trained based on the DistBelief [53] distributed deep network algorithm. This algorithm used a proper quantity of model along with data parallelism. The proposed model used the modified version of GoogleNet. The objective of the study has been to

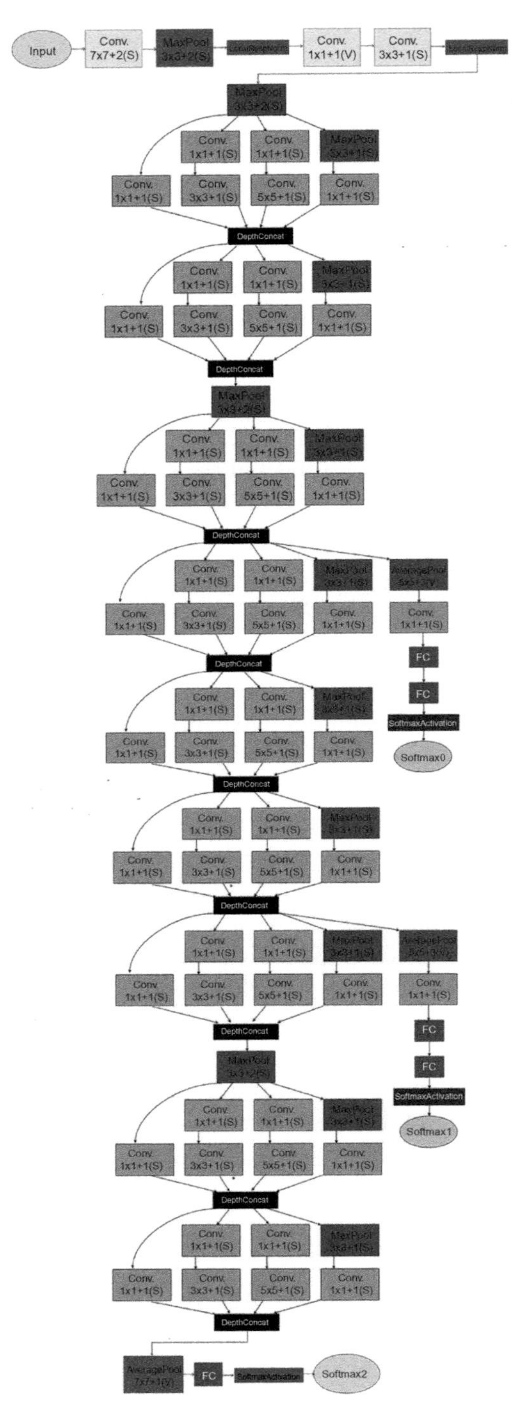

FIG. 6.8 Network of GoogleNet.

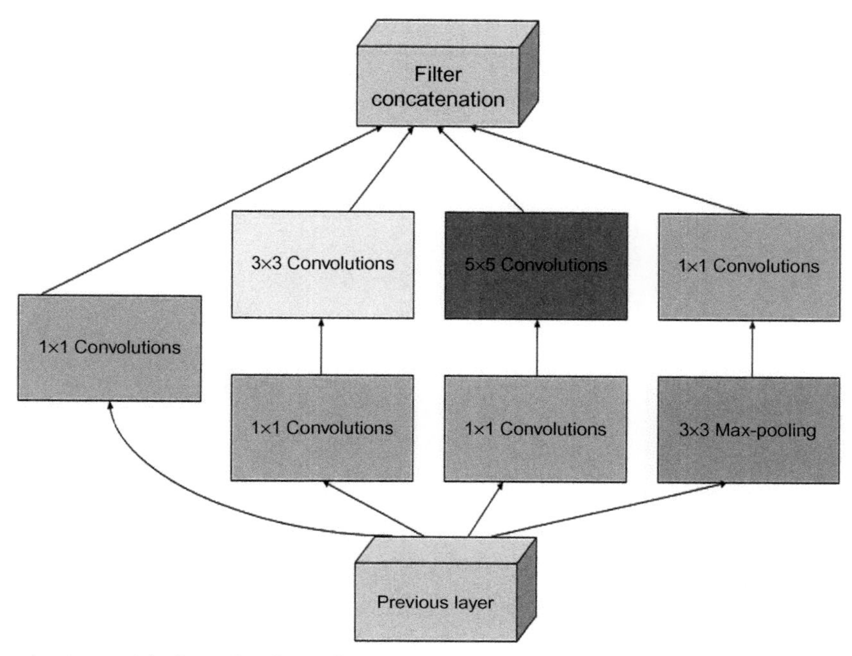

FIG. 6.9 Inception layer with dimensionality reduction.

predict between the two-class labeled (normal or abnormal). Hence, the output of the final layer had been remodeled with two. Using an asynchronous stochastic gradient descent, the proposed model was trained with a momentum of 0.9 and an initial learning rate of 0.0001. The detailed description of the layers using the proposed model is illustrated in Table 6.2 and the setup parameters are as in Table 6.3. The weights of the first convolutional layer are shown in Fig. 6.10. Whereas, Fig. 6.11 illustrates the generation of activation map using a scalogram image.

6.4.3 Performance parameter

Four parameters are estimated to measure the performance of the proposed model based on a deep learning method. The parameters are: True positive (tp), true negative (tn), false positive (fp), and false negative (fn). When a model correctly predicts the healthy subjects, then the term is known as true positive, whereas when a model correctly predicts the pathological subjects, then the term is known as true negative. Similarly, when a model incorrectly predicts the healthy subjects, then the term is known as false positive, whereas when a model incorrectly predicts the pathological subjects, then the term is known as a false negative.

Table 6.2 Details of different layers of proposed model (modified GoogleNet)

Layer	Padding	Stride	Output	Depth	1 × 1	3 × 3 Reduce	3 × 3	5 × 5 Reduce	5 × 5	Pool proj.
Input			224 × 224 × 3							
Convolution 1	7 × 7	3	112 × 112 × 3	1						
Max pool 1	3 × 3	2	56 × 56 × 64	0						
Convolution 2	3 × 3	1	56 × 56 × 192	2		64	192			
Max pool 2	3 × 3	2	28 × 28 × 192	0						
Inception (3a)			28 × 28 × 256	2	64	96	128	16	32	32
Inception (3b)			28 × 28 × 480	2	128	128	192	32	96	64
Max pool 3	3 × 3	2	14 × 14 × 480	0						
Inception (4a)			14 × 14 × 512	2	192	96	208	16	48	64
Inception (4b)			14 × 14 × 512	2	160	112	224	24	64	64
Inception (4c)			14 × 14 × 512	2	128	128	256	24	64	64
Inception (4d)			14 × 14 × 528	2	112	144	288	32	64	64
Inception (4e)			14 × 14 × 832	2	256	160	320	32	128	138
Max pool 4	3 × 3	2	7 × 7 × 832	0						
Inception (5a)			7 × 7 × 832	2	256	160	320	32	128	128
Inception (5b)			7 × 7 × 1024	2	384	192	384	48	128	128
Avg. pool			1 × 1 × 1024	0						
Fully connected 1 (dropout 60%)			1 × 1 × 1024	0						
Fully connected 2 (linear)			1 × 1 × 1000	1						
Fully connected 3 (softmax)			1 × 1 × 2	0						

Table 6.3 Setup parameters using CNN model.

CNN parameters	Values
Initial learning rate	0.0001
Mini-batch size	15
Maximum epochs	10
Validation frequency	10
Validation patience	Infinity
Verbose	1

$$\text{Sensitivity} = \frac{tp}{tp + fn} * 100\% \tag{6.4}$$

$$\text{Specificity} = \frac{tn}{tn + fp} * 100\% \tag{6.5}$$

$$\text{Accuracy} = \frac{tp + tn}{tp + tn + fp + fn} * 100\% \tag{6.6}$$

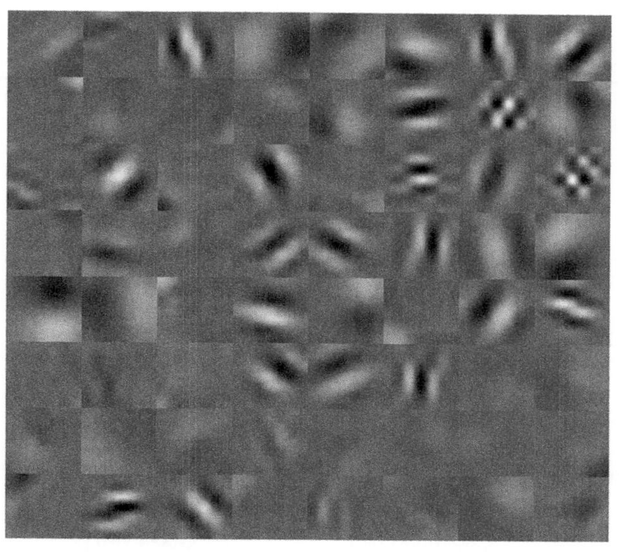

FIG. 6.10 Weights of the first convolutional layer.

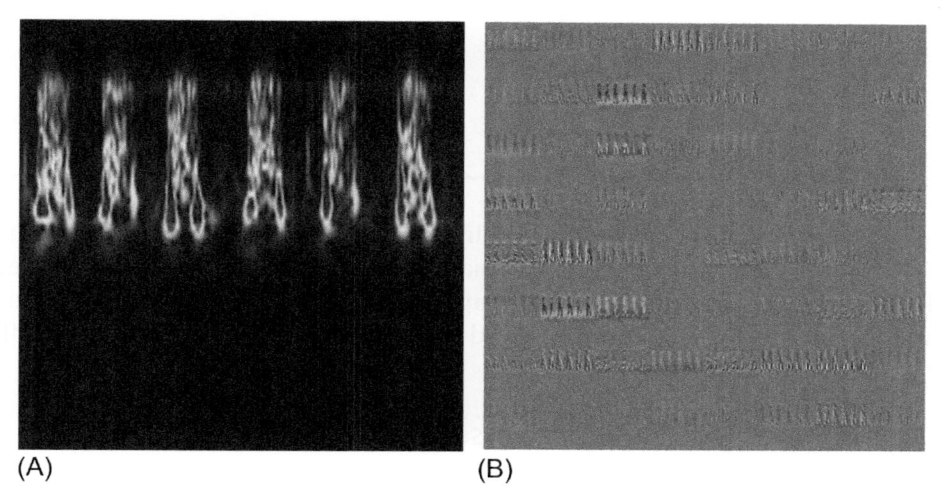

FIG. 6.11 Generation of activation map using first convolutional layer. (A) A scalogram image with annotation c0010 (1–10,000 sample). (B) Corresponding activation map.

6.5 Result

In this section, the performance of the proposed model is computed by randomly selecting 50% of the scalogram images for training and the remaining 50% for validation of the CNN-based model. This section has been divided into two sections. The first section

Table 6.4 A brief description about datasets for training and validating our CNN-based model.

Datasets	Training (50%)			Validating (50%)		
	Total	Normal	Abnormal	Total	Normal	Abnormal
a	205	59	146	204	58	146
b	245	193	52	245	193	52
c	16	4	12	15	3	12
d	28	14	14	27	13	14
e	1071	979	92	1070	979	91
f	57	40	17	57	40	17
Whole (a–f)	1620	1288	332	1620	1287	333

Table 6.5 Performance of our proposed model based on deep learning for different datasets.

Dataset	Sensitivity (%)	Specificity (%)	Accuracy (%)
a	87.70	32.70	72.06
b	21.20	94.30	78.80
c	100.00	66.66	93.33
d	92.90	61.50	77.80
e	84.60	99.50	98.22
f	29.40	87.50	70.18
Whole	88.58	87.80	87.96

involves the performance analysis of the individual datasets (from dataset a to f) and the second section involves the performance analysis of total datasets. The detailed description of the datasets for training and validating the proposed model has been illustrated in Table 6.4. The performance of our proposed model based on deep learning for different datasets has been explained in Table 6.5.

6.5.1 For analyzing individual datasets

The PCG-based scalogram images are used for training and validating the proposed model. CNN based on a deep learning algorithm had been employed for predicting the abnormality in the scalogram images. The performance of CNN for different datasets has been explained in Table 6.5. An accuracy of 72.6% with 87.7% sensitivity was obtained for dataset a. For dataset b, CNN predicted specificity of 94.3% with an accuracy of 78.8%. For dataset c, CNN-based deep learning classifier successfully classified the scalogram images with 93.3% accuracy and sensitivity of 100%. The classification accuracy of 77.7% with a sensitivity of 92.9% had been estimated for dataset d. Furthermore, the performance of the proposed model had improved for dataset e with an accuracy of 98.22% with a sensitivity and specificity of 84.6% and 99.5%, respectively. For dataset f, the CNN-

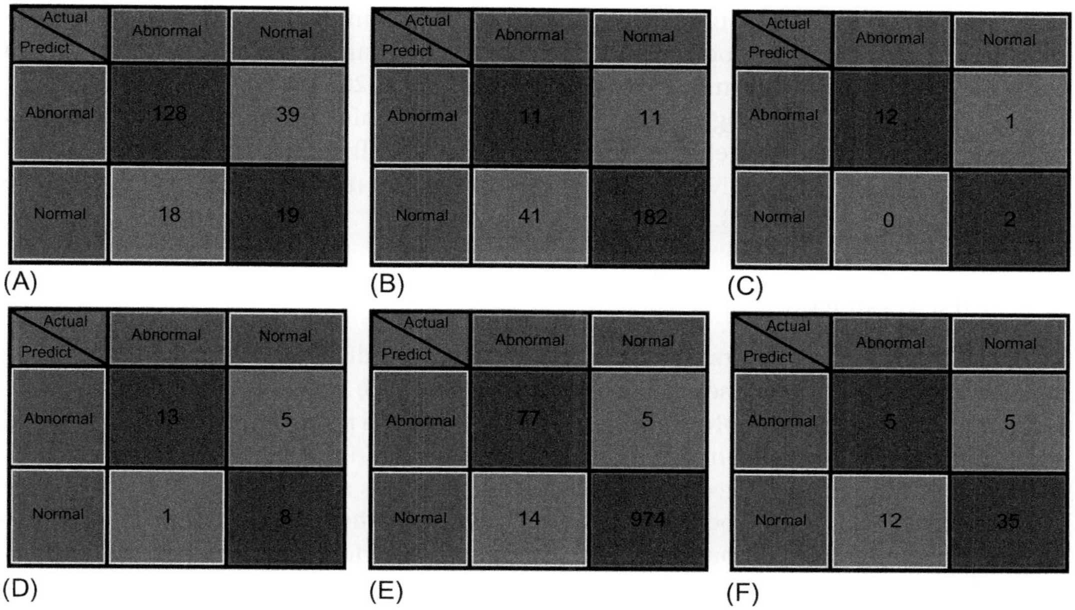

(A)

Actual / Predict	Abnormal	Normal
Abnormal	128	39
Normal	18	19

(B)

Actual / Predict	Abnormal	Normal
Abnormal	11	11
Normal	41	182

(C)

Actual / Predict	Abnormal	Normal
Abnormal	12	1
Normal	0	2

(D)

Actual / Predict	Abnormal	Normal
Abnormal	13	5
Normal	1	8

(E)

Actual / Predict	Abnormal	Normal
Abnormal	77	5
Normal	14	974

(F)

Actual / Predict	Abnormal	Normal
Abnormal	5	5
Normal	12	35

FIG. 6.12 Confusion matrix for different datasets using CNN classifier. (A) Dataset a. (B) Dataset b. (C) Dataset c. (D) Dataset d. (E) Dataset e. (F) Dataset f.

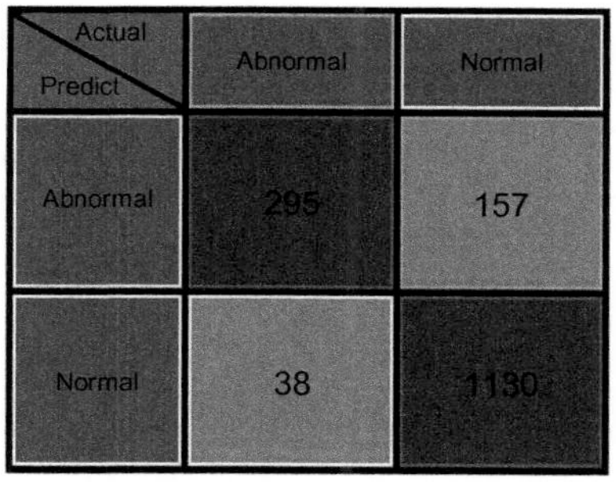

Actual / Predict	Abnormal	Normal
Abnormal	295	157
Normal	38	1130

FIG. 6.13 Confusion matrix for analyzing whole datasets.

based classifier achieved an accuracy of 70.18% with 87.5% specificity. The confusion matrix for analyzing different datasets using the proposed classifier is illustrated in Fig. 6.12.

6.5.2 For analyzing whole datasets

The scalogram images of 5 seconds of PCG recording had been used to train, validate, and estimate the performance of the proposed CNN-based deep learning model. To eliminate the chance of overfitting the model, 50% of the images are randomly split for training and the remaining 50% for validation of the model. Out of 3240 images, 1620 images that comprised of 332 abnormal scalogram images and 1288 normal scalogram images had been employed for training the model and the remaining 1620 images that comprised of another 333 abnormal scalogram images and 1287 normal scalogram images had been used for validating the model. An accuracy of 87.96% with sensitivity and specificity of 88.58% and 87.80%, respectively, was achieved using the model as explained using the confusion matrix in Fig. 6.13.

6.6 Discussion

To date, very few researchers have analyzed PCG for predicting the abnormality of the heart sound based on an unsegmented cardiac cycle [5, 35, 39, 46]. Using the same datasets published by the PhysioNet/CinC 2016 Challenge, the performance of the proposed model was computed by comparing it with different existing methods. In this chapter, using the time-frequency-based scalogram images, it was shown that the performance for anomaly prediction can be enhanced. Table 6.6 shows the comparative analysis of different datasets using traditional methods in which the performance of the model is predicted based on sensitivity, specificity, and accuracy. The performance of the model depends on the quantity of training and validating datasets, which can be observed in Table 6.5. The performance of the CNN-based classifier increases with the increase of

Table 6.6 Comparative analysis of our method with different existing methods using different datasets from PhysioNet 2016.

Dataset	Method	Sensitivity (%)	Specificity (%)	Accuracy (%)
a	**Proposed**	**87.70**	**32.70**	**72.06**
	Homomorphic	96.00	8.00	71.00
	Curve fitting	71.00	32.00	60.00
	Fractal dimension	77.00	43.00	67.00
b	**Proposed**	**21.20**	**94.30**	**78.80**
	Homomorphic	25.00	81.00	70.60
	Curve fitting	23.00	77.00	65.00
	Fractal dimension	32.00	81.00	71.00
c	**Proposed**	**100.00**	**66.67**	**93.33**
	Homomorphic	25.00	100.00	80.00
	Curve fitting	81.00	35.00	69.00
	Fractal dimension	92.00	90.00	92.00
d	**Proposed**	**92.90**	**61.50**	**77.80**
	Homomorphic	25.00	81.00	70.60
	Curve fitting	23.00	77.00	65.00
	Fractal dimension	32.00	81.00	71.00
e	**Proposed**	**84.60**	**99.50**	**98.22**
	Homomorphic	23.00	95.00	89.70
	Curve fitting	36.00	94.00	89.00
	Fractal dimension	93.00	99.00	98.00

Table 6.7 Comparative analysis of our method with state-of-the-art methods using whole datasets from PhysioNet 2016.

Author	Sensitivity (%)	Specificity (%)	Accuracy (%)
Proposed CNN	**88.58**	**87.80**	**87.96**
Dominguez et al. [40]	93.20	95.12	97.00
Wei et al. [45]	98.33	84.67	91.50
Bradley et al. [36]	90.00	88.45	89.26
Tang et al. [38]	88.00	87.00	88.00
Plesinger et al. [44]	89.00	81.60	85.00
Mostafa et al. [43]	76.96	88.31	82.63
Masun et al. [42]	79.60	80.60	80.10
Singh et al. [54]	93.00	90.00	90.00
Philip et al. [46]	77.00	80.00	79.00

training and validating datasets. The proposed model achieved a classification accuracy of 78.80%, 93.33%, and 98.22% from three different datasets.

Comparative analysis of the proposed CNN model with state-of-the-art methods using whole datasets is shown in Table 6.7. The variation of abnormal and normal scalogram images resulted in diminishing the performance of the proposed CNN model. The classification accuracy obtained for the previous methods varied between 79% to 97%. Whereas, their sensitivity and specificity varied between 76.96% to 98.33% and 80% to 95.12%, respectively. The classification accuracy of 97% had been achieved by Dominguez et al. [40], but their approach for classification of PCG recording was complex. Most of the classification approaches in the literature had been based on the segmentation of the heart cycle. However, Philip et al. [46] employed classification by eliminating complex segmentation process and achieved a classification accuracy of 79% with sensitivity and specificity being 77% and 80%, respectively.

The proposed method based on deep learning has defined an alternative approach for analyzing the classification of PCG, that is, recorded signal. This proposed model achieved an overall accuracy of 87.96% with sensitivity and specificity of 88.58% and 87.96%, respectively. It further proved that the proposed method provided comparable performance when compared with other segmented- and unsegmented-based classification methods. Hence, CNN based on deep learning can overcome the hindrances encountered by the existing traditional methods.

6.7 Conclusion

This chapter has introduced a new method for predicting the anomaly of the PCG recordings based on 5-second scalogram images. Using the PCG recordings published in the PhysioNet/CinC 2016 Challenge, the data had been preprocessed followed by conversion to 2D scalogram images using the CWT approach. These images were used to train and validate using CNN-based deep learning algorithm. The proposed method was also evaluated for whole datasets by comparing it with the previous state-of-the-art methods. Hence, the proposed CNN-based classifier achieved high classification performance by elimination of complex segmentation processes, which can aid the cardiologist or physician to analyze heart abnormality more accurately.

References

[1] N. Townsend, M. Nichols, P. Scarborough, M. Rayner, Cardiovascular disease in Europe—epidemiological update 2015, Eur. Heart J. 36 (2015) 2696–2705.

[2] D. Lloyd-Jones, R.J. Adams, T.M. Brown, M. Carnethon, S. Dai, G. De Simone, T.B. Ferguson, E. Ford, K. Furie, C. Gillespie, A. Go, K. Greenlund, N. Haase, S. Hailpern, P.M. Ho, V. Howard, B. Kissela, S. Kittner, D. Lackland, L. Lisabeth, A. Marelli, M.M. McDermott, J. Meigs, D. Mozaffarian, M. Mussolino, G. Nichol, V.L. Roger, W. Rosamond, R. Sacco, P. Sorlie, R. Stafford, T. Thom, S. Wasserthiel-Smoller, N.D. Wong, J. Wylie-Rosett, Heart disease and stroke statistics-2010 update, Circulation 121 (2010) 948–954.

[3] S. Sasayama, Heart disease in Asia, Circulation 118 (2008) 2669–2671.

[4] F.L. de Hedayioglu, Heart Sound Segmentation for Digital Stethoscope Integration, University of Porto, 2011.

[5] S. Yuenyong, A. Nishihara, W. Kongprawechnon, K. Tungpimolrut, A framework for automatic heart sound analysis without segmentation, Biomed. Eng. (Online) 10 (2011) 13.

[6] D. Roy, J. Sargeant, J. Gray, B. Hoyt, M. Allen, M. Fleming, Helping family physicians improve their cardiac auscultation skills with an interactive CD-ROM, J. Contin. Educ. Health Prof. 22 (2002) 152–159.

[7] E. Etchells, C. Bell, K. Robb, Does this patient have an abnormal systolic murmur? J. Am. Med. Assoc. 277 (1997) 564–571.

[8] S. Mangione, L.Z. Nieman, Cardiac auscultatory skills of internal medicine and family practice trainees: a comparison of diagnostic proficiency, J. Am. Med. Assoc. 278 (1997) 717–722.

[9] M.Z.C. Lam, T.J. Lee, P.Y. Boey, W.F. Ng, H.W. Hey, K.Y. Ho, P.Y. Cheong, Factors influencing cardiac auscultation proficiency in physician trainees Singapore, Med. J. 46 (2005) 11.

[10] D.S.M. Amin, B. Fethi, Features for heartbeat sound signal normal and pathological computer, Recent Pat. Comp. Sci. 1 (2008) 1–8.

[11] J. Singh, R.S. Anand, Computer aided analysis of phonocardiogram, J. Med. Eng. Technol. 31 (2007) 319–323.

[12] J. Martinez-Alajarin, R. Ruiz-Merino, Efficient method for events detection in phonocardiographic signals, Bioeng. Bioinspired Syst. II SPIE 5839 (2005) 398–410.

[13] C. Ahlstrom, K. Hglund, P. Hult, J. Hggstrm, C. Kvart, P. Ask, Distinguishing innocent murmurs from murmurs caused by aortic stenosis by recurrence quantification analysis, Int. J. Biol. Life Sci. 1 (2008) 201–206.

[14] L.H. Cherif, S.M. Debbal, F. Bereksi-Reguig, Segmentation of heart sounds and heart murmurs, J. Mech. Med. Biol. 8 (2008) 549–559.

[15] C. Ahlstrm, Nonlinear Phonocardiographic Signal Processing, Doctoral dissertation, Institutionen fr medicinsk teknik, 2008.

[16] D. Kumar, P. Carvalho, M. Antunes, J. Henriques, M. Maldonado, R. Schmidt, J. Habetha, Wavelet transform and simplicity based heart murmur segmentation, Comput. Cardiol. 2006 (2006) 173–176.

[17] G.D. Clifford, C. Liu, B. Moody, D. Springer, I. Silva, Q. Li, R.G. Mark, Classification of normal/abnormal heart sound recordings: the PhysioNet/computing in cardiology challenge 2016, 2016 Computing in Cardiology Conference (CinC), IEEE, 2016, pp. 609–612.

[18] S. Leng, R.S. Tan, K. Tshun, C. Chai, C. Wang, D. Ghista, L. Zhong, The electronic stethoscope, Biomed. Eng. (Online) 14 (2015) 66.

[19] A. Noponen, S. Lukkarinen, A. Angerla, R. Sepponen, Phono-spectrographic analysis of heart murmur in children, BioMed. Cent. 7 (2007) 23.

[20] C. Liu, D. Springer, Q. Li, B. Moody, R.A. Juan, F.J. Chorro, F. Castells, J.M. Roig, I. Silva, A.E.W. Johnson, Z. Syed, S.E. Schmidt, C.D. Papadaniil, L. Hadjileontiadis, H. Naseri, A. Moukadem, A. Dieterlen, C. Brandt, H. Tang, M. Samieinasab, M.R. Samieinasab, R. Sameni, R.G. Mark, G.D. Clifford, An open access database for the evaluation of heart sound algorithms, Physiol. Meas. 37 (2016) 2181–2213.

[21] G. Marascio, P.A. Modesti, Current trends and perspectives for automated screening of cardiac murmurs, Rev. Cardiovasc. Technol. 5 (2013) 213–218.

[22] J. Herzig, A. Bickel, A. Eitan, N. Intrator, Monitoring cardiac stress using features extracted from S1 heart sounds, IEEE Trans. Bio. Eng. 62 (2014) 1169–1178.

[23] C. Ahlstrom, P. Hult, P. Rask, J.E. Karlsson, E. Nylander, U. Dahlstrm, P. Ask, Feature extraction for systolic heart murmur classification, Ann. Biomed. Eng. 34 (2006) 1666–1677.

[24] S. Jabbari, H. Ghassemian, Modeling of heart systolic murmurs based on multivariate matching pursuit for diagnosis of valvular disorders, Comput. Biol. Med. 41 (2011) 802–811.

[25] I. Maglogiannis, E. Loukis, E. Zafiropoulos, A. Stasis, Support vectors machine-based identification of heart valve diseases using heart sounds, Comput. Methods Prog. Biomed. 95 (2009) 47–61.

[26] N.E. Huang, Z. Shen, S.R. Long, M.C. Wu, H.H. Shih, N. Yen, C.C. Tung, H.H. Liu, The empirical mode decomposition and the Hilbert spectrum for nonlinear and non-stationary time series analysis, Proc. R. Soc. A 454 (1998) 903–995.

[27] E. Messner, M. Zhrer, F. Pernkopf, Heart sound segmentation—an event detection approach using deep recurrent neural networks, IEEE Trans. Biomed. Eng. 65 (2018) 1964–1974.

[28] S.E. Schmidt, C. Holst-Hansen, C. Graff, E. Toft, J.J. Struijk, Segmentation of heart sound recordings by a duration-dependent hidden Markov model, Physiol. Meas. 31 (2010) 513–529.

[29] C.N. Gupta, R. Palaniappan, S. Swaminathan, S.M. Krishnan, Neural network classification of homomorphic segmented heart sounds, Appl. Soft Comp. 7 (2007) 286–297.

[30] Y. Soeta, Y. Bito, Detection of features of prosthetic cardiac valve sound by spectrogram analysis, Appl. Acoust. 89 (2015) 28–33.

[31] M.S. Obaidat, Phonocardiogram signal analysis: techniques and performance comparison, J. Med. Eng. Technol. 17 (1993) 221–227.

[32] B. Ergen, Y. Tatar, H.O. Gulcur, Time-frequency analysis of phonocardiogram signals using wavelet transform: a comparative study, Comput. Methods Biomech. Biomed. Eng. 15 (2012) 371–381.

[33] S. Debbal, A. Tani, Heart sounds analysis and murmurs, Int. J. Med. Eng. 8 (2016) 49–62.

[34] S. Chauhan, P. Wang, C.S. Lim, V. Anantharaman, A computer-aided MFCC-based HMM system for automatic auscultation, Comput. Biol. Med. 38 (2008) 221–233.

[35] M. Hamidi, H. Ghassemian, M. Imani, Classification of heart sound signal using curve fitting and fractal dimension, Biomed. Signal Process Control 39 (2018) 351–359.

[36] B.M. Whitaker, P.B. Suresha, C. Liu, G.D. Clifford, D.V. Anderson, Combining sparse coding and time-domain features for heart sound classification, Physiiol. Meas. 38 (2017) 1701.

[37] W. Zhang, J. Han, S. Deng, Heart sound classification based on scaled spectrogram and tensor decomposition, Expert Syst. Appl. 84 (2017) 220–231.

[38] H. Tang, Z. Dai, Y. Jiang, T. Li, C. Liu, PCG classification using multidomain features and SVM classifier, BioMed. Res. Int. (2018).

[39] S.W. Deng, J.Q. Han, Towards heart sound classification without segmentation via autocorrelation feature and diffusion maps, Future Gener. Comput. Syst. 60 (2016) 13–21.

[40] J.P. Dominguez-Morales, A.F. Jimenez-Fernandez, M.J. Dominguez-Morales, G. Jimenez-Moreno, Deep neural networks for the recognition and classification of heart murmurs using neuromorphic auditory sensors, IEEE Trans. Biomed. Circuits Syst. 12 (2018) 24–34.

[41] V. Maknickas, A. Maknickas, Recognition of normal-abnormal phonocardiographic signals using deep convolutional neural networks and Mel-frequency spectral coefficients, Physiol. Meas. 38 (2017) 1671.

[42] M.N. Homsi, P. Warrick, Ensemble methods with outliers for phonocardiogram classification, Physiol. Meas. 38 (2017) 1631.

[43] M. Abdollahpur, A. Ghaffari, S. Ghiasi, M.J. Mollakazemi, Detection of pathological heart sounds, Physiol. Meas. 38 (2017) 1616.

[44] F. Plesinger, I. Viscor, J. Halamek, J. Jurco, P. Jurak, Heart sounds analysis using probability assessment, Physiol. Meas. 38 (2017) 1685.

[45] W. Han, Z. Yang, J. Lu, S. Xie, Supervised threshold-based heart sound classification algorithm, Physiol. Meas. 39 (2018) 115011.

[46] P. Langley, A. Murray, EHeart sound classification from unsegmented phonocardiograms, Physiol. Meas. 38 (2018) 1658–1670.

[47] A.P. Kamson, L.N. Sharma, S. Dandapat, Multi-centroid diastolic duration distribution based HSMM for heart sound segmentation, Biomed. Signal Process Control 48 (2019) 265–272.

[48] H. Naseri, M.R. Homaeinezhad, H. Pourkhajeh, Noise/spike detection in phonocardiogram signal as a cyclic random process with non-stationary period interval, Comput. Biol. Med. 43 (2013) 1205–1213.

[49] S.A. Singh, S. Majumder, A novel approach OSA detection using single-lead ECG scalogram based on deep neural network, J. Mech. Med. Biol. 19 (2019) 1950026.

[50] C.-Y. Lee, P.W. Gallagher, Z. Tu, Generalizing pooling functions in convolutional neural networks: mixed, gated, and tree, Artif. Intell. Stat. 51 (2016) 464–472.

[51] C. Szegedy, W. Liu, Y. Jia, P. Sermanet, S. Reed, D. Anguelov, D. Erhan, A. Rabinovich, Going deeper with convolutions, in: IEEE Conference on Computer Vision and Pattern Recognition (CVPR), 2015, pp. 1–9.

[52] M. Lin, Q. Chen, S. Yan, Network in Network, 2013. ArXiv preprint arXiv:1312.4400.

[53] J. Dean, G. Corrado, R. Monga, K. Chen, M. Devin, M.Z. Mao, M. Ranzato, A. Senior, P. Tucker, K. Yang, Q.V. Le, A.Y. Ng, Large scale distributed deep networks, Advances in Neural Information Processing Systems, 2012, pp. 1223–1231.

[54] S.A. Singh, S. Majumder, Classification of unsegmented heart sound recording using KNN classifier, J. Mech. Med. Biol. 19 (2019) 1950025.

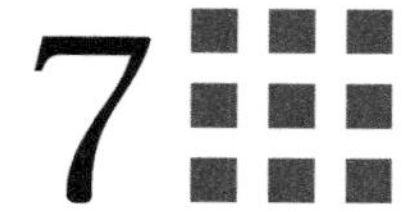

Development of a laboratory medical algorithm for simultaneous detection and counting of erythrocytes and leukocytes in digital images of a blood smear

Ana Carolina Borges Monteiro[a], Yuzo Iano[a], Reinaldo Padilha França[a], Rangel Arthur[b]

[a]SCHOOL OF ELECTRICAL ENGINEERING AND COMPUTING (FEEC), STATE UNIVERSITY OF CAMPINAS (UNICAMP), CAMPINAS, BRAZIL

[b]SCHOOL OF TECHNOLOGY (FT), STATE UNIVERSITY OF CAMPINAS (UNICAMP), LIMEIRA, BRAZIL

7.1 Introduction

Increasingly, medical areas are dependent on automated equipment that provides greater reliability and safety for professionals and patients [1]. Developing methodologies and equipment that solve problems in the health area is as important as developing ways to reduce the cost of existing methodologies to positively affect the quality of life of less-favored populations.

Although the cost of world wealth is estimated at trillions of dollars per year, millions of people from underdeveloped countries live on less than $1 a day and are classified at the "extreme poverty" level. In this context, approximately half of the world's population lives on less than $2 a day. Examples of this reality can be seen in sub-Saharan Africa and southern Asia, where >500 million people live with a monetary amount less than a dollar per day. In the region of Latin America and the Caribbean, 222 million individuals are poor, in which 96 million (18.6%) live in an indigenent situation [2].

It is important to note that the studies of Amartya Sen, winner of the 1998 Nobel Prize in Economics, clearly states that a single strand of analysis of the poverty situation cannot be determined, and categorizing all individuals in the same way does not take into account personal characteristics and conditions. Sen stated that the analysis of the poverty situation of the individuals would also need to consider the individual's ability to

Deep Learning Techniques for Biomedical and Health Informatics. https://doi.org/10.1016/B978-0-12-819061-6.00007-0

access nutrition, education, health, and well being, which directly impacts and reflects on the basic capacity to act in a society [3].

In the current global context, many people still live in underdeveloped or developing countries, where access to healthcare is often not a guaranteed right for all individuals. Removing people from this implies neglecting the early diagnosis of various diseases. Considering that the hemogram is responsible for qualitatively and quantitatively evaluating blood cells, this examination is responsible for the detection of a vast range of diseases such as genetic anemias, anemia, leukemias, thrombosis, parasitic, viral and/or bacterial infections, hemorrhage, and allergic response to external agents, among others [4].

7.2 Blood cells and blood count

Currently, the blood count is a highly requested exam due to its wide applicability [5]. The hemogram is a laboratory-based medical examination dependent on peripheral blood collection through venipuncture. This blood should be deposited in an anticoagulant tube and properly homogenized to prevent clots from forming. Through this biological material, it is possible to analyze the quantity and quality of the blood cells [6].

Human blood is composed of blood cells and plasma. These blood cells originate inside the bone marrow of long bones. Within the medulla is a microenvironment controlled by the presence of cytokines, hormones, stimulating factors, and interleukins, among other substates. When they reach a maturation state, these cells are released into the bloodstream where they exert their biological functions [7]. In this way, it is possible to analyze them by means of an erythrogram, leukogram, and plaquetogram, which evaluate red blood cells (RBCs), leukocytes, and platelets, respectively [5].

The erythrogram is composed of the total RBC count, hemoglobin concentration, hematocrit determination, and hematimetric indices. RBCs may also be referred to as erythrocytes and are characterized by biconcave disk morphology, measuring a certain 7 μm in diameter [8, 9]. The total RBC count is expressed from 4.0 to 6.0×10^6, with values below 4.0×10^6 strongly indicative of the presence of genetic anemias, deficient anemias, or internal or external hemorrhages. Values $>6.0 \times 10^6$ are indicative of polycythemia, which is responsible for severe complications with embolism and stroke [10, 11].

Hemoglobin is classified as a hemeprotein, a member of the globular proteins. These proteins are characterized by having the spherical or globular coiled polypeptide chain with various types of secondary structures. Another important feature that characterizes globular proteins is the positioning of the side amino acid chains, which is reflected in the structure and stability of the hydrophobic interactions. Most of its hydrophobic side groups are located inside the molecules, far from exposure to water, and most of the hydrophilic (polar) side groups are on the surface of the molecules [8, 9].

The hemeproteins are constituted by a prosthetic group, heme, and by a part protein, globin. They are responsible for various types of catalytic activity and perform various functions in biological systems. They exhibit maximum absorption at 400–440 nm (Sorét's band), characteristic of porphyrins. The prosthetic group heme confers to these proteins a characteristic color, and consists of an organic part and an iron atom in the ferrous state [Fe(II)]. The heme consists of a tetrapyrrole structure in which the four pyrrole rings are connected by methane bridges with four methyl radicals at the periphery, two vinyl radicals and two proprionyl radicals in different arrangements, of which protoporphyrin IX is one of the isomers. Fe (II) is in the center of protoporphyrin IX coordinated by four bonds to the nitrogen of each pyrrole ring [12, 13].

The iron atom may be in the ferrous (II) or the ferric (III) oxidation state, and the corresponding forms of both myoglobin and hemoglobin are referred to as ferro- and ferri- (or meta-), respectively. Only those in the oxidation state (II) can bind to oxygen. Thus, each erythrocyte has four iron atoms, which are responsible for the connection with the oxygen and subsequent conduction to the tissues of the organism [12, 13].

In this way, hemoglobin is constituted by the heme group, which is formed by iron and globin chains. These chains can be alpha, beta, gamma, or delta, resulting in the major types of hemoglobin. The three most common types of hemoglobin are HbA1, HbA2, and HbF. The HbA1 is formed by two alpha chains and two beta chains, and is present in higher concentrations in human blood. The **HbA2** is composed basically by two delta chains and two alpha chains. Thus, the **HbF** has two alpha chains and two gamma chains, and is present in a higher concentration in newborns; its concentration decreases according to the development. It is important to note that severe mutations in hemoglobin chains may be incompatible with life [14, 15].

In this context, many methodologies were developed for the determination of hemoglobin. In the hemogram, the technique consists basically of causing hemolysis and subsequently reading the concentration of hemoglobin from spectrophotometry. The values obtained should be between 12.0 and 16.0 g/dL of blood. Values below this reference value are strongly indicative of anemia. In this case, patients commonly present with fatigue, shortness of breath, and pallor. Patients with high hemoglobin concentrations are often due to smoking, dehydration, or the presence of malignant tumoral disease [16].

Hematocrit is the percentage of blood cells present in the plasma. Reference values range from 36% to 48% depending on the sex and age of the patient. Values above that indicate polycythemia, although lower results are indicative of anemia. In turn, through the medical literature, the hematimetric indices are performed by means of mathematical formulas [16]. The hematimetric indexes are the Mean Corpuscular Hemoglobin Concentration (CHCM), Mean Corpuscular Hemoglobin (HCM), and Mean Corpuscular Volume (MCV). MVC is responsible for evaluating the size of RBCs, that is, classifying them as macrocytic or microcytic. In turn, the HCM and CHCM are responsible for assessing the staining of erythrocytes. RBCs with low staining are called hypochromic, whereas those with more staining are called hyperchromic [16, 17].

Only with the association of the values of total RBC count, hemoglobin concentration, and determination of hematocrit and hematocrit indexes are able to provide a safe medical report [5].

Leukocytes are the body's defense cells, acting both on innate and acquired immunity. The leukocytes are round in shape and measure between 8 and 15 μm in diameter. In the leukogram, these cells should be present in the amount of 4000–11,000 leukocytes per mm^3

of blood. Values lower than this are called leukopenia, which is caused by a sedentary lifestyle [14]. The increase in the count of these cells is called leukocytosis and is triggered by the presence of infections and/or inflammations. When the amount of leukocytes exceeds $\geq$90,000 leukocytes/mm^3 of blood associated with the presence of blasts, there is strong indication to the presence of leukemia. Considering that leukocytes are subdivided into five types (monocytes, neutrophils, eosinophils, basophils, and lymphocytes), the leukogram relies on the determination of the absolute and relative values of these cells [18].

The indication of the presence of acute bacterial infections is related to the quantitative increase of neutrophils, called neutrophilia; however, patients with burns, postoperative period, stress, myocardial infarction, use of glucocorticoids, and intense physical exercise are also synonymous with neutrophilia. The decrease in the neutrophil count is termed neutropenia and may be indicative of diseases such as medullary infiltration, immunosuppression due to human immunodeficiency virus (HIV), food deficiency, radiation exposure, alcoholism, or use of cytotoxic drugs [19, 20].

Eosinophilia is referenced by the increase of eosinophils, which have primary action in the intestines, and also are present in the epithelium and the respiratory system. The increase in its quantity in the body may indicate intestinal infections by helminths (ascariasis, schistosomiasis, ancylostomiasis, etc.) and may also involve dermatoses, dermatitis, chronic myeloid leukemia, acute myeloid leukemia, or even respiratory allergic processes or pernicious anemia. Nevertheless, its decrease—eosinophilia—or even the absence of eosinophils in the bloodstream is considered normal due usually to the administration of adrenocorticotrophic hormones or corticoids, related to eosinopenia [19, 20].

Basophils are barely present in the bloodstream, yet the low amount present in the body has an important role in acting against allergic reactions that carry out the release of IgE and histamines, alleviating the symptoms. Whereas basophilia is the process related to a greater presence of circulating basophils, where it also indicates the presence of leukemia and even allergic reactions, in contrast, basopenia is not classified strictly as a pathological process in an organism [19–21].

Related to phagocytosis processes, monocytes are very essential cells in the body, as their increase in volume, production, and release is called monocytosis. Its increase indicates the presence of chronic infections and even suggests tissue repair. In contrast, monopenia is related to the indication of decreased production and release of monocytes in the body, generally associated with the lifestyle of the patient due normally to stress. In this way, all leukocytes and their subtypes according to their variations and amounts show the age of the patient [19, 20].

The increase in the amount of lymphocytes in the bloodstream (lymphocytosis) is directly related to the presence of viral pathologies such as HIV, rubella, hepatitis, etc., as blood cells active in viral infection processes [19, 22]. Already, a decrease in the amount of lymphocytes (lymphopenia) in the body can be understood as acute stress, chronic HIV, or even the presence of liver cirrhosis. It is important to observe that lymphocytosis is a normal condition considered only during childhood. Like lymphocytes, other blood cells undergo changes in their amounts according to the life stage, as shown in Table 7.1 [21].

Table 7.1 Reference values of the leukogram.

Total WBC (4000–11,000/mm^3)	Relative WBC (%)	WBC absolute (leukocytes/mm^3)
Blastos	0.0%	0.0
Myelocytes	0.0%	0.0
Promyelocytes	0.0%	0.0
Metamielocytes	0.0%	0.0
Segmental neutrophils	45.0%–74.0%	1700–8000
Rod neutrophils	0.0%–4.0%	00–440
Eosinophils	1.0%–5.0%	50–550
Basophils	0.0%–2.0%	00–100
Lymphocytes	20.0%–50.0%	900–2900
Atypical lymphocytes	0.0%–0.3%	00–330
Monocytes	2.0%–10.0%	300–900

Its methodology is variable between manual and automated, and the automated methodology does not exclude the use of manual techniques, because it is used as a form of confirmation of altered reports [4].

7.3 Manual hemogram

The manual methodology is dependent on an optical microscope and the making of a blood smear [20]. The microscope has improved over the years and is used as a powerful tool, mainly for analysis in the fields of biology and medicine. Currently, microscopes are more complex instruments; however, they follow the same logic of using associated lenses for image enlargement. Microscopes are made up of two parts: mechanics and optics [8, 23].

The mechanical part is responsible for the support and regulation of the instrument. The optical part of the microscope is formed by a halogen lamp, a condenser, a diaphragm, objective lenses, and ocular lenses. The halogen lamp acts as a light source, whereas the condenser has the purpose of projecting a beam of light onto the sample being analyzed. The diaphragm allows control of the brightness and obtaining the appropriate contrast. After passing through the sample, this light beam penetrates the objective lens. The objective lenses, in turn, are responsible for projecting an enlarged image into the focal plane of the eyepiece, which again magnifies the same image. Thus, the magnification of the image is the product between the increase of the objective lenses associated with the increase of the ocular lenses. Therefore, when using a 40 × objective and a 10 × eyepiece, the result is a magnification of 400 × [16, 23].

The details observed in the image are provided by the resolution limit of the optical microscope and not by its ability to increase the respective size of the images. Thus, the resolution power of an optical system is in its ability to detect details, expressed by the resolution limit, which is determined by the smaller distance between two points. The capacity to increase is only of value when it is accompanied by a parallel increase in resolving power. Thus, the resolution limit is dependent on the objective lens, with the ocular lenses unable to add detail. The function of the ocular lenses is restricted to the enlargement of the image size previously provided by the objective lens [8, 23].

The resolution limit is dependent: from the numerical aperture (NA) of the objective lenses and the wavelength of the applied light. In turn, the resolution limit (RL) of the objective lens is given by Eq. (7.1):

$$\mathrm{RL} = \frac{k \times \lambda}{\mathrm{NA}} \tag{7.1}$$

where k is an estimated constant in seconds, numerically represented by 0.61, and λ (lambda) is the wavelength of the light employed. Due to the greater sensitivity of the human eyes to the bandwidth of green-yellow, the value of 0.55 μm is used in calculating the resolution limit. By means of this formula, it is notable that the resolution limit is directly linked and also proportional to the wavelength of the light in use, which is also inversely proportional to the numerical aperture in relation to the objective lens [24].

Through the microscope, it is possible to analyze the blood smear. For preparation, it is necessary that few microliters of peripheral blood be deposited on a glass slide, followed by application of dyes responsible for coloring the cellular structures, which facilitate the visualization of these cells. This blood is subjected to visual recognition of erythrocytes, leukocytes, and platelets, with consequent application of the quantified values to the formulas present in the literature. All this process takes a considerable time, besides not giving great reliability to the medical reports because the whole process is dependent on the health professional, who is subject to factors such as stress, fatigue, and inattention, among others. In contrast, the hemogram performed by the manual method presents low cost [4].

7.4 Automated hemogram

A greater agility in the realization of the exams, as well as the release of the reports was implied through the automation of the hemogram; however, this brought more expensive methodology compared to manual hemograms. In the 1950s, Coulter Eletronic, Inc. introduced the impedance principle for cell counts [4].

The principle of impedance is based on the fact that electrically conductive cells are diluted in a conducting solution of electricity. This cell suspension is weighed through an orifice with a diameter of about 100 μm, where an electric current passes through. This electric current originates from two electrodes: one located on the inner side of the hole and positively charged, and another located on the outer side of the hole and negatively charged. In this way, each time the cell passes through the hole, it interrupts the electric current, and there is a change in the conductance; consequently, each interruption is counted as a particle [25].

Over the years, the impedance principle was enabled with counters capable of measuring cell volume. Such evolution was the result of the correlation of the proportionality of the magnitude of the interruption of the electric current (pulses) with the cellular volume. Thus, it was observed that small pulses correspond to small volumes, whereas large pulses result from larger volumes [5, 25, 26].

From this correlation between the magnitude of the electric current and the cellular volume, a new concept was created called the threshold concept. The threshold concept is responsible for the classification of cells according to their volume, thus allowing the detection of globular volume. The globular volume corresponds to the hematocrit performed in the manual blood count; however, it received this name because it is performed without the need for microcentrifugation. The threshold concept and impedance principle both have the responsibility for the introduction of multiparameter devices present on the market, which are able to realize simultaneous cell counts using separate channels [4].

In the 1960s, the conductivity technique was developed based on a high-frequency electromagnetic current, which is responsible for providing information about cell volume, nucleus size, and cytoplasmic content of granulations. Later, in 1970, laser light scatters and hydrodynamic fluid techniques were introduced. Both techniques preserve nuclei and granulation of leukocytes, retracting only the cytoplasmic membrane. These techniques are based on the principles of diffraction, refraction, and reflection of emitted light [5, 25, 26].

The RBCs are undetectable in these techniques, which leads to the solution of this problem that the erythrocytes are counted through hydrodynamic focus and flow cytometry, one by one by means of an extremely fine capillary. These cells are then subjected to a laser beam, where the light scattering is analyzed at different angles of deviation, zero degrees indicates the cell size, 10 degrees is indicative of the internal structure, and 90 degrees indicative of leukocytes and its content of granulations and characteristics of looseness [5, 25, 26].

Nowadays, there are many multiparameter devices using dispersion techniques of emitted light, impedance, and conductivity where these technologies are associated with the cytochemical characteristics of cells (the myeloperoxidase) and the respective use of reagents, which analyze certain cell types. From another point of view, before effectively buying a hematology device, it is necessary to consider the number of hemograms per day × samples/h of the apparatus; quality control; technical assistance; automation device × types of patient attended; employee training; the cost of each blood count; and interfacification [4].

7.5 Digital image processing

In view of this, it is necessary that there is a reduction in the cost of the methodology in use to perform the hemogram without loss of time and reliability of the exam. In this context, it is important to note that, since the creation of Matlab software, many tools were created to solve problems in medical areas. Matlab is defined as a powerful computational environment technique; an interactive environment for numerical computation, visualization, and programming; and a high-level programming language. Matlab is able to perform the data analysis, the development of algorithms, creation of models and applications, and also build mathematical tools capable of exploring multiple approaches by searching for a faster solution than traditional programming languages like Visual Basic, Java, or C/C ++. It can be used for video and image processing, tests and measurements, control systems, computational finance, and computational biology, among others. In addition, it is an easy-to-use tool [11].

Based on this, many developed methodologies are based on image processing techniques such as ultrasound, X-ray, and magnetic resonance, among others [11]. For this, it is admitted that a digital image represents a two-dimensional function $f(x, y)$, that x and y are spatial coordinates and, respectively, the amplitude of f for whatever pair of coordinates are then called and considered as image intensity. The term grayscale is used, relative to the intensity of monochrome images. The intensity of the level of each pixel is fundamental information that has the role of forming the image. The values of pixel intensity are also used in performing operations such as filtering or segmentation [27, 28].

The intensity of the grayscale level is responsible for describing how dark or bright the pixel corresponds to its presented position. There are many ways to record the intensity of each pixel. A pixel can be represented by 0 or 1, where 0 represents the pixel with low intensity (black in many cases) and 1 shows the pixel at high intensity (or white) [29, 30].

Colored images are designed by individual combination images. In color system RGB (Red, Green, and Blue), a color image is comprised of three individual monochrome images with respect to primary images or red, green, and blue components. Due to this reason, several techniques developed use monochrome images, which can be extended to color images through processing three individual image components [31].

In addition, an image can have continuous aspect with relation to the x and y coordinates and, in the amplitude, its conversion to a digital format demands that the coordinates are digitized, as well as the amplitude. Sampling is the name given to the digitization of the coordinate values, and quantization is the name for the amplitude scanning. Thus, a digital image is when x, y, and the values of the amplitude of f are all finite, characterizing these quantities [29–31].

The resultant of the quantization and sampling is an array of real numbers. It is declared that the image $f(x, y)$ is a sample that results in an image that has M rows and N columns. The original image is defined as $(x, y) = (0,0)$. The following coordinate values along the first line of the image are $(x, y) = (0,1)$. The notation (0,1) is usually used to denote the second sample along the first line [29, 31].

In this way, the process of image segmentation is the first step in general object recognition in diverse applications as the identification of regions of interest (ROIs) in a scene or annotating the data. Image segmentation is an essential process for an image analysis task. The segmentation process consists of the act of homogeneously fractionating an image into groups of spatially connected pixels [29–31].

There are many techniques for describing and recognizing correct image viewing, with image comprehension a process highly dependent on the outcome of segmentation processing. There are many methodologies of image segmentation. However, the focus of this study is the Hough Transform (HT).

7.6 Hough transform

The HT was presented by Paul V.C. Hough in 1962. Its application was based on particle physics, aimed at the detection of lines and arcs in photographs acquired in cloud chambers. The HT is classified in the middle range of image processing hierarchy. This methodology is applied to previously treated images and free of irrelevant details. Thus, this method is dependent on the processes of filtering, and boundary and edge detection. The method assigns a logical label to a selected object that, until then, existed exclusively as a collection of pixels; in this way, it made classification as a segmentation procedure possible [32, 33].

The logic by behind of the method is based on the concept that parametric shapes in an image may be detected through points accumulated in the parameter space. Thus, if a particular shape is located and present in a desired image, mapping is performed with respect to all points in the parameter space that must be grouped around the corresponding parameter values to that shape. This mapping approach effects the distribution and disjointment of the elements contained in the image to a localized accumulation point [32, 33].

The detection of objects in images is always one of the essential objectives of areas related to digital image processing (DIP). In simpler cases, the detection of straight lines may come from several black dots arranged discreetly in the image, which contains a white background. This problem can be solved through several techniques with different degrees of precision and degree of computation but are capable of detecting groups of collinear or quasicollinear figure points. For this, it is necessary to test the established lines with respect to all pairs of points, and the calculation used should consider that n points are proportional to $n2$ and may be prohibitive, approximately [34].

To solve the problem of finding collinear points, Rosenfeld [35] described a method for Hough [36], where the solution was to substitute the search for collinear points with equivalent mathematical solutions of search and detection of simultaneous lines. In this methodology, each of the points is transformed into a straight line parameter. This space has, by definition, a parametric representation that describes lines in an image plane. Thus, Hough opted for the choice of using tilt intersection parameters. Thus, its space parameter was a two-dimensional tilt intercept plane.

Thus, the foundation of the methodology developed by Hough is based on the set of all the lines (straight) that are present in the plane of a certain image, thus giving rise to two parameters. In this context, a straight line, being arbitrary, can be assumed in a parameter space as a single point, and is called the normal parameterization. In turn, the specific parameterization in a straight line, with respect to an angle 0, is normal, and an algebraic distance p from its origin [34, 36]. Based on this, the mathematical representation of a line corresponding to this described geometric is given by Eq. (7.2):

$$x \cos \theta + y \sin \theta = p \tag{7.2}$$

Thus, if 0 is restricted to the interval $[0, \pi]$, then the parameters of a row are consequently deleted, causing each row in the coordinate plane x, y to correspond to a single point in the plane $\theta - p$.

In this context, we can explain from the assumption of the existence of a set defined by $\{(x_1, y_1), \dots (x_n, y_n)\}$, where there are n points of the figure and the goal to be achieved is to find a set of specific straight lines and adjust into it [34, 36]. For this, it is necessary to transform the points (x_i, y_i) in sine curves on the plane $\theta - p$. This plane is then mathematically defined in Eq. (7.3):

$$p = x_i \cos\theta + y_i \sin\theta \tag{7.3}$$

Thus, it becomes easier to demonstrate that the curves with respect to the points of collinear figures have a general intersection point. Being such a point in the plane is given by $\theta - p$, said (θ_0, p_0). Such a point is able to pass by the collinear points. Based on this, the solution of the collinear point detection problem can be extended and converted to solve the problem of finding concurrent curves [34, 36].

With respect to point-to-curve transformation, the dual property can also be established; for this, it is possible to exemplify from a set of points $\{(\theta_1, \cdot p_1), \dots (\theta_K, p_K)\}$ present in the plane $\theta - p$, all supported in the curve through Eq. (7.4):

$$p = x_0 \cos\theta + y_0 \sin\theta \tag{7.4}$$

In this way, it is easy to demonstrate that all points corresponding to lines in the plane of x, y are coordinates of the digital image. It is possible to summarize the transformation properties of a curve point through four properties: (1) a point in the plane of the image has correspondence to the sinusoidal curve present in the plane of the parameter; (2) a point in the plane of the parameter has correspondence to a straight line located in the plane of the image; (3) points residing in a same line (straight) in the plane of the image corresponds to curves through a common point located in the plane of the parameter; and (4) points supported on the same curve in the parameter plane correspond to the lines present with respect to the same point in the plane of the image [34, 36].

The computational load of this method can quickly lead to increasing the number of parameters that define and detect the shape. The number of parameters is distinct according to the form: rows have two parameters, circles have three parameters, and ellipses have five parameters. The Hough method was applied to all those described, but the ellipse is in practical terms and probably at its maximum limit. Its greatest advantage is supported in specialized insight as aerial photo analysis, manufacturing quality control, and also particle physics data analysis [32].

One of the greatest benefits of using the HT is that the detection of partially occluded forms is still detected in the evidence of its visible parts, that is, all segments contained in same circle collaborate with the detection of this circle regardless of the gaps possibly present between them [32, 33].

7.7 Review

Based on this scientific theoretical basis, many authors have developed algorithms based on different techniques of DIP, aimed at the detection and counting of erythrocytes and leukocytes present in digital blood smear images. **Walsh and Raftery** have developed a work related to precise and efficient detection of curves in digital images, emphasizing the relevance of the HT. In this paper, they investigated the use of specific grouping techniques with the aim at the simultaneous identification of multiple curves present in the digital image. They also used probabilistic arguments for the creation of the stopping conditions of the developed algorithm. A blood smear image containing 26 erythrocytes counted 23 cells using the methodology developed by the authors [37].

Song stated that the HT is admittedly assumed as a powerful tool for extracting graphic elements of images due to its overall vision and robustness in noise or environmental degradation. However, the application of the HT was restricted to small-sized digital images because of the peak detection of the technique, and the scan line has a computational consumption much longer for large-size digital images [38]. In turn, Smereka modified the HT, proposing an improvement in the detection of circular objects present in the image with low contrast. The original HT and its various modifications were then discussed, tested, and compared, concluding that both can be improved in the criteria of efficiency and performance, as well as computational complexity of the algorithm [39].

Wu et al. developed their study using the circular histogram based on Otsu Transform for the purpose of segmenting white blood cells (leukocytes), thus stating that, for detecting blood cells, the methods of edge detection are weak because not all boundaries are well determined and delimited, which makes it difficult to acquire all edge information and thus locating cells in precise way [40].

Moallem et al. created a new algorithm capable of segmenting and detecting overlapping RBCs in digital blood smear imaging. Initially, the binarization of the image is performed with the posterior location of the cell center by means of clustering adaptable to medium changes. Finally, the Gradient Vector Flow (GVF) technique was used to detect erythrocytes. The method was then systemically tested on a dataset acquired through Chittagong Medical College Hospital located in the port city of Chittagong in Bangladesh [41].

Shahin et al. developed a cross-sectional study of neutrophil-based multiscale similarity measures for the detection of leukocytes. For this, a score was used with the purpose of measuring the similarity of the staining components of the blood smear images [42]. In this same context, Banerjee developed an algorithm based on the techniques of morphological operations, thresholding, and labeling of objects for the detection of leukocytes in digital images of blood smear [43].

Therefore, the present study aims to develop an algorithm of segmentation of images capable of simultaneously detecting and counting the erythrocytes and leukocytes present in blood smear digital images. This methodology should meet the low-cost criteria and standards without loss of quality and reliability required for medical environments.

7.8 Materials and methods

The algorithm of detection and simultaneous counting of erythrocytes and leukocytes was developed based on the union of two DIP techniques: the HT and the detection of objects by coloring. In this way, the algorithm was named HT-DC (Hough Transform + Color Detection). This methodology was developed following the steps described in Fig. 7.1.

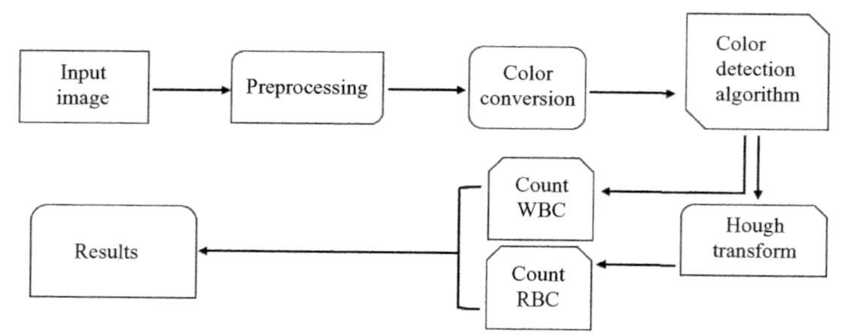

FIG. 7.1 Logic used in the development of the HT-DC algorithm for the simultaneous detection and counting of erythrocytes and leukocytes in digital images of blood smear.

The image input is based on the acquisition of digital images of blood cells obtained through open access to virtual platforms. The selected images were in the formats "jpg," "jpeg," and "png," and represented cells in a nonpathological state. These images were transferred to the virtual development environment of the Matlab software, in its version 8.3 of 64-bit (2014a). Then they went through the preprocessing stage, conversion of images in RGB pattern to grayscale, and then to binary scale. These processes are responsible for correcting certain problems that may arise from the moment of image capture such as poor quality camera, poor brightness, and poor microscope condenser aperture, resulting in images with low brightness and/or sharpness.

In terms of morphology, leukocytes and erythrocytes are distinct from each other. The leukocytes are rounded cells with nuclei in different forms of lobulations and granulations, characterized by blue coloration [5]. Taking advantage of these natural characteristics of the body's defense cells, the algorithm performs a process of separation of coloring RGB, avoiding that leucocytes are counted as erythrocytes, marking the leukocytes with the "+" sign, which indicates the center of each detected form, as shown in Fig. 7.2. The detection of objects by coloring can be performed through the Matlab software's Image Processing Toolbox.

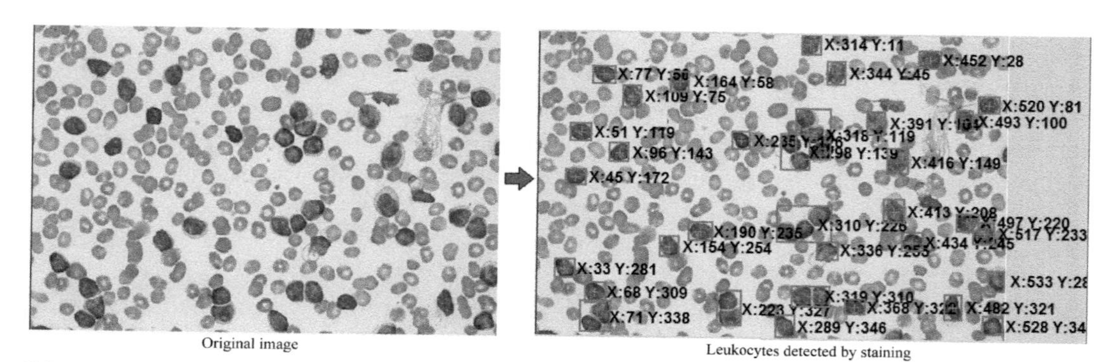

FIG. 7.2 Detection and counting of leukocytes by the HT-DC algorithm.

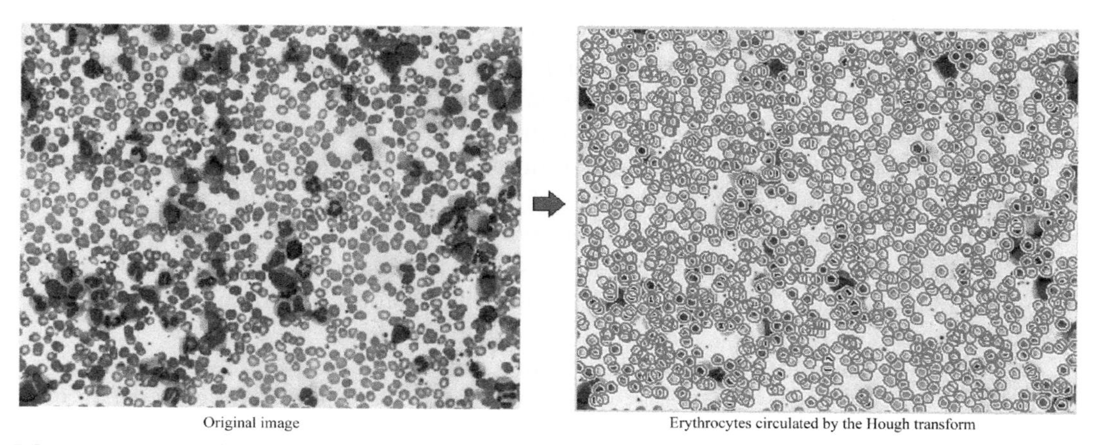

FIG. 7.3 Detection and counting of erythrocytes by the HT-DC algorithm.

The erythrocytes, in turn, are detected and quantified through the HT. The HT was presented by Paul Hough in 1962 by means of a patented technology based on application on particle physics, where it aimed at the detection of lines and arcs contained in the photographs achieved in cloud chambers. The HT is classified in the middle range of the image processing hierarchy. The method assigns a logical label to an object in particular, existing until then only as a collection of digital pixels; thus, it can be identified as a segmentation procedure [32, 33].

The parametric shapes are the principle of this method, which by means of an image can be detected as the points accumulated in a parameter space. What results in, if a particular shape is present in the digital image, all the points of this form can be mapped and likewise grouped around the corresponding parameter values with respect to the parameter space. Such approach maps disjoint and distribute the elements of a certain image to a localized accumulation point, in evidence of their visible parts occur of partially occluded forms still be detected, that is, all segments contained in a same circle cooperate in the detection of this independently of the gaps that may exist between them [32, 33].

In this present context, the HT is responsible for detecting edge points on each cell object and drawing a circle with that point as the origin and radius. It also uses a three-dimensional matrix, the first two dimensions responsible for representing the coordinates of the matrix, which increases each time the circle is drawn around the rays over each edge point. An accumulator has been implemented that maintains the proper count [19], as shown in Fig. 7.3.

7.9 Results and discussion

The experiments were administered through the selection of 10 digital images of blood smear fields containing erythrocytes and leukocytes in a nonpathological state. These images were transferred to the Matlab software and followed all the steps mentioned in the previous session. The results of field counts of erythrocytes and leukocytes were released separately. These values were then compared to those found during manual counting performed by a biomedical professional. The comparisons between the manual methodology and the HT-DC algorithm are presented in Figs. 7.4 and 7.5.

It is important to highlight that the differences in terms of quantity in blood cell counts are due to the physiological characteristics of the human organism, because of the bone marrow releases around 4.0 to 6.0×10^6 of erythrocytes and 4.0 to 11.0×10^3 leukocytes into the bloodstream [44]. In this context, 2146 blood cells were analyzed in the digital images of blood smear. The values expressed in Figs. 7.4 and 7.5 were then used to determine the accuracy, sensitivity, and specificity of the HT-DC methodology.

Two points as important as developing techniques and systems that facilitate daily laboratory and hospital tasks (1) is to classify the methodology developed as effective or ineffective according to the type of analysis to which it is submitted, and (2) to perform the validation of results to quantify their discriminative power. These actions are of extreme importance because the simple fact of quantifying the hits in a test group is incapable of adequately reflecting the efficiency of a system, because the quantification of a system must consider the quality and distribution of the data in test groups [45].

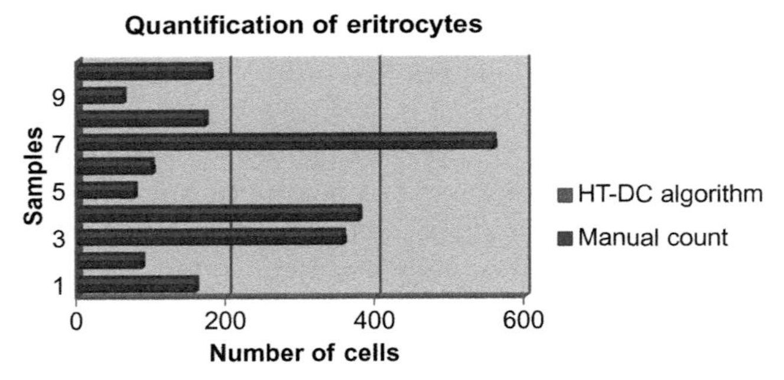

FIG. 7.4 Comparison of erythrocytes by manual methodology and HT-DC algorithm.

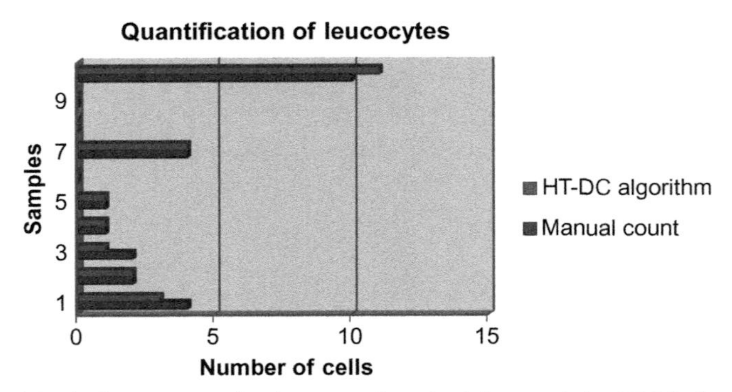

FIG. 7.5 Comparison of the leukocyte count by the manual methodology and the HT-DC algorithm.

Sensitivity, specificity, and accuracy analysis consider the test variables with binary elements (1 and 0, which classify as event or nonevent, respectively). For this, it is necessary to determine a cut-off point, that is, a rule of prediction that determines where the individual should be framed. One of the ways to determine the cut-off point is through the Receiver Operating Characteristic (ROC) Curve. This curve plots $P\left(\hat{Y}=1|\,Y=1\right)$ (called sensitivity) versus $1-P\left(\hat{Y}=0|\,Y=0\right)$ (called specificity) for all possible cut-off points between 0 and 1 [46, 47].

The second stage of the process consists of discriminating the events (1) from nonevents (0). For this, the following metrics are used: Accuracy, Sensitivity, Specificity, True Predictive Positive, and True Predictive Negative. All these metrics should be applied to the confusion matrix, which is represented by contingency Table 7.2, which is the predicted value in the line and the observed value in the column (true value) [45–47].

So, the metrics and their matches are

- True Positive: Cells are counted correctly by HT-DC algorithms compared to manual counts.
- True Negative: Digital blood smear images with leukocyte absence in both manual counts and counts by the HT-DC algorithm.
- False Positive: The test is blood cells counted further by the HT-DC algorithm in relation to manual counting.
- False Negative: Leukocytes are counted by the algorithm HT-DC and, in manual counting, these leukocytes are absent.

Later, through the values obtained in the manual counts and counts by the HT-DC algorithm associated to the values present in the confusion matrix, two binary matrices were created: (1) matrix for the quantification of erythrocytes and (2) a matrix for leukocytes. These matrices were transferred to Matlab software, where the following values were obtained: Sensitivity, Specificity, AROC (area of the ROC curve), VPP—True Positive Predictive (probability that an evaluated individual with a positive result is actually sick), and NPV—True Negative Predictive (probability that an evaluated individual with a negative result is actually normal) [45]. The ROC curve of the erythrocytes is shown in Fig. 7.6 and is complemented by Table 7.3, where the other values provided by the function used are presented. The same is true in Fig. 7.7, where the ROC curve of leukocytes is expressed and complemented by Table 7.4.

Considering the HT-DC algorithm obtained the values of 80% and 85% accuracy in erythrocytes and leukocytes, respectively, the methodology developed in the present

Table 7.2 Matrix of confusion.

True value (observed value)		
Predictive value	$Y=1$	$Y=0$
	TP (true positive)	FP (false positive)
	FN (false negative)	TN (true negative)

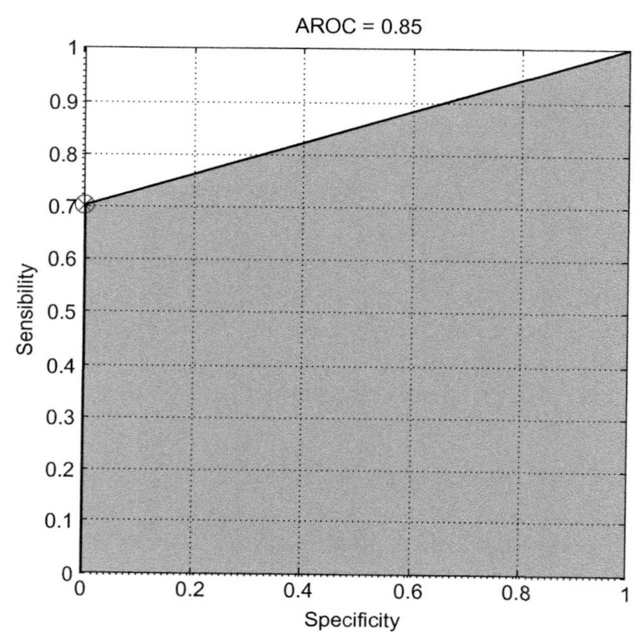

FIG. 7.6 ROC curve of erythrocyte count by HT-DC methodology.

Table 7.3 ROC curve parameters to erythrocyte counts quantified by the HT-DC algorithm.

Erythrocytes by HT-DC	Parameters of the ROC curve
Distance	0.3
Threshold	0.5
Sensitivity	70%
Specificity	99%
AROC	0.85
Accuracy	85%
PPV	99%
NPV	76.92%

study demonstrates a 6.25% improvement in the detection of these blood cells when compared to the results obtained in the works of Arivu et al. [48].

The accuracy, sensitivity, and specificity assessment demonstrate that the HT-DC algorithm can be used as a reliable tool for detecting and counting blood cells, as these values range from 80% to 99%. The reliability of exams is a primary variable both for the health professional and for the patient, as it is able to avoid the issuance of false positive or false negative medical reports, preventing the individual from being excluded or referred to inadequate treatment of anemia, leukemia, platelet disorders, or infections and/or inflammation [49].

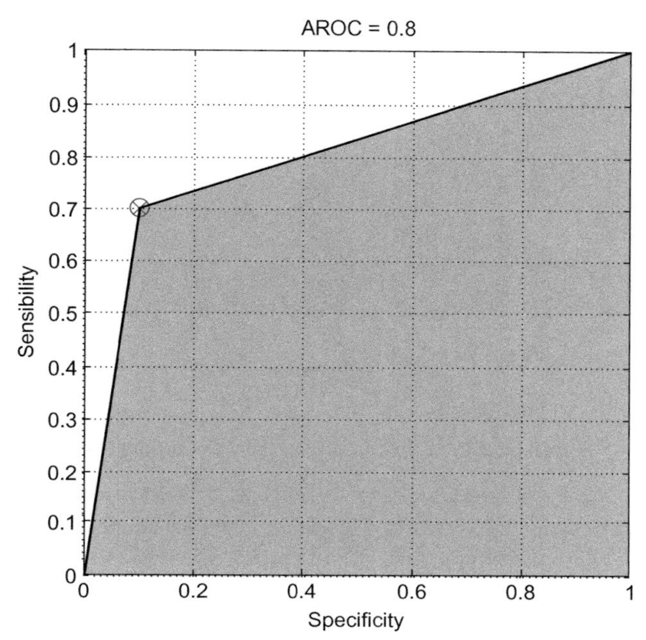

FIG. 7.7 ROC curve of leukocyte count by the HT-DC methodology.

Table 7.4 ROC curve parameters for leukocyte count analysis quantified by the HT-DC algorithm.

Leucocytes by HT-DC	Parameters of the ROC curve
Distance	0.31
Threshold	0.5
Sensitivity	70%
Specificity	90%
AROC	0.8
Accuracy	80%
PPV	87.5%
NPV	75%

The extreme importance of analyzing the reliability of a methodology through the determination of accuracy, sensitivity, and specificity, and making the methodology accessible to all is of enormous relevance. In this context, the HT-DC algorithm was evaluated for processing time. The processing time was measured using Matlab software using the "cputime" function. The "cputime" command is responsible for returning the total CPU time (in seconds) used by the application in use in Matlab from the moment it was started [28–30].

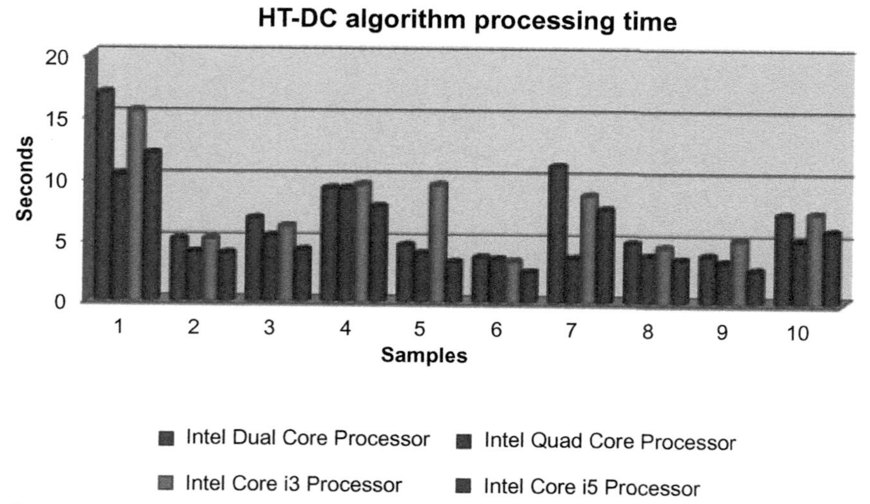

FIG. 7.8 Analysis of the processing time of the HT-DC algorithm in different hardware.

In this way, the HT-DC algorithm was tested in four different hardware: (1) Dual Core processor with 2GB RAM, (2) Quad-Core processor with 6GB RAM, (3) Intel Core i3 processor with 4GB RAM, and (4) Intel Core i5 processor with 8GB RAM. The values obtained, in seconds, in each of the 10 samples evaluated are presented in Fig. 7.8.

The processing time analyzes were performed through descriptive statistics by means of the determination of the mean time in seconds, as presented in Table 7.5. According to the processing time found in Thejashwini and Padma studies, it is 20 to 40 s by sample [49]. Thus, the HT-DC algorithm shows an improvement of 18.27%, 13.0%, 18.51%, and 11.47% when compared to the values obtained by the simulations performed on Intel Dual Core, Quad-Core, i3, and i5, respectively.

The performance of tests in different hardware has the purpose of unlinking the methodology of specific equipment and can be performed on different physical platforms without loss of accuracy, sensitivity, specificity, and processing time. In this way, the HT-DC algorithm can be seen as the first step to the complete accomplishment of a hemogram by means of a simple personal computer or notebook, making this important exam more accessible to remote and/or less favored places of underdeveloped countries and in

Table 7.5 Average processing time of the HT-DC algorithm in different hardware.

Hardware	Average processing time Time
Intel Dual Core Processor—2GB RAM memory	7.31″
Intel Quad Core Processor—6GB RAM memory	5.20″
Intel Core i3 Processor—4GB RAM memory	7.46″
Intel Core i5 Processor 8GB RAM memory	4.59″

development. This accessibility also aims to help professionals create their own laboratory who are undertaking the field of clinical analysis.

7.10 Future research directions

The potential of this research lies in the development of software to detect and count blood cells that presents high reliability, accuracy, and agility, as well as reduced cost. For this, it is necessary to develop more algorithms capable of performing the other counts performed during the hemogram.

Thus, the next steps of this research will be the development of a medical algorithm capable of detecting and counting platelets and blood cells in Neubauer's chamber. These deep learning techniques can be seen as strong allies for such tasks. Consequently, there will be the creation of a graphical interface for easy usability.

Through these future studies, the populations of countries with lower per capita income may be able to waive the use of high-cost hematological equipment and its reagents. On the other hand, people from developed countries may see this new methodology as an ally in the confirmation of hematological reports with alterations.

7.11 Conclusion

The HT-DC blood cell counting and detection algorithm developed in the present study presented satisfactory results in accuracy, sensitivity, and specificity, demonstrating that the methodology is capable of transmitting reliability in the medical reports issued. The reliability requirement is as important for patients as it is for professionals, because false-negative or false-positive results cause great damage to the lives of both.

The processing time analyzes of the HT-DC algorithm in hardware with different configurations presented results lower than 8 mean seconds. This suggests that the methodology presents one of the indispensable criteria for laboratories of clinical analysis: the speed of the methodology and the emission of the results. Fast methodologies allow the patient to be referred more quickly for treatment or to the necessary complementary tests, thus increasing the chance of cure.

Therefore, the proposal of this chapter in which the counting of RBCs and leukocytes in blood smear images can be seen as one of the great fruits of the use of computer engineering techniques for the solution of problems in medical areas. In this way, the HT-DC algorithm can be seen as the first step toward the reduction of hemogram costs and its consequent accessibility to less favored populations to a fast, efficient, and reliable methodology.

Acknowledgments

We thank the National Council for Scientific and Technological Development (CNPq)—Brazil, for all financial support for the development of this research, and the State University of Campinas (UNICAMP), Brazil, for providing all necessary structure for the development of this study.

References

[1] E.M. Suplicy, Renda de cidadania: a saída é pela porta, Cortez Editora, São Paulo, 2004.

[2] A. Sen, Development as Freedom, Anchor Books, New York, 2000.

[3] J.D. Bronzino, Medical Devices and Systems: The Biomedical Engineering Handbook, third ed., Taylor & Francis Group, United States, 2006.

[4] R. Failace, F. Fernandes, Hemograma, manual de interpretação, fifth ed., Artemed, Porto Alegre, 2009.

[5] F. Bernadette, G.A. Rodak, K.D. Fristma, Hematology—Clinical Principles and Applications, Elsevier, New York, 2015.

[6] Abbot laboratorie de México S.A, Atlas Com Interpretacion Histogramas Y Escatergramas, E.G, Buenos Aires, 2002.

[7] M.L. Turgeon, Clinical Hematology Theory and Procedures, fourth ed., Lippincott Williams and Wilkins, Philadelphia, 2004.

[8] E. De Robertis, J. Hib, De Robertis Bases da Biologia Celular e Molecular, fourth ed., Guanabara Koogan, São Paulo, 2006.

[9] H. Carr, F.B. Rodak, Clinical Hematology Atlas, third ed., Artmed, Porto Alegre, 2009.

[10] R.M. Xavier, J.M. Dora, E. Barros, Laboratório na Prática Clínica: Consulta Rápida, third ed., Artmed, Porto Alegre, 2016.

[11] R. Hoofman, E.J. Benz, L.E. Silberstein, H.E. Heslop, J.I. Weitz, J. Anastasi, Hematology: Basic Principles and Practice, sixth ed., Elsevier, Canadá, 2013.

[12] J.W. Baynes, M.H. Dominiczak, Bioquímica Médica, fourth ed., Elsevier, Rio de Janeiro, 2015.

[13] D.L. Nelson, M.M. Cox, Príncipios de Bioquímica de Lehninger, seventh ed., Artmed, Porto Alegre, Brasil, 2018.

[14] R.I. Handin, S.E. Lux, T.P. Stossel, Blood: Principles and Pratice of Hematology, second ed., Lippincott Williams and Wilkins, Philadelphia, Pennsylvania, United States, 2003.

[15] M. Melo, C.M. Silveira, Laboratório de Hematologia: Teorias, Técnicas e Atlas, Rubio, 2015.

[16] E. Molinaro, L. Caputo, R. Amendoeira, Métodos Para a Formação de Profissionais em Laboratórios de Saúde, Fio Cruz, Rio de Janeiro, 2009.

[17] A.H. Schmaier, L.M. Petruzzelli, Hematology for the Medical Student, Lippincott Williams and Wilkins, 2003.

[18] T. Verrastro, T.F. Lorenzi, S.W. Neto, Hematologia e Hemoterapia: Fundamentos de Morfologia, Patologia e Clínica, Atheneu, Rio de Janeiro, 2005.

[19] A.S.A. Nemer, F.J. Das Neves, J.E.S. Ferreira, Manual de solicitação de exames laboratoriais, Revinter, 2010.

[20] J.B. Bain, Blood Cells: A Practical Guide, fourth ed., Blackweel, Australia, 2006.

[21] A.K. Abbas, A.H. Lichtman, S. Pillai, Cellular and Molecular Immunology, seventh ed., Elsevier, Amsterdã, 2012.

[22] D. Provan, C.R.J. Singer, T. Baglin, I. Dokal, Oxford Handbook of Clinical Haematology, second ed., Oxford University Press, United Kingdom, 2004.

[23] L.G. Koss, C. Gompel, Introdução a Citopatologia Ginecológica Com Correlações Histológicas e Clínicas, sixth ed., Roca, São Paulo, 2006.

[24] L.C. Junqueira, J. Carneiro, Biologia Celular e Molecular, ninth ed., Guanabara Koogan, São Paulo, 2012.

[25] D. Chabot-Richards, T.I. George, White blood cell counts: reference methodology, Clin. Lab. Med. 35 (1) (2015) 11–24.

[26] R. Green, S. Wachsmann-Hogiu, Development, history, and future of automated cell counters, Clin. Lab. Med. 35 (2015) 1–10, https://doi.org/10.1016/j.cll.2014.11.003.

[27] C.C. Reyes-Aldosoro, Biomedical Image Analysis Recipes in Matlab: For Life and Engineers, Wiley Blackwell, London, 2015.

[28] W.K. Pratt, Introduction to Digital Image Processing, second ed., CRC Press, Boca Raton, United States, 2014.

[29] R.C. Gonzalez, R.E. Woods, S.L. Eddins, Digital Image Processing Using Matlab, second ed., McGraw-Hill Education, New York, 2009.

[30] S.R. Devasahayam, Signals and Systems Engineering: Signals Processing and Physiological Systems Modeling, Springer, New York, 2013.

[31] T. Morris, Image Processing With Matlab, Pearson/Prentice Hall, Upper Saddle River, United States, 2005.

[32] A. Danko, Review of the Hough Transform Method, With an Implementation of the Fast Hough Variant for Line Detection, School of Informatics, Computing and Engineering: Indiana University Bloomington, Bloomington, Indiana, 2008.

[33] F.A. Nava, The intersective Hough transform for geophysical application, Geofis. Int. 53 (3) (2014) 321–323.

[34] O.D. Richard, P.E. Hart, Use of the Hough transformation to detect lines and curves in pictures, Commun. ACM 15 (1972) 11–15, https://doi.org/10.1145/361237.361242.

[35] A. Rosenfeld, Picture Processing by Computer, Academic Press, New York, 1969.

[36] P.V.C. Hough, Method and Means for Recognizing Complex Patterns, U.S. Patent 3,069,654, Dec. 18, 1962.

[37] D. Walsh, A.E. Raftery, Accurate and Efficient curve detection in images: the importance sampling Hough transform, Pergamon J. Pattern Recogn. Soc. 35 (7) (2001) 1421–1431.

[38] J. Song, R.M. Lyu, A Hough Transform Based Line Recognition Method Utilizing Both Parameter Space and Image Space, Department of Computer Science & Engineering, The Chinese University of Hong Kong, Shatin, N.T., Hong Kong SAR, P.R. China, 2004.

[39] M. Smereka, I. Duleba, Circular object detection using a modified Hough transform, Int. J. Appl. Math. Comput. Sci. 18 (1) (2008) 85–91, https://doi.org/10.2478/v10006-008-0008-9.

[40] J. Wu, P. Zeng, Y. Zhou, C.A. Oliver, Novel color image segmentation method and its application to white blood cell image analysis, in: 8th International Conference on Signal Processing, 2006.

[41] G. Moallem, H. Sari-Sarraf, M. Poostchi, R.J. Maude, K. Silamut, M.A. Hossain, S. Antani, S. Jaeger, G. Thoma, Detecting and segmenting overlapping red blood cells in microscopic images of thin blood smears, Proc, SPIE 10581, Medical Imaging 2018: Digital Pathology, 105811F (6 March), 2018.

[42] A.I. Shahin, Y. Guo, K.M. Amin, et al., A novel white blood cells segmentation algorithm based on adaptive neutrosophic similarity score, Health Inf. Sci. Syst. 6 (1) (2018). https://doi.org/10.1007/s13755- 017-0038-5.

[43] S. Banerjee, B.R. Ghosh, S. Giri, D. Ghosh, Automated system for detection of white blood cells in human blood sample, in: S.C. Satapathy et al., (Eds.), Smart Computing and Informatics, Smart Innovation, Systems and Technologies, vol. 77, 2018. https://doi.org/10.1007/978-981-10-5544-7_2.

[44] B. Ciesla, Hematology in Practice, second ed., F.A Davis Company, Philadelphia, United States, 2012.

[45] T. Kawamura, Interpretação de um teste sob a visão epidemiológica: eficiência de um teste, Arq. Bras. Cardiol. 79 (4) (2002) 437–441.

[46] R.M.E. Sabattini, Um Programa para o Cálculo da Acurácia, Especificidade e Sensibilidade de Testes Médicos, Rev. Informédica 2 (12) (1995) 19–21.

[47] C. Shein-Chung, Statistical Design and Analysis of Stability Studies, CRC Press, Boca Raton, United States, 2007.

[48] S.K. Arivu, M. Sathiya, Analyzing blood cell images to differentiate WBC and counting of linear & non-linear overlapping RBC based on morphological features, Elixir Comp. Sci. Engg. 48 (2012) 9410–9413.

[49] M. Thejashwini, M.C. Padma, Counting of RBC's and WBC's using image processing technique, Int. J. Recent Innov. Trends Comput. Commun. 3 (5) ISSN: 2321-8169, 2015.

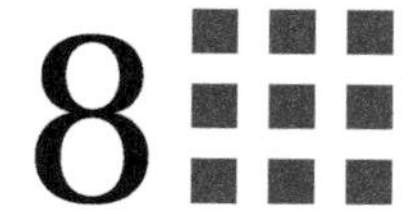

Deep learning techniques for optimizing medical big data

Muhammad Imran Tariq[a], Shahzadi Tayyaba[b], Muhammad Waseem Ashraf[c], Valentina Emilia Balas[d]

[a]THE SUPERIOR UNIVERSITY, LAHORE, PAKISTAN
[b]THE UNIVERSITY OF LAHORE, LAHORE, PAKISTAN
[c]GC UNIVERSITY LAHORE, LAHORE, PAKISTAN
[d]DEPARTMENT OF AUTOMATION AND APPLIED INFORMATICS, AUREL VLAICU UNIVERSITY OF ARAD, ARAD, ROMANIA

8.1 Relationship between deep learning and big data

Big data and deep learning (DL) are different technologies that have gained high focus in the areas of machine learning and artificial intelligence. Big data has become very significant as a number of pharmaceutical organizations, both private and public, have gathered a lot of medicine-related information, which can contain helpful data about various issues, for example, national intelligence, digital security, and medical informatics. Enterprises like Microsoft, Google, and Amazon analyze a lot of data for analysis and business decisions, influencing present and future innovations (Fig. 8.1).

DL algorithms determine mind-boggling abstractions as portrayals of data through a various leveled learning process. Complex abstractions are found at a specific dimension dependent on moderately basic deliberations that were determined at a previous dimension in the progression. The core advantage of DL is the examination and learning of a lot of uncensored data, making it a significant tool for big data analysis where essential data aren't sorted or ordered.

Big data may be distinct from its fundamental features, that is, large size, great variety, high speed, and high reliability [1–3]. The most obvious feature of big data is the large volume that contains an explosive amount of data. For example, Flickr generates approximately 3.6 terabytes of data and Google treats about 20,000 terabytes per day. The National Security Agency (NSA) reports that almost 1.8 PB of data are collected on the Internet every day [4]. Many people say that big data technology is produced rapidly and needs ongoing handling. The continuous investigation of big data is vital for the electronic trade for online administration. Another significant element of big data is that it shows an extensive number of noisy items, insufficient and uncertain objects, erroneous objects, and indistinguishable items [5]. The volume of big data keeps increasing at an extraordinary rate (Fig. 8.2).

Deep Learning Techniques for Biomedical and Health Informatics. https://doi.org/10.1016/B978-0-12-819061-6.00008-2

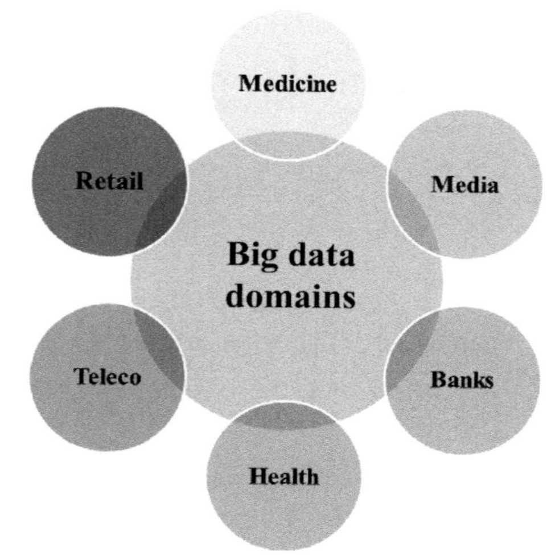

FIG. 8.1 Application of big data.

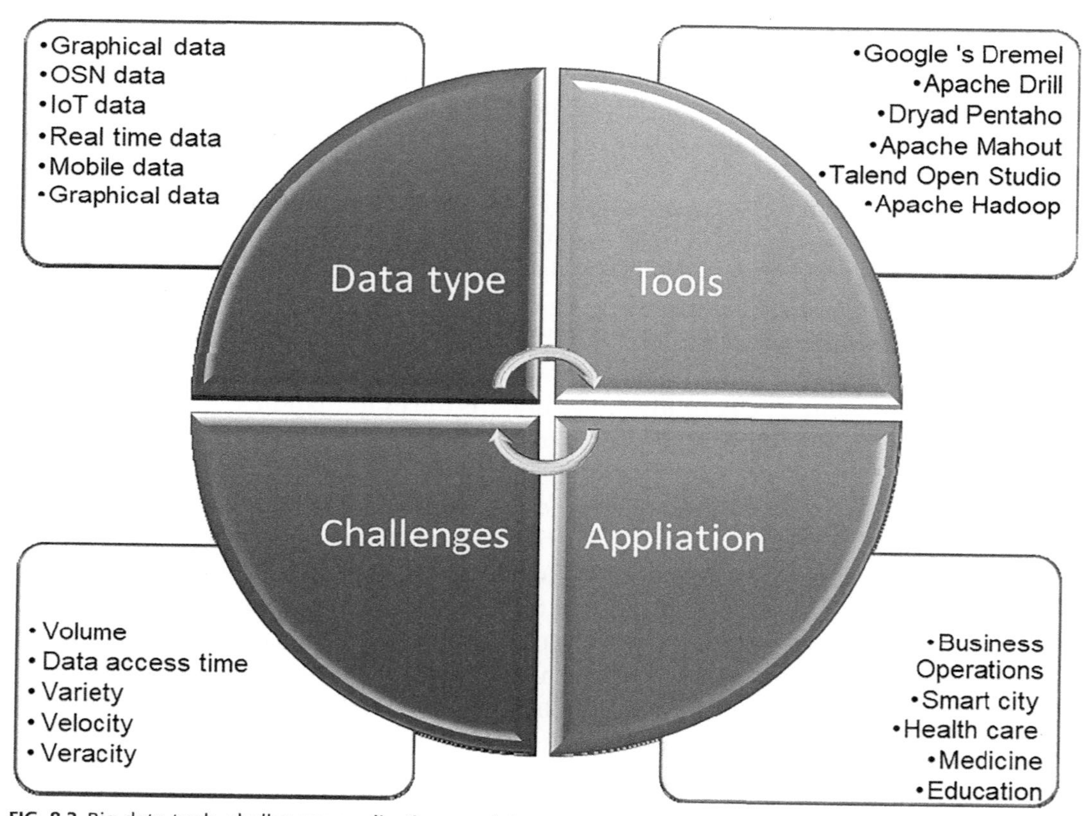

FIG. 8.2 Big data tools, challenges, applications, and data types.

In any case, solely using just big data is insufficient. For most applications, for example, industry and medicine, the key is to discover important data and extract it from big data to help prediction services. Take, for instance, the physical gadgets that endure mechanical disappointments in modern manufacturing. If we can analyze the values gathered for the gadgets in a viable manner, then we can take prompt measures to stay away from these debacles. Although big data offer critical open doors for a wide scope of fields, including e-commerce, mechanical control, and shrewd medicine, they present numerous troublesome issues in data extraction and data handling. Truth be told, it is hard for traditional methods to successfully analyze and control big data, and to accomplish an incredible assorted variety of extraordinary trustworthiness.

DL assumes a significant job in big data arrangements because it can collect important information about complex systems. Specifically, DL has become a standout among the most dynamic investigations in machine learning networks since its presentation in 2006 [6–8]. In any case, traditional training techniques for multilayer neural network systems must dependably lead to a perfect locally else they can't ensure a convergence solution. Subsequently, multilayer neural networks have not received broad applications in spite of the fact that multilayer neural networks may work better for learning qualities and exhibitions.

8.2 Roles of deep learning and big data in medicine

Artificial intelligence and DL are reclassifying the manner in which organizations work in various businesses around the globe. With the beginning of DL, organizations understood that they would now be able to process and utilize every one of the data they have gathered throughout the years. These data could now be analyzed and conclusions utilized to develop effective and efficient business decisions. Hence, DL and its conjunction with big data can be used effectively and actively in the medical field to give precise findings and conveyance of supermedical medications.

According to a recent survey, a majority of medical organizations collect patient data from healthcare organizations for cognitive analytics and to develop expressive insights. In addition, researchers have expected a dramatic expansion of DL and big data in the medical industry, which will obviously help reduce the costs of medicine and medical treatment by almost half.

Let's assume that when a patient visits a specialist for medical treatment and is told about their medical problems, the doctor starts maintaining the record of the patient in his computer. This encourages the specialist to understand the tests that are necessary before giving medicine effectively to the patient, to better understand the health of the patient, and to effectively diagnose the exact problem and diseases of the patient. For instance, a radiologist takes the assistance of a PC to perform magnetic resonance imaging (MRI) to investigate the organs and structure of the body and precisely analyze a patient's medical issue. The medical history is checked constantly. In view of past tests, the patients at long last are given a treatment plan or medical advice to improve their health.

8.2.1 What makes deep learning and big data necessary in medicine?

Machine learning algorithms rapidly process huge datasets and give helpful insights into knowledge that permits awesome healthcare services. Despite the fact that the business was moderate in embracing this innovation, it is now quickly getting up to speed and is giving effective preventive and prescriptive healthcare solutions.

Organizations in the healthcare division are currently progressively utilizing computational capacity to dissect voluminous datasets and recognize designs that give valuable experience from the current patient data to make a precise determination and give better patient consideration.

DL and big data algorithms process large datasets quickly and provide useful information to manufacture high quality medicine. Although the adoption ratio of the medicine industry toward DL and big data is not appreciable, it is now rapidly growing to provide successful medical solutions.

The data of medical associations has expanded definitely and needs the computational capacity to examine extensive datasets to distinguish patterns from existing patient data for precise medical advancement.

8.2.2 How are deep learning and big data changing the medicine industry?

After combining DL, big data, and artificial intelligence, the medicine industry can take more precise decisions in the manufacturing of new medicine, dramatically improving operational efficiency and eliminating unwanted costs.

With the help of improved technology previously discussed, the medical industry can save more lives. Medicine industry can formulate new medicine by developing new formulas and testing those new medicines. Likewise, diagnosis won't require serious time, and patients will quickly learn about the deceases, what they are suffering from, and what they should do next.

8.2.3 Examples and application of machine learning in medicine

A number of medical organizations are using machine learning to formulate new medicine formulas and imaging programs. DL and big data fields are useful for medicine because big data and DL algorithms work well to understand and learn patterns, which is very important in medicine. The application of machine learning in medicine/pharma can be found in the following fields.

8.2.3.1 Disease identification/diagnosis

Disease identification and its diagnosis have a key role in machine learning research in pharma. Machine learning and big data are playing vital roles in the development of more than 1000 medicines and cancer cures. Currently, medical organizations are facing the biggest challenge of dealing with a large number of results. The giant organizations quickly jumped to DL algorithms to handle their thousands of patients.

It is not surprising that the older players were first to jump into the car, especially in the areas where they are most needed, such as recognizing and treating cancer. In October 2016, IBM Watson Health announced IBM Watson Genomics, a partnership with Quest Diagnostics, which aims to make significant advances in the field of microbiology by integrating knowledge computation and sequencing of gene sequences.

8.2.3.2 Personalized treatment

Personalized medicine is based on individual health data associated with predictive analysis, which is a hot research area narrowly related to better evaluation of disease. The field is currently subject to supervised learning, allowing physicians to choose between limited sets of diagnoses, for example, to assess a patient's risk based on signs and genetic data.

8.2.3.3 Drug discovery

The use of machine learning in the early detection of drugs has potential for multiple uses, from the initial testing of drug compounds to predicting the expected success rate based on biological agents. This includes research and development of discovery techniques, such as sequencing the next generation.

It seems that determining the exact medication, which includes identification of disease mechanisms of "multiple agents" and, in turn, alternative routes of treatment, is limited in this space. Much of this research involves unsurprised learning, which is still largely limited to identification of data patterns without predictions.

8.2.3.4 Clinical trial research

DL contains many potential applications that are useful to help carry out pilot research and guidance. The advanced predictive analytics application can identify candidates for clinical trials in a wider range of data than currently done, including visits to social and medical media, for example, as well as genetic information when looking at specific groups; this would generate results for smaller, faster, and cheaper trials in general.

8.2.3.5 Smart electronic health records

Classifying documents through support vector devices and optical character recognition (OCR) is a key technique based on machine learning to help advance e-health and numbered information. Two examples of innovations in this area are in handwriting recognition techniques in Matlab and the Google Cloud API for visual recognition.

8.3 Medical big data promise and challenges

Despite the fact that the capability of big data analytics is promising, currently the use of big data analytics is generally promissory. Thus, it is necessary to explain a few big data challenges in medicine.

8.3.1 Promises and challenges

During the literature review, it was revealed that there are many challenges in data protection, data gathering, and exchange of medical information in terms of big data [9]. Big data analytics techniques using advanced technologies have the ability to transform data repositories and make informative decisions. Big data in terms of medicine also faces issues such as confidentiality, integrity, availability, nondisclosure, security threats, vulnerabilities, and governance issues that must be addressed [10]. Information like the treatment of nanoparticles in cancer treatment may be included in big data to deliver an overview of the best cure for cancer, particularly when nanotechnology is essential to administer drugs in the treatment of cancer [11]. Medical organizations are facing a number of challenges with medical big data. We explored these and discuss them in the following:

8.3.1.1 Data aggregation challenges

First of all, patients and financial data are often disseminated through many creditors, whether in hospitals, administrative offices of private and public hospitals, government organizations, servers, and workstations. Collecting and arranging all these data from different locations to produce new data requires a lot of planning. In addition, each participating organization must understand and agree upon the big data types and formats it intends to analyze. Looking beyond the coordination issues between organizations in which it is stored, the accuracy and quality of this data must be proven. This needs not only data cleansing but also a review of data governance issues and many more data aggregation challenges.

8.3.1.2 Policy and process challenges

After the validation and aggregation of the data, several policy and process issues must be addressed. Health Insurance Portability and Accountability Act (HIPAA) regulations require that policies and procedures of health and medical organizations that use patient information must protect this information. Access control, authentication, authorizations, contingency planning, risk management, and other rules must be deployed to secure the information, and these rules and regulations complicate the task.

8.3.1.3 Management challenges

Another challenge of big data analysis in medicine and health requires organizations to adjust their ways of doing trade. It is likely that data scientists will be needed along with the IT staff that has the necessary skills to execute the analysis. Some organizations may struggle with the idea of having to rebuild a large part of their IT infrastructure, although cloud service providers mitigate a majority of these concerns. Pharmacists and administrators of medical organizations may need some time before they can trust the invisible advice that big data can provide.

8.3.1.4 Cloud storage

Cloud computing is the most well-known technology of this world. It provides many services to its users, and cloud storage is one the key features of cloud computing. We may use cloud storage to store data in the cloud network. Hence, the cloud will require enough storage space and enough speed to load the data at the same time [12]. Storage is also used to store heavy-sized images like X-ray and MRI apart from Word documents. The cloud system must enable the user to create graphical presentations of the existing data in such a way that clients can visualize them, quickly understand them, and make quick decisions.

8.3.1.5 Data accommodation

A substantial and condensed data framework is required to house all data, which must be predictable and disentangled. This is to ensure that clients can recover data without issues. It is a difficult assignment to make every single important system interconnected. There is a culture of dispute inside individual organizations where a few parts can control the data for their own necessities rather than the organizations in general [13].

8.3.1.6 Data personnel

At present, it is a very difficult to teach computer science and pharmaceuticals to data scientists who have experience in the field of statistics. There must be a standard data entry protocol that consolidates all information entered by the person who enters the data, even though there are many variations in the data entry employees. All these measures should be taken to ensure continuity and uniform format for data entry [14].

8.3.1.7 Data nature

Data integration won't just incorporate data inside a medical framework; it must also include external data. Although it offers potential advantages, it is a test as far as protection, security, and legal issues. Medical data are generally comprised of patients looking for treatment in hospitals or clinics, yet there is no data about healthy individuals. By incorporating healthy people in the database, it will give superior comprehension to the idea of sickness and medication. At the point when data turns out to be later, it is basic that data is exchanged to clients quickly to settle on clinical choices and improve health results.

8.3.1.8 Technology incorporation

The deficiency of data to help basic leadership, rules planning, or strategy making are one of the issues with big data. The redefinition and selection of new innovations are very slow, and it can influence medicine formulation and research. Without an appropriate selection of technology, big data can't create and distribute information. The data is frequently divided and scattered among various stakeholders, for example, service providers, retailers, medical clinics, organizations, and patients. The resolution of this issue is to gather all the most relevant data at a central point like a database.

8.4 Medical big data techniques and tools

To handle medical big data, pharmaceutical associations require explicit devices that may deal with a complex type of data effectively and successfully. Because traditional instruments can't deal with the massive amounts of medical data, analysis tools are discussed in the next sections. Big data tools are grouped into batch processing, stream processing, and interactive analysis, and are illustrated in Fig. 8.3.

8.4.1 Batch processing technique and tools

Apache Hadoop was basically designed for batch processing, is generally utilized for big data, and the vast majority of the medical big data applications use Hadoop as a platform. Hadoop is essentially and broadly utilized in different fields including medicine, for example, machine learning, data mining, and so on. Hadoop disperses a bundle of big data through a few machines. It functions admirably in the handling of medical big data, as it is explicitly planned to process big data bunches. Some of the big data technique tools are clarified here:

8.4.1.1 Apache Hadoop

Apache Hadoop is a software-based platform created for data-concentrated applications for distributed data. It utilizes Map Reduce as a model [15]. It was developed by Google in collaboration with other organizations to critically organize and develop large datasets.

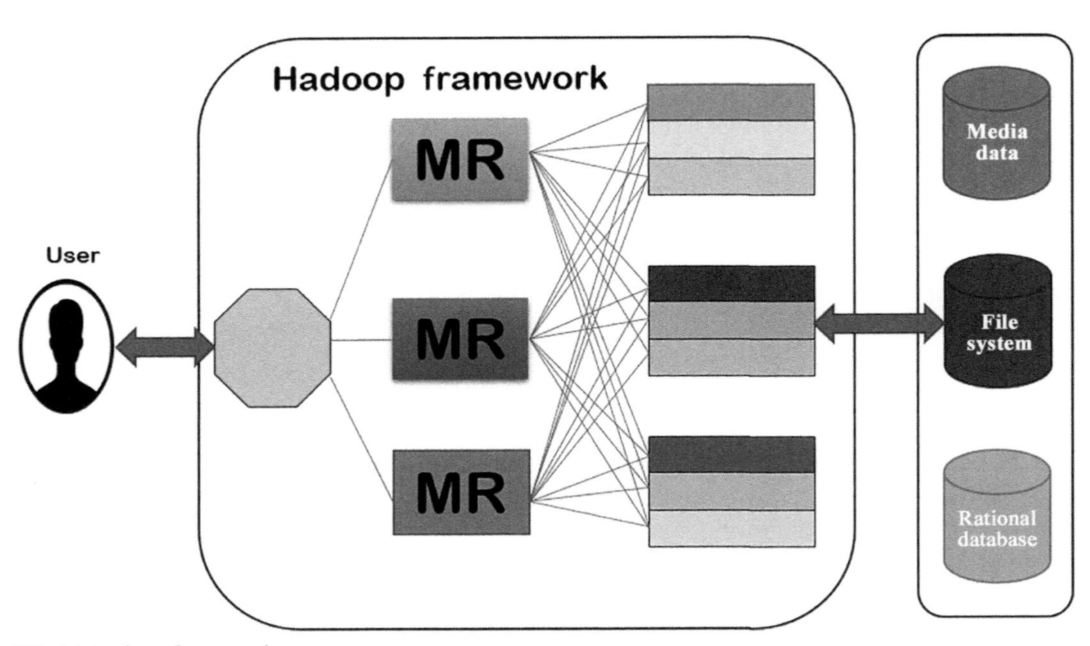

FIG. 8.3 Hadoop framework.

Map Reduce divides the complex into small issues, and then again divides these small chunks into further small chunks, repeating this process until the subproblem can be addressed directly and easily. The services of processing clusters will be utilized to resolve these subproblems; these processing clusters work in parallel, and finally the results of these processing clusters are integrated to formulate a concrete solution of the actual problem. Fig. 8.3 explains the Hadoop framework.

This technique works in two stages known as Map and Reduce. Likewise, the structure of Hadoop comprises of two sorts of modes called master and slave nodes. The function of the Map stage is to divide the main problem into subproblems, and then it distributes these subproblems to slave nodes to further process them. The slave nodes work on them until it reaches the outcome of the solutions of the subnodes. Finally, these slave nodes integrate the results and send them to the master node that accordingly integrates the results of all slave nodes. The inclusion of the Map Reduce model in the Hadoop structure made it an incredible system for the parallel processing of big datasets in extensive accumulations in an increasingly dependable manner. The Map Reduce structure is clarified thoroughly together with other kinds of data evaluation and examination tools [16].

8.4.1.2 Dryad

Dryad is a programming model that handles different processes in a parallel and distributed way [17]. Microsoft initially started Dryad purely as research. It can deal with a small level of data with a small number of nodes. The idea of mass is utilized to actualize and process programs in a distributed way [18]. Moreover, one of the benefits of the Dryad structure is that software engineers don't need to know parallel programming [1].

Applications that use this structure work in a guided way, comprising of headers and borders. It is a general-purpose execution environment for the application, which uses a distributed execution graph and parallel data applications. Most of its concentration is on the throughput of the application rather than the latency rate. The main goal of Dryad is the automatic management of the scheduling algorithms, distribution of data and fault tolerance, etc. Dryad is like a middleware abstraction layer that runs distributed execution graphs for the users [19].

8.4.1.3 Talend Open Studio

Talend Open Studio is freely available for developers and users. The program was launched in 2006 after intensive research for many years [20]. Talend is based on Eclipse, which primarily supports ETL-oriented applications provided for local sites, publication, and software as a service (SaaS) [21]. Talend Open Studio is used to integrate operating systems. In addition to the ETL tool for data storage, data processing, and data migration, the company breaks the traditional property model, providing open, innovative, attractive, and powerful software solutions with the flexibility to satisfy the data integration needs for all environments. Today Talend Open Studio is the most innovative and open source data integration tool in the market. Talend Open Data Integration Studio helps

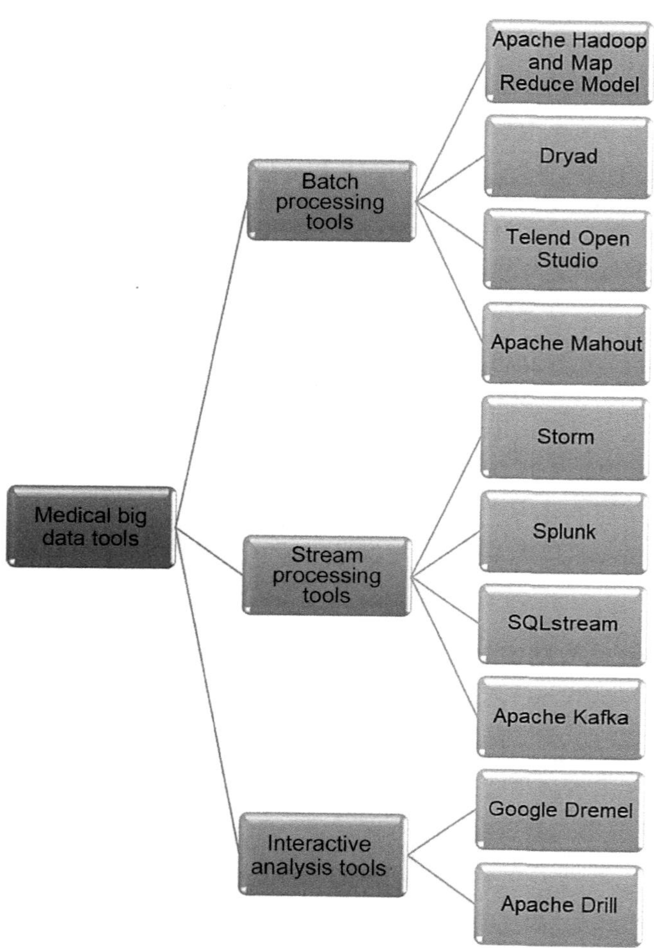

FIG. 8.4 Medical big data tools.

you connect your data to the objective, the desired location, in a specified way, in a timely manner [22].

Talend Open Studio is created by Apache Hadoop. It is not like Hadoop, as it facilitates the user to solve their difficulties without coding in Java. Furthermore, it also offers the facility of drag and drop in respect of specific tasks. The visible illustration of the user's components facilitates the understanding of their work, but at the same time it does not provide many details to understand the mechanism in depth [23] (Fig. 8.4).

8.4.1.4 Apache Mahout

Apache Mahout is a data gathering technique and tool that addresses difficulties related to ML systems for data evaluation. It is usable for a large scale of applications that need smart data investigation [24]. Organizations like Microsoft, Amazon, LinkedIn, and Google have machine learning algorithms to tackle big data issues. Apache Mahout utilizes the Hadoop and Map Reduce platforms. It utilizes many effective, efficient, and well-structured algorithms, for example, aggregation, classification, dimension decrease, pattern analysis, and so forth.

8.4.1.5 Pentaho

Pentaho is a platform used to deal with decisions relating to business, and it also manages enormous and unstructured data. It provides the facility to its users to easily work through simplified access, good integration with other tools, and excellent data exploration through graphical visualizations [25]. Moreover, it enables its client to make very efficient decisions based on the data of their organizations, which increase the performance of the organization [26]. Like different projects, for example, JasperSoft, a few Pentaho tools have been created and are connected to databases, for example, Cassandra and MongoDB.

8.4.2 Stream processing tools

Stream processing deals with real-time immense amounts of data. Applications, for example, industry sensors, document preparing, and online communications require continuous handling of huge amounts of data. Data on a large scale that also requires real-time processing needs an extremely low transfer of data during its processing [2]. The Map Reduction model gives a high transition period, where the data of the Map stage must be saved on disks before the decrease stage starts, which makes a significant interim and makes it an inconceivable procedure to process the data in real time [27]. Big data in streaming processing face fewer difficulties with regards to extensive data sizes, high data rates, and time lags. To overpower the problems faced by the Map Reduce, constant stages, for example, Storm, Splunk, and Apache Kafka are exhibited [28]. A few of them are discussed here:

8.4.2.1 Storm

The most prominent platform to execute streaming in real time is known as Storm. Storm was grown explicitly for a data stream that is difficult to work with and guarantees the total handling of the data. Billions of medical records every second were prepared in a node, which makes Storm a compelling platform for the development of data [29]. On the Storm platform, an alternate topology is utilized for various jobs [3]. A topology (Fig. 8.5) is the mathematical diagram of computation, and it may be developed in any computer programming languages. It has two sorts of nodes, namely key and a screw [30].

The beginning stage on the graph that reestablishes the source of the stream is the key, whereas the screw is responsible for handling the input stream and, after that, creating the output streams. It has two sorts of nodes as outlined in Fig. 8.6. It has just a single master node and the remainder of the nodes are worker nodes. The primary function of these nodes is to implement two further types of daemons called Nimbus and Supervisor. Both

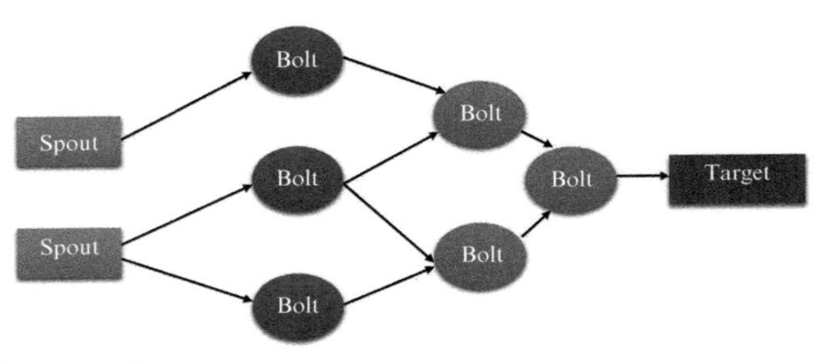

FIG. 8.5 A Storm topology example.

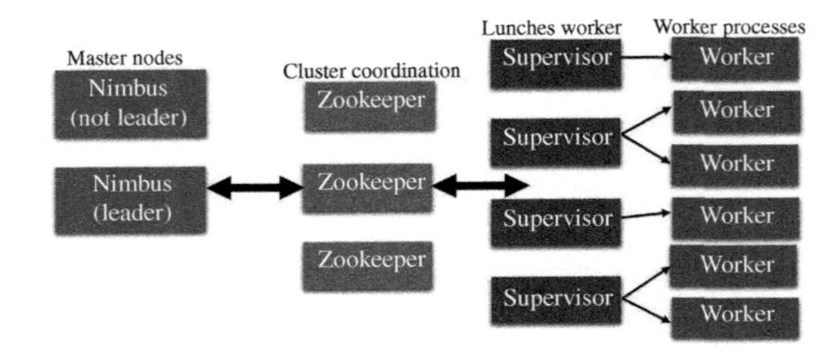

FIG. 8.6 A Storm cluster.

daemons have different functions like function in the Map Reduce framework. The function of the daemon Nimbus is to distribute load across the Storm cluster, scheduling assigned load, and monitoring the whole processes of the system. In case of failure of the system at any stage in the cluster, the Nimbus shall act and detect it.

The Supervisor nodes execute and comply with the tasks assigned by the Nimbus nodes, and it also acts based upon the instructions of the Nimbus. The Storm cluster topology partitions the workload and communicates to the worker nodes, and every worker executes the assigned topology workload.

8.4.2.2 Splunk

Splunk is an intelligent system that uses a large amount of data in real time to exploit the information of large data generated by machines. Cloud technologies and extensive data help users search, monitor, and analyze the data generated by the device through the Web interface [31], and show results intuitively, such as graphs, reports, and alerts. Splunk is developed to provide standards for numerous applications to analyze system infrastructure difficulties and information technology, as well as give intellect to corporate processes [32]. It is a superlative type of Splunk data examination. It is completely dissimilar to other batch processing tools. Its features comprise indexing created or unstructured data generated by the machine, real-time searches, analytical results reports, and information panels [33].

8.4.2.3 Apache Kafka

At first, Apache Kafka was created by LinkedIn to deal with the stream and playback data. The processing of memory is utilized to choose data in real time [34]. It is a distributed informing framework that conveys reliable messages with high efficiency, distributed processing, and data stacking in parallel [35]. Activity data are essentially the activities that somebody performs on a site, for example, duplicating substance, clicks, and looking for keywords [36].

8.4.3 Interactive analysis tools

This is the time of big data, and engineers has created open source frameworks to address the issues, not just for batch processing, and furthermore to control the stream, notwithstanding for interactive processing. It enables medical stakeholders to legitimately communicate with the data and evaluate the data as per their requirements.

8.4.3.1 Google Dremel

In 2010, Google presented architecture called Dremel. Dremel's plan is unique from Apache Hadoop, which was created to deal with batch processing [37]. In addition to this, Google Dremel has the capability to play out a lot of inquiries in a second in a table containing a billion rows with the assistance of trees and data from multilevel columns. Additionally, Google Dremel supports a huge number of computation processors, and it has also had the capability to process a billion bytes of data from a huge number of Google clients [38].

8.4.3.2 Apache Drilling

The interactive analysis supported by distributed platform deals with big medical data is called Apache Drilling. Apache Drilling is more adaptable than Google Dremel to help diverse inquiry dialects, data types, and sources [39]. The Apache Drilling's purpose is to support a huge number of servers, processing data in the size of bytes, and dealing with billions of client records without any loss of time. It is intended to successfully discover overlapping data. Google Dremel and Google Apache have some expertise in broad intuitive analysis to answer customized inquiries, for example, stockpiling utilized by HDFS and bunch analysis, utilizing the Map Reduce model [40]. You can check data through units of the GB in light of questions in almost no time, either in a distributed files system or a column.

8.5 Existing optimization techniques for medical big data

Data is the most important aspect of various activities in the world, and it is being generated every second. The traditional system cannot calculate and analyze big data types that use a huge data tool like Hadoop. Hadoop always saves, computes, and analyzes data in a parallel environment. Actually, every medical organization relies on information obtained from a large amount of pharmaceutical data. To solve the huge problem of complex data, a lot of work has been done. As a result, different types of tools have been developed. The aims of this section are to explore existing big data optimization techniques.

The material following Table 8.1 gives concise thoughts regarding big data advancement procedures and its analysis. The table additionally examines the properties of optimizing techniques and furthermore educates how these techniques are valuable for the improvement in big data basic decision-making processes.

Table 8.1 additionally discusses the destinations of this segment, such as different authors' research on the optimization of medical big data tools, the techniques utilized for enhancement as Flex Allocation Scheduler (FAS) and Recursive Chunk Division for Optimal Pipelining (RCD-OP), merger of various small files, and grouping of numerous files.

Table 8.1 Optimization implemented.

Author	Objective	Technique	Application
Wolf et al. [41]	FLEX scheduler for optimizing schedules and a slot apportionment scheduling optimizer for Map Reduce jobs	Flex allocation scheduler	Job scheduling
Dokeroglu et al. [42]	Improve the overall performance of the Hadoop Hive and shorter execution time	Multiple query optimization framework and SharedHive	Map Reduce
Okcan and Riedewald [43]	Decrease the amount of data transferred from mappers to reducers	Anticombining	Map Reduce
Dong et al. [44]	Improve the cloud storage and access efficiencies of small files of big data	The merger of small files	Hadoop distributed file system
Yildirim et al. [45]	Optimization of big data transfers. Pipelining, parallelism, and concurrency	Recursive chunk division for optimal pipelining	Map Reduce optimization
Kim et al. [46]	Application layer throughput optimization model to achieve higher accuracy and lower overhead predictions	Parallel stream number prediction	Throughput optimization
Arslan et al. [47]	End-to-end data transfer throughput with very low overhead	Predictive end-to-end data transfer optimization algorithms	HARP
Roy et al. [48]	Improve the capability to cope up with the problems of great dimensionality and scatteredness of data	Increasing the range of search procedure	Ensuring constructive population diversity
Kolomvatsos et al. [49]	Effective time optimized arrangement for advanced analytics in big data	Conduct the entering partial results	Improve the performance of querying big data clusters

8.5.1 Big data optimization tools for medicine

Numerous big data advancements have serious performance needs like analysis of big data in real time. The Hadoop Map Reduce still faces big data challenges to optimize a huge amount of data at different places in a distributed environment, and that data is gradually increasing day by day. To quickly streamline the enormous amounts of information, clients ordinarily depend on nitty-gritty execution investigation to recognize potential execution bottlenecks. To mitigate the difficulties of performance, numerous types of performance evaluation tools have been suggested and developed to exactly estimate the performance of big data during runtime. In this section, we will discuss several big data optimization tools that optimize big data in every aspect.

8.5.1.1 Sonata

SONATA is a performance evaluation tool, and IBM data products like Symphony are already using it. It was basically designed to tune the optimization of big data and Map Reduce. SONATA is mostly used by application developers, system analysisists, designers, and architects. SONATA has the capability to provide different types of analysis techniques to effectively look inside the execution of big data. For example, an overall overview of the system includes the number of running tasks, how much CPU and memory the tasks are consuming, how much network bandwidth is consumed by each task, usage of the storage against each master and slave node, timeline, and reduction of task in slave nodes. SONATA also identifies abnormal usage of resources to its user and also identifies respective nodes. Another feature of SONATA is to show a breakdown of the map and tasks. The previously discussed features of SIONATA are correlated with each other to identity the critical phases of the system. The key feature/characteristic of SONATA is the graphical representation of all the output and also providing a statistics overview to its users.

A concern of the architecture of SONATA is that it is comprised of basically four types of the layers: data gathering, data loading, performance visualization, and then recommending optimizations to its user. All these layers are connected with each other and work step-by-step after the collection of data. The data collection layer is comprised of two components, namely monitor and aggregator. SONATA stores gathered data into history files of the database through a data loading layer. The data loading layer is critical, and the next layers depend upon the successful task of this layer. It dynamically loads all the gathered data through the aggregator and stores it into a database. The third stage of SONATA is the visualization of the performance. It gives an overall overview of the Map Reduce, a detailed overview against each node, and also statistical output. The last layer of SONATA is the recommendations for the optimization of big data. SONATA effectively outlines the overall performance of the system to the practitioners and exactly informs the key areas that require optimization. Guo et al. [50] evaluated and tested the performance overhead of SONATA and found that it only generates less than 5% overhead on the system.

8.5.1.2 ECL-Watch

Xu et al. [51] state that tuning performance of big data is a very challenging and critical problem due to its extraordinary complexity and high number of distributed systems. Due to an immense number of layers in big data, it is very hard or nearly impossible to identity an exact layer that has performance issues. Analyzing a big data system without an appropriate performance evaluation tool is monotonous, time wasting, and boring work. The analyzers always demand different types of functionalities in performance optimization tools.

To address these issues, the authors introduced ECL-Watch as a performance analysis tool that gives an interface to the HPCC systems; it also permits analysists to view information about the system and each master and slave node of the big data that correlate with each other. It displays all nodes currently executing a task in the system, their respective topologies, and logs. Furthermore, it allows users to watch the status of the tasks and files. Similar to SONATA, it also has four different features, and these features have further subfeatures in deep analysis insight into the system.

The activity features of the ECL-Watch provide overall activities of the clusters in the system, tasks in queue, set their priority, pause the task, terminate the task, and stop the queue. The second feature is called Workunit, which provides more details about the task; users may analyze each task in depth. Similar to SONATA, the third feature of the ECL-Watch is to give graphical visualization of the task and also generate statistics against each node.

8.5.1.3 Turbo

Turbo is also a performance analysis tool, but the distinct feature of Turbo is that it automatically performs all the performance analysis tests without any interaction of the user. It enables inexperienced beginners to get experience. It is the opposite of SONATA with respect to its handling. It was built to enable beginners to understand the details about big data. Users who have no expertise to deploy big data can get the benefits from Turbo software. It automatically calls the performance tuning process and quickly completes it. It is pertinent to add here that the performance gain must be greater than baseline, and this is automatically achieved by this tool. Furthermore, it automatically tunes Map Reduce, Virtualization, JVM, OS, and hardware as well. The functionality of Turbo starts with the optimization of the underlying hardware layer, which comprises network adopters, switches, routers, and servers, etc. In the second state, it tunes the Java Virtual Machine and then optimizes the runtime engine, which is comprised by the optimizer. It also has features for experienced analysists. Turbo-experienced users can manually optimize performance of big data.

8.5.1.4 Other big data optimization tools

Ananthanarayanan et al. [52] effectively monitored Map Reduce and mitigated an outlier. After a deep understanding of the Map Reduce, it classified the problems of the Map Reduce into three types, namely machine, network, and work. The authors proposed

Mantri as a system that can analyze and monitor the activities of the tasks and cull outliers. It uses cause and resource awareness techniques in the Mantri. It effectively evaluates the cost of each task after identifying the actual problem with the task. The authors proved that, if the Mantri were deployed in a Microsoft Bing search engine, then it can reduce completion time by 32%.

Dai et al. [53] developed a technique named HiTune, which is an accessible, light-weight, and efficient performance analyzer, particularly for Hadoop. The purpose of HiTune is the fine-tuning of cloud-based big data. HiTune is based upon distributed tools and dataflow-driven execution analysis. All the more explicitly, HiTune performs correlated performance activities simultaneous on different tasks and nodes, reproduces the dataflow-based procedure of the Map Reduce application, and relates the asset used to the dataflow model. HiTune is, for the most part, utilized for postexecution analysis and improvement.

The authors developed a Theia visualization utility to show the performance of Map Reduce tasks. It provides comprehensive details about each task, whether it is suspended, running, or idle, even though it also provides the duration of the task, present status, and resources allocated to the task. These visualizations can support the analyzers in optimizing the performance of the tasks and also finding out the exact node, hardware, OS, JVM, and server where performance is decreasing and also supporting the analyzer to troubleshoot problems.

PerfXPlain is a complete framework that allows users to complete the optimization process through the wizard by asking different types of performance-related questions [54]. After examining the questions and answers, the PerfXPlain automatically generates a detailed report. PerfXPlain uses query language to develop a question, process the answer of the questions through concerned algorithms, and finally produce a report based on the previous history of Map Reduce.

8.6 Analyzing big data in precision medicine

In this section, we offer an outline of recent growth in the analysis of big data with respect to medicine. Medical and biology data provide many opportunities for the development of precision medicine [55]. The main challenges in the development of precision medicine are also highlighted, and the authors also provide recent developments.

Precision medicine, which is otherwise called predictive and participatory medicine, is a way to customize the practice of medicine. Treatment and anticipation techniques like blood tests that consider individual inconsistency are not new. Likewise, sex, race, and time of cytotoxic cytotoxicity and serotonin are considered to reduce the danger of organ transplant rejection [56].

The challenge of applying the concept of precision medicine to clinical and omic datasets is due to patient characteristics that are available and physicians who cannot directly explain due to big size and complexity [57].

Big data is a wide term for large or complex datasets, so traditional methods of data processing are not enough. The previously discussed characteristics of big data apply to the existing biological and medical datasets.

Subsequently, medical business confronts a developing gap between our capacity to deliver big biomedical data and our ability to analyze and translate it.

The development of precision medicine faces four related problems that are discussed as follows:

8.6.1.1 Subtyping and biomarker discovery

Also called stratification of patients, this process identifies subgroups of patients who may be utilized for a treatment methodology and a particular subgroup predicts results. Subtype describes inner standards, which allude to subtypes in which patients are classified with likenesses in their basic components of infection, and primary standards, which narrate with genuine groups [58]. Nonetheless, which are explicit patterns and prototypes as well as how they are found, remain open. In spite of various meanings, subtyping is a grouping task and remain an active, viz-a-viz developing, piece of the AI. The larger part of the deceases like cancer, autism, and Parkinson's have been inspected through the sub-lens of this subtyping [59].

8.6.1.2 Drug repurposing and personalized treatment

Its objectives are identification proof and advancement of another equation for abandoned pharmacology. Capitalizing on existing medications takes into account limiting the advancement cost of pharmacology when compared with new findings and expansions [60, 61]. Drug repurposing is additionally going to preclinical assessments, which incorporate anticipating helpful regimens and security of the treatment. Patient subtyping and an accurate expectation of mending treatment results are key for beginning customized medications [60].

8.6.2 Biomedical data

In biological data, the complexity increased in two ways, one with respect to the number of samples and second with respect to heterogeneity.

8.6.2.1 The increasing number of samples

Because the capture of new innovation is quicker and less expensive, the number of entities with data is expanding rapidly. Assume that the quantity of accessible human genomes has expanded significantly in the most recent decade. Right now, the exome total group comprises 60,706 dissimilar human exomes [61]. The nature of the joined genomes depends on a vast degree on the connection between the short all-out perusing lengths and the length of the objective quality grouping. Furthermore, for a similar individual, an expanding number of tests is caught, and the data can be gathered through various tissues by utilizing a solitary cell genome or in various conditions [62].

8.6.2.2 Increasing heterogeneity of captured data

The amount of diverse biological objects through which information can be gathered is expanding. The collected data is very large in volume, and even the administration of basic data has become difficult.

8.6.3 Deep learning in medicine

As stated in previous sections, big data deals with large-scale variety of data and complications, and it requires capable algorithms to extract hidden knowledge in data. At present, big data uses statistical and machine learning methods as computational techniques. These methods have already proven that these can act to cover the differences between production and interpretation, but these methods still need improvement [63].

Machine learning methods have attracted the attention of big data analysis because of their outstanding ability to mine a huge amount of data, the diverse nature of data, and heterogeneous data types, which are a big principal issue in the precision of medicine and informatics [64].

8.6.4 Computational methods

There are four types of computational models and their details are given as follows:

8.6.4.1 Disease subtyping and biomarker discovery

Grouping patients into further subgroups on the basis of various criteria like genomics and clinical data is called subtyping. The principal target of subtyping is to accomplish progressively exact expectations of the anticipated aftereffects of people that can be utilized to advance treatment decisions. Numerous medicines have profited by subtyping, including Parkinson's disease as well as cancer. The exploration uncovered that cancer growth is a considerable disease among most considered diseases by subtyping. It is a disease where the deviations of the genome accumulate and finally malfunction in the regulation of the cellular system [65]. In view of the particular arrangement of qualities that are differentially communicated, they ascertain the separation among patients and make progressive gatherings. Explicit subnets are expected as working units that can act as new foci for drugs and their combinations.

8.6.4.2 Drug repurposing and personalized treatments

Various techniques for computation have been proposed for medication repurposing that can be grouped by various criteria. For instance, from a data perspective, it is suggested classification in medication and disease-based methodologies. The first set of techniques uses some similarities between drugs and the drug group, and concludes that a new drug candidate be reconfigured from the cluster that is able to do the same as other drugs in the cluster. The second group of techniques utilizes similarities among diseases of the groups and the advancement of another medication to be reconfigured by growing the known links between the drug and a few individuals from the group to the rest of the group. Different techniques use target-based likenesses to induce new drugs. These methodologies might be ordered dependent of similarities. Medical organizations frequently use artificial intelligence, machine learning, and Web-based technologies to identity drugs. Notwithstanding, the best limitation of these strategies is the absence of learning of these structures of numerous protein targets and the general computational expenses of testing a solitary pharmacological response.

8.7 Conclusion

Development continuously moves forward in newly invented technologies like cloud computing, Internet, and DL; now, all types of organizations have stepped into the field of big data. The medicine industry also uses big data and DL for its medical data. Although the medical industry adoption of big data and DL is relatively late, it is developing rapidly. It is a fact that medical big data is increasing, and the types of medical data are also gradually increasing, hence, the structure of medical big data will become more complex. The visual analysis of medical big data can play a vital role in the medical industry, pharmaceutical industry, public healthcare, and life sciences. In this chapter, we analyzed the relationship between DL and big data, their roles in medicine, their effects on the development of the medicine industry, applications of DL and big data in medicine, challenges and promises of both DL and big data with respect to medicine, prevailing techniques and tools, and existing big data optimization techniques that can be used in the medical industry.

There are a number of performance optimization of medical big data. We found that SONATA is the best big data performance optimization tool that can only be used by experts. The second tool is EC-Watch, which is also used for the same purpose; this tool has unique features that other tools do not. The third tool that is explained in detail is Turbo. It automatically optimizes Map Reduce and medical big data without human intervention. It permits big data inexperienced and beginners to optimize performance. The others identified, including these three tools, will significantly improve the efficiency of medical big data.

References

[1] M. Ahmed, S. Choudhury, F. Al-Turjman, Big data analytics for intelligent internet of things, in: R. Bera, S.K. Sarkar, O.P. Singh, H. Saikia (Eds.), Advances in Communication, Devices and Networking, Springer, Singapore, 2019, pp. 107–127, https://doi.org/10.1007/978-3-030-04110-6_6.

[2] A.A. Alani, F.D. Ahmed, M.A. Majid, M.S. Ahmad, Big data analytics for healthcare organizations a case study of the Iraqi healthcare sector, Adv. Sci. Lett. 24 (2018) 7783–7789, https://doi.org/10.1166/asl.2018.13017.

[3] T. Alhussain, Medical big data analysis using big data tools and methods, J. Med. Imaging Health Inform. 8 (2018) 793–795, https://doi.org/10.1166/jmihi.2018.2400.

[4] A. Samuel, M.I. Sarfraz, H. Haseeb, S. Basalamah, A. Ghafoor, A framework for composition and enforcement of privacy-aware and context-driven authorization mechanism for multimedia big data, IEEE Trans. Multimedia 17 (2015) 1484–1494, https://doi.org/10.1109/TMM.2015.2458299.

[5] B. Saha, D. Srivastava, Data quality: the other face of big data, in: 2014 IEEE 30th International Conference on Data Engineering. Presented at the 2014 IEEE 30th International Conference on Data Engineering (ICDE), IEEE, Chicago, IL, USA, 2014, pp. 1294–1297, https://doi.org/10.1109/ICDE.2014.6816764.

[6] Y. Bengio, A. Courville, P. Vincent, Representation learning: a review and new perspectives, IEEE Trans. Pattern Anal. Mach. Intell. 35 (2013) 1798–1828, https://doi.org/10.1109/TPAMI.2013.50.

[7] Y. LeCun, Y. Bengio, G. Hinton, Deep learning, Nature 521 (2015) 436–444, https://doi.org/10.1038/nature14539.

[8] J. Schmidhuber, Deep learning in neural networks: an overview, Neural Netw. 61 (2015) 85–117, https://doi.org/10.1016/j.neunet.2014.09.003.

[9] Q.K. Fatt, A. Ramadas, The usefulness and challenges of big data in healthcare, J. Health Commun. 03 (2018)https://doi.org/10.4172/2472-1654.100131.

[10] W. Raghupathi, V. Raghupathi, Big data analytics in healthcare: promise and potential, Health Inform. Sci. Syst. 2 (2014)https://doi.org/10.1186/2047-2501-2-3.

[11] R.R. Wakaskar, Brief overview of nanoparticulate therapy in cancer, J. Drug Target. 26 (2018) 123–126, https://doi.org/10.1080/1061186X.2017.1347175.

[12] S. Rallapalli, R.R. Gondkar, U.P.K. Ketavarapu, Impact of processing and analyzing healthcare big data on cloud computing environment by implementing Hadoop cluster, Proc. Comput. Sci. 85 (2016) 16–22, https://doi.org/10.1016/j.procs.2016.05.171.

[13] K.R. Malik, T. Ahmad, M. Farhan, M. Aslam, S. Jabbar, S. Khalid, M. Kim, Big-data: transformation from heterogeneous data to semantically-enriched simplified data, Multimed. Tools Appl. 75 (2016) 12727–12747.

[14] C.S. Kruse, B. Smith, H. Vanderlinden, A. Nealand, Security techniques for the electronic health records, J. Med. Syst. 41 (2017) https://doi.org/10.1007/s10916-017-0778-4.

[15] J. Dean, S. Ghemawat, MapReduce: simplified data processing on large clusters, Commun. ACM 51 (2008) 107–113.

[16] A. Pavlo, E. Paulson, A. Rasin, D.J. Abadi, D.J. DeWitt, S. Madden, M. Stonebraker, A comparison of approaches to large-scale data analysis, in: Proceedings of the 35th SIGMOD International Conference on Management of Data—SIGMOD '09. Presented at the 35th SIGMOD International Conference, ACM Press, Providence, Rhode Island, USA, 2009, p. 165, https://doi.org/10.1145/1559845.1559865.

[17] A.Y. Zomaya, S. Sakr (Eds.), Handbook of Big Data Technologies, Springer International Publishing, Cham, 2017 https://doi.org/10.1007/978-3-319-49340-4.

[18] S.P. Menon, N.P. Hegde, A survey of tools and applications in big data, in: 2015 IEEE 9th International Conference on Intelligent Systems and Control (ISCO). Presented at the 2015 IEEE 9th International Conference on Intelligent Systems and Control (ISCO), IEEE, Coimbatore, India, 2015, pp. 1–7, https://doi.org/10.1109/ISCO.2015.7282364.

[19] R. Bruno, P. Ferreira, A study on garbage collection algorithms for big data environments, ACM Comput. Surv. 51 (2018) 1–35, https://doi.org/10.1145/3156818.

[20] C. Stergiou, K.E. Psannis, T. Xifilidis, A.P. Plageras, B.B. Gupta, Security and privacy of big data for social networking services in cloud, in: IEEE INFOCOM 2018—IEEE Conference on Computer Communications Workshops (INFOCOM WKSHPS). Presented at the IEEE INFOCOM 2018—IEEE Conference on Computer Communications Workshops (INFOCOM WKSHPS), IEEE, Honolulu, HI, 2018, pp. 438–443, https://doi.org/10.1109/INFCOMW.2018.8406831.

[21] S. Shamim, J. Zeng, S.M. Shariq, Z. Khan, Role of big data management in enhancing big data decision-making capability and quality among Chinese firms: a dynamic capabilities view, Inf. Manag. 56 (6) (2018) 103135, https://doi.org/10.1016/j.im.2018.12.003 S0378720618302854.

[22] F. Prasser, H. Spengler, R. Bild, J. Eicher, K.A. Kuhn, Privacy-enhancing ETL-processes for biomedical data, Int. J. Med. Inform. 126 (2019) 72–81, https://doi.org/10.1016/j.ijmedinf.2019.03.006.

[23] H. Sebei, M.A. Hadj Taieb, M. Ben Aouicha, Review of social media analytics process and big data pipeline, Soc. Netw. Anal. Min. 8 (2018) 30, https://doi.org/10.1007/s13278-018-0507-0.

[24] J. Veiga, R.R. Expósito, J. Touriño, Performance evaluation of big data analysis, in: S. Sakr, A. Zomaya (Eds.), Encyclopedia of Big Data Technologies, Springer International Publishing, Cham, 2018, pp. 1–6, https://doi.org/10.1007/978-3-319-63962-8_143-1.

[25] N. Dey, A.E. Hassanien, C. Bhatt, A.S. Ashour, S.C. Satapathy (Eds.), Internet of Things and Big Data Analytics Toward Next-Generation Intelligence, Studies in Big Data, Springer International Publishing, Cham, 2018https://doi.org/10.1007/978-3-319-60435-0.

[26] A. Villar, M.T. Zarrabeitia, P. Fdez-Arroyabe, A. Santurtún, Integrating and analyzing medical and environmental data using ETL and business intelligence tools, Int. J. Biometeorol. 62 (2018) 1085–1095, https://doi.org/10.1007/s00484-018-1511-9.

[27] A. Oussous, F.-Z. Benjelloun, A. Ait Lahcen, S. Belfkih, Big data technologies: a survey, J. King Saud Univ. Comput. Inf. Sci. 30 (2018) 431–448, https://doi.org/10.1016/j.jksuci.2017.06.001.

[28] M.O. Gökalp, K. Kayabay, M.A. Akyol, A. Koçyiğit, P.E. Eren, Big data in mHealth, in: E. Sezgin, S. Yildirim, S. Özkan-Yildirim, E. Sumuer (Eds.), Current and Emerging MHealth Technologies, Springer International Publishing, Cham, 2018, pp. 241–256, https://doi.org/10.1007/978-3-319-73135-3_15.

[29] D. Palanikkumar, S. Priya, Brain storm optimization graph theory (BSOGT) and energy resource aware virtual network mapping (ERVNM) for medical image system in cloud, J. Med. Syst. 43 (2019) 37, https://doi.org/10.1007/s10916-018-1155-7.

[30] N. Sudhakar Yadav, B. Eswara Reddy, K.G. Srinivasa, Cloud-based healthcare monitoring system using storm and Kafka, in: S. Chakraverty, A. Goel, S. Misra (Eds.), Towards Extensible and Adaptable Methods in Computing, Springer, Singapore, 2018, pp. 99–106, https://doi.org/10.1007/978-981-13-2348-5_8.

[31] I. Ndukwe, B. Daniel, R. Butson, Data science approach for simulating educational data: towards the development of teaching outcome model (TOM), Big Data Cogn. Comput. 2 (2018) 24, https://doi.org/10.3390/bdcc2030024.

[32] M. Singh, M.N. Halgamuge, G. Ekici, C.S. Jayasekara, A review on security and privacy challenges of big data, in: A.K. Sangaiah, A. Thangavelu, V. Meenakshi Sundaram (Eds.), Cognitive Computing for Big Data Systems Over IoT, Springer International Publishing, Cham, 2018, pp. 175–200, https://doi.org/10.1007/978-3-319-70688-7_8.

[33] A. Mohamed, M.K. Najafabadi, Y.B. Wah, E.A.K. Zaman, R. Maskat, The state of the art and taxonomy of big data analytics: view from new big data framework, Artif. Intell. Rev. (2019) https://doi.org/10.1007/s10462-019-09685-9.

[34] S. Vitabile, M. Marks, D. Stojanovic, S. Pllana, J.M. Molina, M. Krzyszton, A. Sikora, A. Jarynowski, F. Hosseinpour, A. Jakobik, A. Stojnev Ilic, A. Respicio, D. Moldovan, C. Pop, I. Salomie, Medical data processing and analysis for remote health and activities monitoring, in: J. Moll, S. Carotta (Eds.), Target Identification and Validation in Drug Discovery, Springer, New York, NY, 2019, pp. 186–220, https://doi.org/10.1007/978-3-030-16272-6_7.

[35] T.-H.-Y. Le, T.-C. Phan, A.-C. Phan, Big data driven architecture for medical knowledge management systems in intracranial hemorrhage diagnosis, in: V.-N. Huynh, M. Inuiguchi, D.H. Tran, T. Denoeux (Eds.), Integrated Uncertainty in Knowledge Modelling and Decision Making, Springer International Publishing, Cham, 2018, pp. 214–225, https://doi.org/10.1007/978-3-319-75429-1_18.

[36] B. Leang, S. Ean, G.-A. Ryu, K.-H. Yoo, Improvement of Kafka streaming using partition and multi-threading in big data environment, Sensors 19 (2019) 134, https://doi.org/10.3390/s19010134.

[37] T.-M. Choi, S.W. Wallace, Y. Wang, Big data analytics in operations management, Prod. Oper. Manag. 27 (2018) 1868–1883, https://doi.org/10.1111/poms.12838.

[38] K. Rani, R.K. Sagar, A rigorous investigation on big data analytics, in: M. Pant, K. Ray, T.K. Sharma, S. Rawat, A. Bandyopadhyay (Eds.), Soft Computing: Theories and Applications, Springer, Singapore, 2018, pp. 637–650, https://doi.org/10.1007/978-981-10-5699-4_60.

[39] E. Begoli, J. Camacho-Rodríguez, J. Hyde, M.J. Mior, D. Lemire, Apache calcite: a foundational framework for optimized query processing over heterogeneous data sources, in: Proceedings of the 2018 International Conference on Management of Data—SIGMOD '18. Presented at the 2018 International Conference, ACM Press, Houston, TX, USA, 2018, pp. 221–230, https://doi.org/10.1145/3183713.3190662.

[40] X. Yu, V. Gadepally, S. Zdonik, T. Kraska, M. Stonebraker, FastDAWG: improving data migration in the BigDAWG polystore system, in: V. Gadepally, T. Mattson, M. Stonebraker, F. Wang, G. Luo, G. Teodoro (Eds.), Heterogeneous Data Management, Polystores, and Analytics for Healthcare, Springer International Publishing, Cham, 2019, pp. 3–15, https://doi.org/10.1007/978-3-030-14177-6_1.

[41] J. Wolf, A. Balmin, D. Rajan, K. Hildrum, R. Khandekar, S. Parekh, K.-L. Wu, R. Vernica, On the optimization of schedules for MapReduce workloads in the presence of shared scans, VLDB J. 21 (2012) 589–609, https://doi.org/10.1007/s00778-012-0279-5.

[42] T. Dokeroglu, S. Ozal, M.A. Bayir, M.S. Cinar, A. Cosar, Improving the performance of Hadoop Hive by sharing scan and computation tasks, J. Cloud Comput. 3 (2014) https://doi.org/10.1186/s13677-014-0012-6.

[43] A. Okcan, M. Riedewald, Anti-combining for MapReduce, in: Proceedings of the 2014 ACM SIGMOD International Conference on Management of Data—SIGMOD '14. Presented at the 2014 ACM SIGMOD International Conference, ACM Press, Snowbird, Utah, USA, 2014, pp. 839–850, https://doi.org/10.1145/2588555.2610499.

[44] B. Dong, Q. Zheng, F. Tian, K.-M. Chao, R. Ma, R. Anane, An optimized approach for storing and accessing small files on cloud storage, J. Netw. Comput. Appl. 35 (2012) 1847–1862, https://doi.org/10.1016/j.jnca.2012.07.009.

[45] E. Yildirim, E. Arslan, J. Kim, T. Kosar, Application-level optimization of big data transfers through pipelining, parallelism and concurrency, IEEE Trans. Cloud Comput. 4 (2016) 63–75, https://doi.org/10.1109/TCC.2015.2415804.

[46] J. Kim, E. Yildirim, T. Kosar, A highly-accurate and low-overhead prediction model for transfer throughput optimization, Clust. Comput. 18 (2015) 41–59, https://doi.org/10.1007/s10586-013-0305-4.

[47] E. Arslan, K. Guner, T. Kosar, HARP: predictive transfer optimization based on historical analysis and real-time probing, in: SC16: International Conference for High Performance Computing, Networking, Storage and Analysis. Presented at the SC16: International Conference for High Performance Computing, Networking, Storage and Analysis, IEEE, Salt Lake City, UT, USA, 2016, pp. 288–299, https://doi.org/10.1109/SC.2016.24.

[48] C. Roy, M. Pandey, S. SwarupRautaray, A proposal for optimization of data node by horizontal scaling of name node using big data tools, in: 2018 3rd International Conference for Convergence in Technology (I2CT). Presented at the 2018 3rd International Conference for Convergence in Technology (I2CT), IEEE, Pune, 2018, pp. 1–6, https://doi.org/10.1109/I2CT.2018.8529795.

[49] K. Kolomvatsos, P. Oikonomou, M. Koziri, T. Loukopoulos, A distributed data allocation scheme for autonomous nodes, in: 2018 IEEE SmartWorld, Ubiquitous Intelligence & Computing, Advanced & Trusted Computing, Scalable Computing & Communications, Cloud & Big Data Computing, Internet of People and Smart City Innovation (SmartWorld/SCALCOM/UIC/ATC/CBDCom/IOP/SCI), IEEE, Guangzhou, China, 2018, pp. 1651–1658, https://doi.org/10.1109/SmartWorld.2018.00282 Presented

at the 2018 IEEE SmartWorld, Ubiquitous Intelligence & Computing, Advanced & Trusted Computing, Scalable Computing & Communications, Cloud & Big Data Computing, Internet of People and Smart City Innovation (SmartWorld/SCALCOM/UIC/ATC/CBDCom/IOP/SCI),.

[50] Q. Guo, Y. Li, T. Liu, K. Wang, G. Chen, X. Bao, W. Tang, Correlation-based performance analysis for full-system MapReduce optimization, in: 2013 IEEE International Conference on Big Data. Presented at the 2013 IEEE International Conference on Big Data, 2013, pp. 753–761, https://doi.org/10.1109/BigData.2013.6691648.

[51] L. Xu, E. Muharemagic, A. Apon, ECL-watch: a big data application performance tuning tool in the HPCC systems platform, in: 2017 IEEE International Conference on Big Data (Big Data). Presented at the 2017 IEEE International Conference on Big Data (Big Data), 2017, pp. 2941–2950, https://doi.org/10.1109/BigData.2017.8258263.

[52] G. Ananthanarayanan, S. Kandula, A.G. Greenberg, I. Stoica, Y. Lu, B. Saha, E. Harris, Reining in the Outliers in Map-Reduce Clusters Using Mantri, in: 9th USENIX Symposium on Operating Systems Design and Implementation (OSDI '10). Presented at the 2010 Operating Systems Design and Implementation, 2010, pp. 265–278.

[53] J. Dai, J. Huang, S. Huang, B. Huang, Y. Liu, HiTune: Dataflow-Based Performance Analysis for Big Data Cloud, in: 2011 USENIX Annual Technical Conference (USENIX ATC'11), 2011, pp. 87–100.

[54] N. Khoussainova, M. Balazinska, D. Suciu, Perfxplain: debugging MapReduce job performance, Proc. VLDB Endow. 5 (2012) 598–609.

[55] A. Alyass, M. Turcotte, D. Meyre, From big data analysis to personalized medicine for all: challenges and opportunities, BMC Med. Genet. 8 (2015) 33, https://doi.org/10.1186/s12920-015-0108-y.

[56] K. He, D. Ge, M. He, Big data analytics for genomic medicine, Int. J. Mol. Sci. 18 (2017) 412, https://doi.org/10.3390/ijms18020412.

[57] J. Roski, G.W. Bo-Linn, T.A. Andrews, Creating value in healthcare through big data: opportunities and policy implications, Health Aff. 33 (2014) 1115–1122, https://doi.org/10.1377/hlthaff.2014.0147.

[58] I.M. Baytas, C. Xiao, X. Zhang, F. Wang, A.K. Jain, J. Zhou, Patient subtyping via time-aware LSTM networks, in: Proceedings of the 23rd ACM SIGKDD International Conference on Knowledge Discovery and Data Mining—KDD '17. Presented at the 23rd ACM SIGKDD International Conference, ACM Press, Halifax, NS, Canada, 2017, pp. 65–74, https://doi.org/10.1145/3097983.3097997.

[59] R. Higdon, R.K. Earl, L. Stanberry, C.M. Hudac, E. Montague, E. Stewart, I. Janko, J. Choiniere, W. Broomall, N. Kolker, R.A. Bernier, E. Kolker, The promise of multi-omics and clinical data integration to identify and target personalized healthcare approaches in autism spectrum disorders, OMICS 19 (2015) 197–208, https://doi.org/10.1089/omi.2015.0020.

[60] J.C. Denny, Surveying recent themes in translational bioinformatics: big data in EHRs, omics for drugs, and personal genomics, Yearb. Med. Inform. 23 (2014) 199–205, https://doi.org/10.15265/IY-2014-0015.

[61] R.E. Hewitt, Biobanking: the foundation of personalized medicine, Curr. Opin. Oncol. 23 (2011) 112–119, https://doi.org/10.1097/CCO.0b013e32834161b8.

[62] M. Naesens, D. Anglicheau, Precision transplant medicine: biomarkers to the rescue, J. Am. Soc. Nephrol. 29 (2018) 24–34, https://doi.org/10.1681/ASN.2017010004.

[63] T. Ching, D.S. Himmelstein, B.K. Beaulieu-Jones, A.A. Kalinin, B.T. Do, G.P. Way, E. Ferrero, P.-M. Agapow, M. Zietz, M.M. Hoffman, W. Xie, G.L. Rosen, B.J. Lengerich, J. Israeli, J. Lanchantin, S. Woloszynek, A.E. Carpenter, A. Shrikumar, J. Xu, E.M. Cofer, C.A. Lavender, S.C. Turaga, A. M. Alexandari, Z. Lu, D.J. Harris, D. DeCaprio, Y. Qi, A. Kundaje, Y. Peng, L.K. Wiley, M.H.S. Segler, S.M. Boca, S.J. Swamidass, A. Huang, A. Gitter, C.S. Greene, Opportunities and obstacles for deep learning in biology and medicine, J. R. Soc. Interface 15 (2018) https://doi.org/10.1098/rsif.2017.0387 20170387.

[64] D. Grapov, J. Fahrmann, K. Wanichthanarak, S. Khoomrung, Rise of deep learning for genomic, proteomic, and metabolomic data integration in precision medicine, OMICS 22 (2018) 630–636, https://doi.org/10.1089/omi.2018.0097.

[65] J.I. Castrillo, S. Lista, H. Hampel, C.W. Ritchie, Systems biology methods for Alzheimer's disease research toward molecular signatures, subtypes, and stages and precision medicine: application in cohort studies and trials, in: R. Perneczky (Ed.), Biomarkers for Alzheimer's Disease Drug Development, Springer, New York, NY, 2018, pp. 31–66, https://doi.org/10.1007/978-1-4939-7704-8_3.

Further reading

J. Andreu-Perez, C.C.Y. Poon, R.D. Merrifield, S.T.C. Wong, G.-Z. Yang, Big data for health, IEEE J. Biomed. Health Inform. 19 (2015) 1193–1208, https://doi.org/10.1109/JBHI.2015.2450362.

G. Bello-Orgaz, J.J. Jung, D. Camacho, Social big data: recent achievements and new challenges, Inform. Fusion 28 (2016) 45–59.

C.P. Chen, C.-Y. Zhang, Data-intensive applications, challenges, techniques and technologies: a survey on big data, Inf. Sci. 275 (2014) 314–347.

Y. Chen, J. Elenee Argentinis, G. Weber, IBM Watson: how cognitive computing can be applied to big data challenges in life sciences research, Clin. Ther. 38 (2016) 688–701, https://doi.org/10.1016/j.clinthera.2015.12.001.

X. Wu, X. Zhu, G.-Q. Wu, W. Ding, Data mining with big data, IEEE Trans. Knowl. Data Eng. 26 (2014) 97–107.

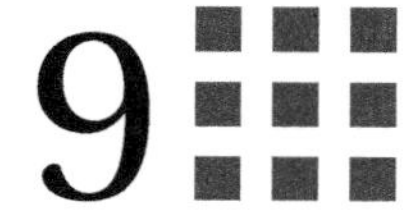

Simulation of biomedical signals and images using Monte Carlo methods for training of deep learning networks

Navid Mavaddat, Selam Ahderom, Valentina Tiporlini, Kamal Alameh
ELECTRON SCIENCE RESEARCH INSTITUTE, EDITH COWAN UNIVERSITY, JOONDALUP, WA, AUSTRALIA

Biomedical signals and images are critical sources of diagnostic information for clinicians and researchers. Recently developed deep learning methods have been proven effective in both clinical decision making [1–4] and their improvement [5, 6]. For supervised deep learning methods, massive data with accurate labels or ground truth is essential for high accuracy [7–9]. However, in biomedical applications, exact measurement of ground truth is often impractical or even impossible. An important avenue to generate data with ground truth is simulation. In this chapter, we explore the use of Monte Carlo (MC) simulation to generate biomedical imaging data and ground truth.

9.1 Introduction to simulation for biomedical signals and images

Biomedical images and signals are widely relied-on sources of anatomical information for diagnostic use and for understanding both normal anatomy and disease. Imaging methods such as X-ray radiography, magnetic resonance imaging (MRI), and ultrasound have become established as critical diagnosis tools in modern medicine. These imaging techniques seek to provide visual representations of hidden structures and systems within the body.

All biomedical imaging and signal technologies are engineered to exploit known physical phenomena, such as electromagnetic radiation (EMR) and sound waves, to interact with biological structures and return signals that can then be reconstructed into human interpretable signals or images. These technologies result in complex systems of interactions between the engineered parts of the system and the biological system being analyzed.

Deep Learning Techniques for Biomedical and Health Informatics. https://doi.org/10.1016/B978-0-12-819061-6.00009-4

9.1.1 Deep learning for classification of biomedical signals and images

9.1.1.1 Supervised machine learning

Machine learning is a field of statistical analysis techniques and algorithms that use experience and computational methods to make predictions [10]. In recent years, machine learning approaches have been successfully applied to a wide variety of applications. Machine learning has been an effective approach to developing many biomedical imaging applications. For example, machine learning-derived predictive models have been shown to perform as well as or better than skilled clinicians in melanoma detection from images [3, 4] and proficient in the diagnosis of retinal disease [2].

Machine learning approaches typically have three main aspects, namely, input data—that is, the *experience* provided to the algorithm; a *predictive model*; and the *predictions* themselves. Input data can be considered as a collection of quantitative measurements or observations of the object properties of interest. This collection of data is often termed training data and is used to train models that will be the basis of predictions to be made from new unseen data [11]. This is essentially the *experience* that the algorithm learns from.

Training data can be considered in the form of an observation or sample together with the desired output, such as a category it belongs to [12]. For the training data, if a label exists for each sample and all of this data is used to train a prediction model, it is a considered a supervised learning technique [11].

Supervised learning can be used to train a model to assign previously unseen examples into predetermined categories, which is referred to as classification. Alternatively, when supervised learning is used to assign a value along a continuum, it is known as regression [11].

9.1.1.2 Deep learning

Deep learning refers to machine learning techniques that have multiple layers of units, where the output of one layer feeds into the next through a nonlinear process [13]. Deep learning algorithms have proven to be a highly effective approach in creating useful predictive models in many fields, including computer vision, speech recognition, and bioinformatics [14]. One reason for the success of deep learning approaches to predictive modeling in recent years is the increase in cost-effective computational resources [13, 15].

Artificial neural networks

Artificial neural networks (ANNs) take inspiration from the neural networks within the brain as an abstract concept but in practice can differ significantly in function [16]. The basic concept is that units known as neurons, which are loosely analogous to neurons in a biological brain, are stimulated by an input to generate an output. This output is then passed, via a synapse, as an analogue to the structure of the biological neuron, which amplifies or attenuates the output (with a parameter known as a weight), to a subsequent neuron unit.

Large numbers of these artificial neurons are connected to form a complex network resulting in a system where inputs at the start of the network are modified by the interconnected neurons until they reach the final outputs at the end of the network. The learning component of this system comes into play through the process of adjustment of the weights until the point that, when inputs with known expected outputs are provided to the network, its outputs match with the expected outputs [16].

Convolutional neural networks

Convolutional neural networks are commonly used form of ANN that add a type of processing that has been inspired by functioning of cortical neurons in animal vision. Input signals are convolved with filters such that the inputs from neighboring neurons are incorporated into each other [14].

9.1.2 Deep learning requires good data

One of the biggest challenges for producing successful predictive models from deep learning is providing enough labeled training data to learn an accurate predictive model. This difficulty exists due to the fact that deep learning requires large numbers of parameters to be adjusted in the learning process [17]. Without large and varied datasets, deep learning techniques risk overfitting models, whereas training results in poor generalization when deployed. For supervised training, this then necessitates a correspondingly large number of data labels to avoid these issues.

As manual labeling of large amounts of data for machine learning can be expensive and error prone, many of the successful applications of deep learning have been possible due to the assembly of large datasets and labeling drawn from public or "crowdsourced" origins, where the effort for manual labeling has already done for other purposes [18].

9.1.3 Labeled biomedical image data is difficult to obtain

Assembling large datasets for biomedical imaging remains challenging due to a number of constraints that make approaches used in other domains, such as crowdsourcing, impractical. Some of these constraints include health risk due to invasive procedures and radiation exposure during imaging, cost of equipment use and the skilled labor required to operate the equipment and provide accurate labeling, confidentiality requirements, and other ethical considerations.

Synthetic data generated from simulation is therefore a useful approach to addressing the lack of availability of large labeled datasets.

9.2 Simulation of biological images and signals

9.2.1 Simulation can generate large amounts of labeled data

The motivating purpose for the development of simulation models for deep learning applications is to generate large, varied, and realistic synthetic datasets and their ground truth labeling.

Simulation as has been described as "imitation of a system" [19] or an "abstraction of reality" [20]. Thus, simulations can be characterized as a working representation of a system or, more specifically for the purposes of biomedical imaging, an imitation or representation of biomedical image or signal.

The development of simulations requires an understanding of the system being modeled. We will refer to this representation as the "system model."

Approaches to system modeling are numerous, and the field of system modeling is itself an active interdisciplinary field of research. One approach involves steps such as breaking down the system into subsystems, appraising the relevance of the parts and processes being modeled, analyzing the accuracy of the model, aggregating the components of the system, and finally validation of the system model [21].

Although simulations can be invaluable for providing insight into the workings of systems of interest, they have a fundamental limitation in that they are essentially simplifications that involve assumptions of the actual process being modeled. We will discuss a number of approaches that have been used to utilize simulation in biomedical imaging despite this limitation, and how this limitation has been addressed and, in some cases, exploited beneficially. Some common motivations for simplification are a lack of knowledge about the system, practical constraints such as limited time to observe many states, lack of resources to observe a working system, hazardous system states, and lack of current insight into the workings of the system that need to be understood [22]. For simulation of physical systems, simplification of the model is necessary as physical systems cannot be completely modeled, even with vast computational resources.

A number of studies have reported promising results by integrating synthetic features of interest into data samples [6, 23, 24]. These systems seek to increase detection rates of features that are of interest but do not occur in sufficient quantities in the original training set to allow for an accurate predictive model to be produced.

In the field of general object detection in images, Dwibedi et al. have implemented a concept of "cutting and pasting" images of objects into different scenes. They have reported that using this simple concept can produce compelling detection rates especially for tasks where recognition of specific instances of objects is required [23].

In the biomedical imaging field, Tom et al. have demonstrated a system to synthesize ultrasound images for the purpose of producing training data for supervised learning or learning aid for clinicians to visualize pathologies that would rarely occur in clinical practice [24]. Their proposed system used a simulation to produce an initial ultrasound image, and then they followed this by passing the simulated image through two stages of generative adversarial networks to produce the images that they term as "patho-realistic." It was proposed that these synthesized images can then be used for predictive model training as well as training clinicians.

9.2.1.1 System modeling

The availability of a priori information about a system can determine the modeling choices made for creating simulations. "White box modeling" refers to systems where all the components and parameters are known a priori. With this form of modeling, the process that the inputs of the system go through to reach resultant output is explicitly modeled and thus intermediate states can be observed and analyzed. In contrast, in "black box" modeling, only the inputs and outputs of the system together with how they are mapped to each other are known. This implies that inner workings cannot be observed or analyzed. In some cases, it may be possible to make inferences about a black box system from the observed relationships between inputs and outputs. Similarly, "grey box modeling" refers to system models where some partial knowledge of the inner states is known.

Monte Carlo methods

MC methods are a class of approximate inference techniques that allow for the mathematical modeling of systems that have a vast number of states, and/or those that are influenced by events of a random nature, to be modeled. MC methods are based on the concept of numeric sampling and can assist in developing probabilistic models [11].

To create an MC model for a system, representative samples of the processes that typically occur within that system are simulated. Due to the random nature of some of the components of these processes, each outcome could possibly be unique. Probability density functions are employed to model the various stochastic processes within the greater system. To build a comprehensive system model, multiple instances of the process being modeled must be observed to capture enough data and effectively model a representative probabilistic distribution.

MC methods for simulation are a promising way data can be generated for deep learning applications, where the ground truth of interest can be more precisely determined while retaining the ability to generate large datasets critical for training high accuracy models using deep learning.

Markov chain

A Markov chain is a method for describing a stochastic model of a complicated sequence of events, where the probability that an event occurs depends only on the state of the previous event [25]. In terms of biological imaging, such as optical coherence tomography (OCT) and X-ray, radiation propagation through material can be described as a Markov Chain. As radiation propagates through tissue, the sequence of events that influence the penetration and path of the radiation can be seen as a chain of events. At each point where an interaction occurs, the next propagation path segment is determined only by the current state of radiation path and the circumstances of the interaction, which sets its trajectory until the next interaction point [11]. An MC simulation can generate a model of the propagation of radiation by taking each chain as an instance that contributes to the greater system model being generated.

9.2.2 Differences between synthetic data and real data

There are many challenges when it comes to system modeling and simulating complex biomedical systems. These include, but are not limited to, inability to define the system, lack of data, ethical considerations, limits of current science and technology, high levels of complexity, lack of appropriate modeling tools, and limited resources.

In the biomedical sciences, some of the previously mentioned challenges are widespread, for example, ethical considerations can impose significant constraints on the process of system modeling. Data may be limited for reasons for confidentiality. Obtaining some types of data may involve invasive procedures that could pose a risk to the health or even the life of the patient. For these reasons and a multitude of others, the system model may only be a limited representation of that actual system and thus produce imaging that is not fully representative of the system it is imitating. Thus simulations, although having the benefit of providing ground truth labeling, will often produce a synthesized data sample, which is only a limited imitation of real data.

9.2.3 Addressing differences between simulated and real data

9.2.3.1 Reality gap bridging techniques

The "reality gap" or "domain gap" can be described as the discrepancy between outputs of simulation and the actual outputs of the real system. For a system model to be a useful generator of data, it is desirable for this discrepancy, or error, to be minimized.

The fundamental way to do this would be to design and tune a system model such that the difference between the model and system is minimal. However, there are a number of reasons why this may not be possible, and hence, it can be valuable to have alternative approaches to bridging this gap. We have covered many of the challenges to accurate system modeling earlier in this chapter, but we will consider some challenges that are being addressed with deep learning in this section.

System identification

"System identification" is a critical part of the system modeling process. Simulation systems need to be tuned to become good representations of physical phenomena. System identification involves tuning the parameters of the system model to match the behavior of the physical system that is modeled [26, 27].

Typically, the main limitations of the system identification techniques are (i) unmodeled physical effects and (ii) unknown material types and dimensions. In some cases, the reality gap can be due to deliberate modeling decisions, such as using existing simulations from a different domain to exploit a predeveloped system model from a different domain [28, 29].

Importance sampling for Monte Carlo simulations For MC simulations, a variance reduction technique known as importance sampling can be used to increase the accuracy of the simulation. Importance sampling works by increasing simulation instances known to be more likely to impact the final probability distribution while compensating this change with a bias function. For biological imaging applications, where it is typical for the radiation to be lost in tissue through known phenomena, importance sampling allows for simulations to be sped up or made more accurate by only focusing on the radiation that is more likely to reach the sensing elements.

Adversarial networks

Adversarial networks are a promising area of machine learning research, useful for bridging the reality gap left by simulations. Adversarial networks can be conceptually simplified as a way to generate a model that creates "synthetic" samples that are convincing enough such that they cannot be distinguished from "real data" when judged by another model known as a critical investigator. The generative model is improved through a process of iteratively checking if the investigator can distinguish if a synthesized sample seems "real," and if not, it adjusts the parameters of the generator until differences between real and synthetic samples can no longer be distinguished [30, 31].

Once such a system is trained, the generative part of the network can be fed inputs, which will produce synthetic outputs that should be indistinguishable from real samples. This allows for the generation of virtually unlimited amount of realistic synthetic samples that can be fed into another training network.

9.2.3.2 Data augmentation

In machine learning, data augmentation refers to methods used to increase the size and variability of training datasets. This term can apply to many methods that seek to generate more data for training; however, in imaging and signals application, it usually refers to a number of simple transformations applied to images and signals to add variability to the training set. These transformations include flipping, jittering, scaling, rotating, and shearing the training images, as well as Gaussian blurring of images and adding Gaussian noise [32, 33]. Although some of these methods model physical changes to create additional synthetic training images, as they don't directly utilize a physical system model in the generation of the images, these type of augmentation techniques may result in widening of the reality gap by introducing variations that would not occur in the real system [32].

9.3 Classification of optical coherence tomography images in heart tissues

9.3.1 Prediction model example: Classification of OCT images from heart tissues

OCT is a nondestructive technique commonly used to conduct surface imaging of biological tissues and other turbid materials. OCT uses the light reflected off a medium target and its interference with the input broadband light to enable a computed reconstruction of the internal structure of that target.

Beyond biomedical imaging [34, 35], OCT has found applications in the semiconductor [36], water treatment [37], and paper industries [38]. OCT is suited to imaging turbid media, that is, cloudy or opaque materials. Effective OCT imaging requires some penetration of EMR, but the material must possess light scattering and reflection properties so the radiation interacting with the biological medium returns to the optical sensor [39].

In comparison to other commonly used medical imaging techniques, such as ultrasonography (ultrasound) and MRI [39], OCT typically enables higher resolution imaging resolving details in biological tissue in the micrometer range. In terms of imaging depth, OCT is typically limited to an imaging depth of 1 to 3 mm for biological tissues [40] but can extend to up to 10 mm for tissues with lower scattering coefficients [41]. These properties as well as some probe and sensor packaging advantages make OCT imaging a compelling solution for heart tissue imaging during tissue ablation procedures.

In this section, we propose a system for distinguishing ablated heart tissues from unablated heart tissues from OCT images. "Real world" OCT images, such as data gathered in vivo from human or animal myocardial tissues, are difficult to obtain due to stringent requirements, such as the need for a skilled clinician, expensive single-use instruments, and regulatory procedures. In vitro experiments, although much less costly, still require a trained technician and are time consuming. To address this lack of real-world data, we will discuss a simulation method for generating data suitable for a deep learning approach to the classification problem. Figs. 9.1 and 9.2 show a hybrid OCT imaging and ablation catheter that can be used to conduct OCT imaging of heart tissues during ablation procedures.

9.3.1.1 Simulation to generate synthetic OCT images

The simulation algorithm for OCT image synthesis can be broken down into a number of logical parts, each of which can be tested and verified before being combined. The two

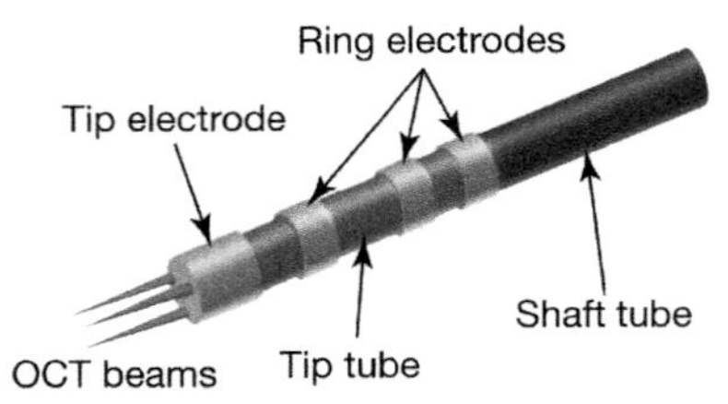

FIG. 9.1 A hybrid RF-OCT catheter structure, consisting of electrode used for RF ablation and three embedded optical fibers extending to the tip of the catheter.

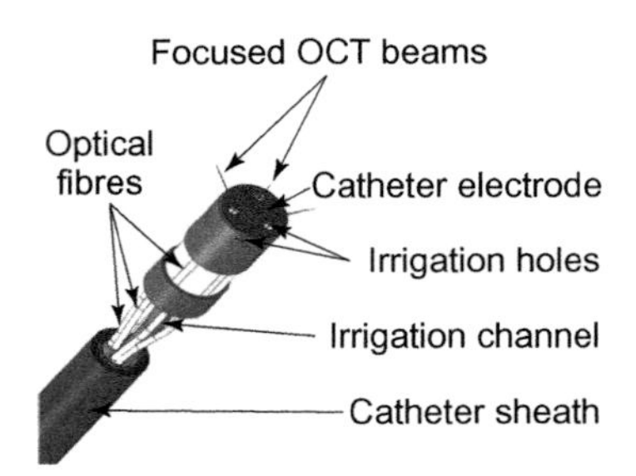

FIG. 9.2 Illustration of the integration of the OCT fiber with an RF catheter and the use of the irrigation holes for OCT illumination and optical back reflection.

main components of the simulation are the interferometry and light propagation simulation algorithms. Together they will be combined to form an OCT simulation algorithm.

System model example: Interferometry system for OCT

The OCT imaging technique makes use of the interference of reflected EMR from a target, with reference radiation from the same source. An understanding of this part of the system is required to effectively model it.

To produce the OCT image, an interferometry setup known as the Michelson Interferometer is used to measure the path length of the light that enters the tissue and returns back to an optical detector (or sensor), as illustrated in Fig. 9.3. The light source, usually a laser or a light-emitting diode (LED) of narrow spectrum, a beam-splitter, two mirrors, and a sensor are used in this setup. As the light moves through the beam-splitter, the two separate paths reflect off the mirrors on their respective arms and return to recombine as they pass back through the beam-splitter. This recombined light signal is then recorded by the sensor. If paths between the beam-splitter and the mirrors have the same length, an interference pattern will be recorded by the sensor [42].

A coherent light source, such as a laser, is a light source that emits photons of similar wavelengths that are in phase with one another, whereas the photons emitted by an incoherent light source do not have the same wavelength and are not in phase with one another, that is, the phases of the waves are completely random [42]. With an incoherent light, an interference pattern will be observed at the sensor of the Michelson Interferometer only when the path difference is less than the coherence length.

OCT is an imaging technique based on light interference, which enables the visualization of the cross-section of a biological tissue with a spatial resolution as low as 10 μm and a penetration depth of around 1 mm. There are two common interferometry approaches to record an OCT signal: Time Domain OCT (TD-OCT) and Frequency Domain OCT (FD-OCT). The main difference between these two methods relates to how the interferometer is altered to measure the differing path lengths within the material being analyzed. As we will be using FD-OCT simulation examples, we will only describe FD-OCT in more detail.

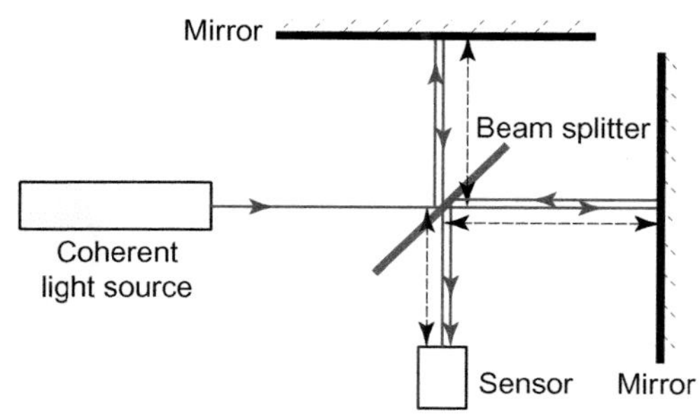

FIG. 9.3 Layout of a Michelson Interferometer.

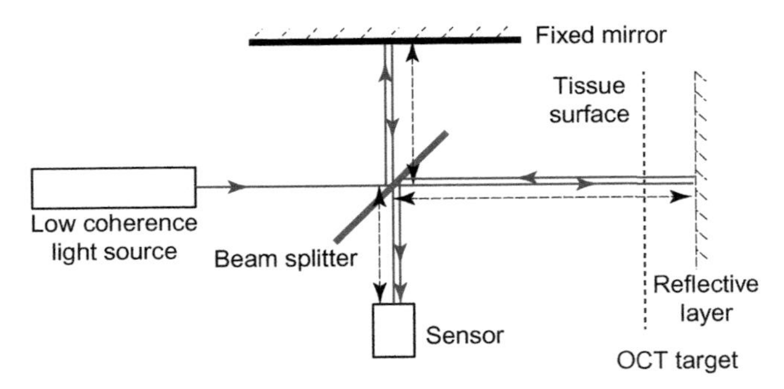

FIG. 9.4 Interferometry set up for a Frequency Domain OCT system.

In the FD-OCT, a single fixed mirror and the surface of the object to be imaged are set at equal distances from the beam-splitter. This layout is illustrated in Fig. 9.4. FD-OCT requires light that is coherent for the length of the imaging depth to produce the interference pattern, so the wavelength of light that is passed through the interferometer is gradually altered over time. This is usually done in a sawtooth pattern, as shown in a study by Zhao [43], to provide some persistence of the pattern at the sensor. As the light sweeps through the different wavelengths, a series of peaks and troughs will be recorded at the sensor. The frequency of this pattern correlates to the depths where light has reflected. To this end, a Fourier Transform is used to return this to the spatial domain [43] producing what is known as an A-scan (Fig. 9.5).

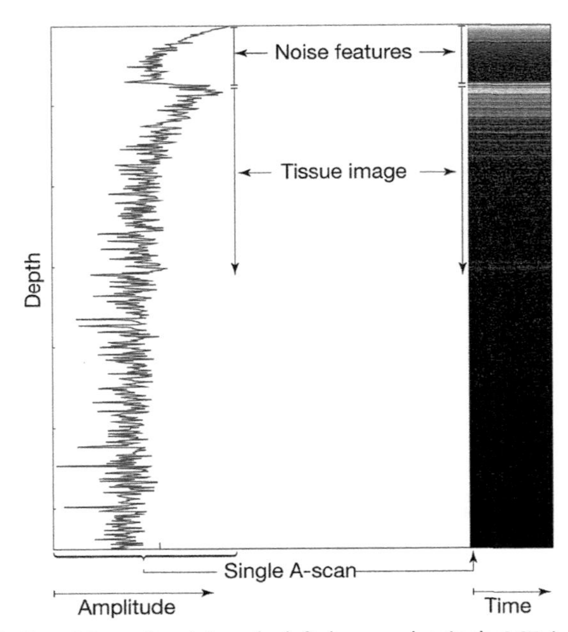

FIG. 9.5 Typical OCT signal of heart tissue the plot on the left shows a plot single OCT signal "A-scan" against depth, whereas the image on the right shows a typical representation of an OCT signal as false-color bundle of 100 A-scans taken within short succession of each other.

If the target is a single fixed fully reflective mirror and d is the path distance from the emitter/sensor, the path length would be $2d$ as it travels from the beam-splitter to the mirror and the same distance back to be recombined. Given the assumption that the mirror is fully reflective and there is no loss along the path, at the point of interference the amplitude of the wave will be doubled. In biological tissue, the light is split into many paths that travel different distances of which only some return to the sensor, and of those that return, due to the interactions that occur within the tissues, the amplitudes of those waves are attenuated by varying amounts.

To simulate the OCT signal, the interferometry module requires the input of path lengths and their associated attenuated amplitudes (or weights). Each of these path lengths z_n can be considered to be fixed mirrors at distance $^{z_n}/_2$ and the amplitude of the return wave to be w_n. The laser sweep can be paired with the interference model to give

$$I(\mathbf{k}) \propto \sum_{n=1}^{N} \sqrt{w_n} \cos(\mathbf{k} z_n) \tag{9.1}$$

where,

$k = \frac{2\pi}{\lambda}$,

$\mathbf{k} = [k_1, \ldots, k_1 + (m-1)\Delta k, \ldots, k_1 + (M-1)\Delta k]$

n is the photon packet number,

N is the total number of photon packets

m is the mth value of k

M is the total number of discrete wavenumbers

The output, $I(\mathbf{k})$, now represents the sensor intensities accumulated during the wavelength sweep. The pattern frequency of this output correlates to the depths where light has reflected. Applying a Fourier Transform will return this to the spatial domain producing a simulated A-scan. Fig. 9.6 illustrates the steps of this process by using a single target as a simple example (the target being a fully reflective mirror).

9.3.1.2 System model example: Physics of light propagation through biological tissue

While each individual photon of light travels through the material, it follows a different path as a result of the various interaction events that act on how it propagates. The nature of these events are influenced by the properties of the media it passes through, affecting its path. Fig. 9.7 illustrates a typical path of light through a turbid material. As some of these events determining the light path are influenced by random processes, each path will be unique. Due to this reason, inferring a model of light propagation through a turbid material is, for practical purposes, an intractable problem, and therefore, an approximation must be used. We will describe an effective means of approximating this system for practical simulation.

9.3.1.3 Optical characterization of biological tissue

Given the complex nature of biological tissue and the limited information available about the optical properties of the tissue ultrastructure, some decisions regarding an

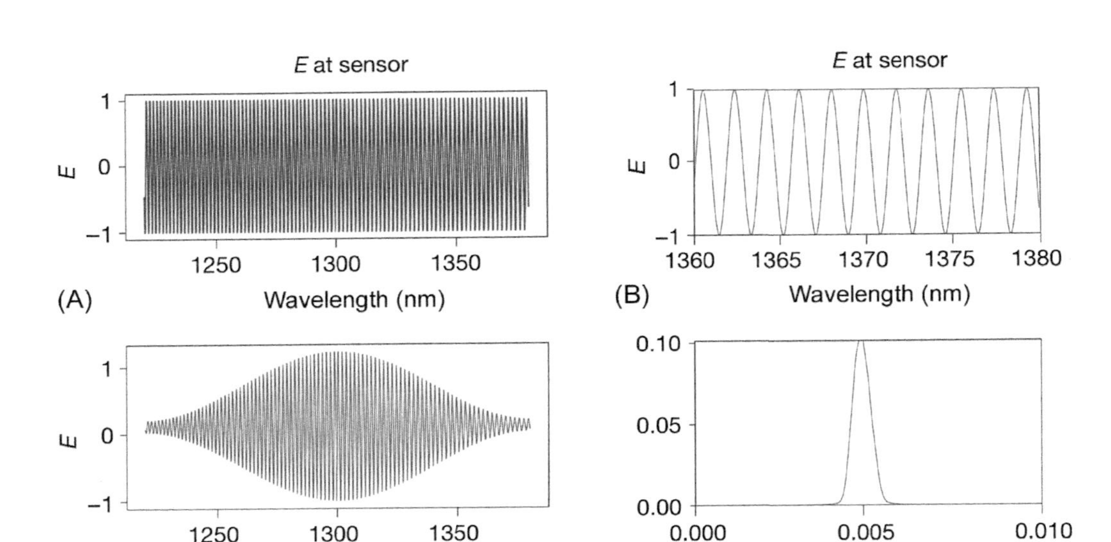

FIG. 9.6 Results from the simulation of an FD-OCT system with the target being a single fully reflective mirror. (A) Sensor response versus wavelength. (B) Zoom in of (A) for more details. (C) Response with a Hamming window applied to reduce the effect of the abrupt cut off window of wavelengths. (D) Response at 0.005 cm obtained by applying a Fourier Transform to recover the depth of the imaged mirror.

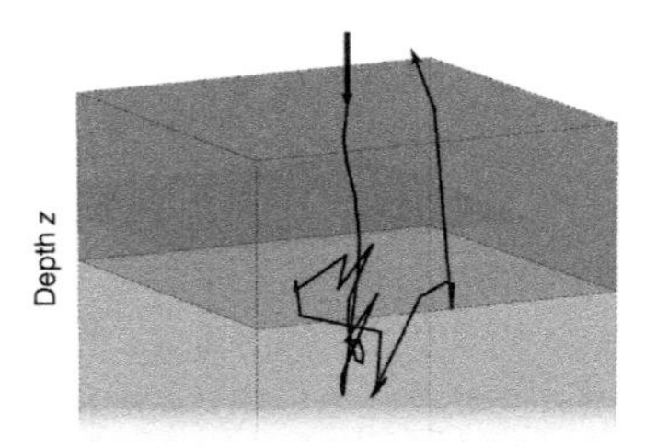

FIG. 9.7 Example of light propagation through a turbid medium. Its path can be altered by many interactions with the media with some paths eventually returning back to the source.

appropriate system model will need to be made. In this section, we will describe the interactions between tissue and light that have been identified by the available literature as significant for forming an OCT image.

For biological tissue characterization, the light propagation model is typically simplified to only take into account the interactions likely to occur in the tissue [44, 45]. Fig. 9.8 is a flow chart detailing a typical light propagation model for a turbid media such as a biological tissue [45]. As light travels through a tissue, it interacts in a number of ways that affect its path. Light can either travel uninterrupted for some distance, be reflected or transmitted through optical boundaries, or be scattered or completely absorbed, and these interactions can also cause it to lose intensity. Fig. 9.9 illustrates some of these interactions.

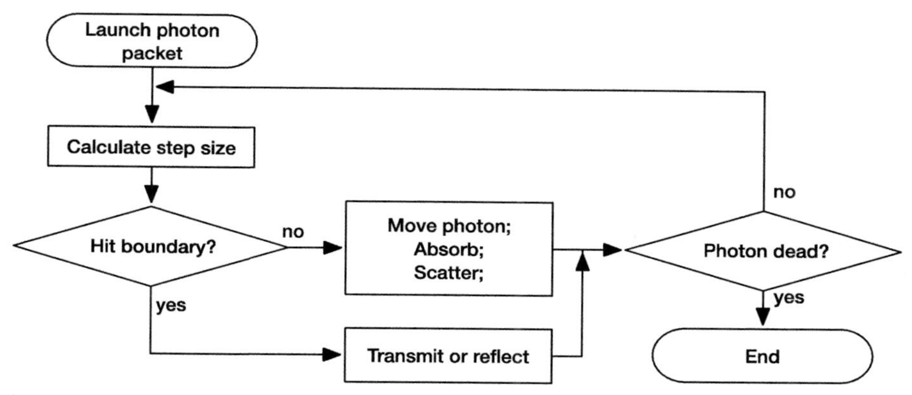

FIG. 9.8 Flowchart of a simple model of the propagation of light. Here the path of a single photon packet is traced until its energy has been absorbed.

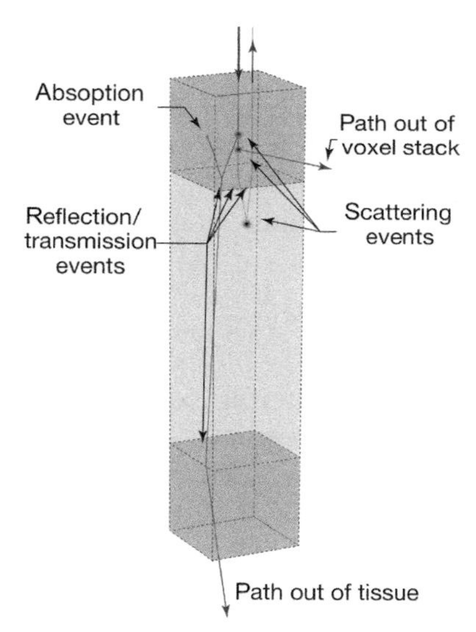

FIG. 9.9 Light-tissues interaction within the 3D stack layout. The different shading colors of the layers indicate differing optical properties.

Biological tissue is often a heterogeneous material consisting of many substructures that affect the propagation of light through it. A biological tissue is often characterized by the key parameters of absorption coefficient, μ_a; scattering coefficient, μ_s; anisotropy factor, g; and refractive index, n. However, in some models, the scattering coefficient μ_s and anisotropy factor g are replaced with the reduced scattering coefficient, μ_s' (the relationship between the parameters is shown in Table 9.1). A number of studies have been conducted to determine the optical parameter values for different tissue types at different wavelengths [41, 44, 46–50].

Table 9.1 Parameters used to describe optical characteristics of biological tissue.

Optical parameter	Notation	Influenced by	Typical units
Absorption coefficient	μ_a	—	cm^{-1}
Scattering coefficient	μ_s	—	cm^{-1}
Scattering function	$p(\theta, \psi)$	(g, μ_s) or μ'_s	sr^{-1}
Anisotropy factor	g	—	(dimensionless)
Refractive index	n'	—	(dimensionless)
Reduced scattering	μ'_s	g, μ_s	cm^{-1}

The scattering coefficient μ_s together with the scattering function $p(\theta, \psi)$ and the anisotropy factor g influence the reduced scattering coefficient μ'_s [44].

It is typically difficult to obtain accurate estimates of these optical parameters. This can be due to the variability of biological tissue, complexity in the models, and other challenges. In addition, another significant restriction of the utility of these estimates is that they are often reported for only a narrow band of wavelengths of light [44]. To help generalize for wavelength, Jacques et al. [44] have attempted to unify a number of different reports by proposing a relationship for determining these coefficients at a particular wavelength. For example, for the reduced scattering coefficient μ'_s, the following function has been proposed:

$$\mu'_s = a\left(\frac{\lambda}{500\,\text{nm}}\right)^{-b} \tag{9.2}$$

which gives standardized reduced scattering coefficient parameters, nominally at the wavelength $\lambda = 500\,\text{nm}$, in terms of two new coefficients a and b [44]. Table 9.2 shows the reported values of a and b for selected tissues types relevant to this study.

Free path

The free path (sometimes referred to as the photon mean free path) is the average distance a photon travels in a particular direction before it collides with another particle causing it to change direction. In a vacuum, this distance may be extremely long, but

Table 9.2 Parameters a and b specifying reduced scattering coefficient of different tissues.

Tissue type	a (cm^{-1})	b	References
Heart	8.3	1.260	[57]
Heart wall	14.6	1.430	[58]
Muscle	9.8	2.829	[58]
Muscle	13.0	0.926	[59]

Data from S.L. Jacques, Optical properties of biological tissues: a review, Phys. Med. Biol. 58(11) (2013) R37.

within a turbid material using near infrared (NIR) light, this distance is often less than 100 μm [44].

Modeling individual photons is infeasible, due to the huge number of photons emitted by a typical OCT light source. To address this issue the process can be simplified by considering large groups of photons as "photon packets" with an associated total energy or "weight." For some interactions, there is a probability that individual photons might be absorbed. This interaction that results in photon absorption can be modeled as a proportional loss of packet weight, and simulation can then be continued with the remaining photons. The free path covered by a photon packet between the points of interaction within a medium is determined by the following equation [41]:

$$s = -\frac{\ln \xi}{\mu_a + \mu_s} \tag{9.3}$$

where

s is the free distance covered by a photon package.
μ_s is the scattering coefficient.
μ_a is the absorption coefficient.
ξ is discrete character energy absorption—a random number uniformly distributed between 0 and 1.

As the free path is traversed, photons are absorbed by the material it passes through, and to model this, the change in weight of the photon packet can be calculated by the following equation [43]:

$$\Delta W = W \frac{\mu_a}{\mu_a + \mu_s} \tag{9.4}$$

where

ΔW is the incremental change in the statistical weight of the photon packet at each point.
W is the statistical weight of the photon packet.
μ_s is the scattering coefficient.
μ_a is the absorption coefficient.

Scattering

Scattering occurs when photons of light interact with particles and change their trajectory [51]. Within a heterogeneous biological material, it is prohibitively complicated to model the individual interaction between photons and a vast array of molecules within the tissue. To create a practical simulation, an estimation is required. Henyey and Greenstein [52] formulated such an estimation function to model this effect within scattering media in terms of the probability of the photon trajectory being scattered by a given angle θ, which is given by

$$p(\cos\theta) = \frac{1-g^2}{2(1+g^2-2g\cos\theta)^{3/2}} \tag{9.5}$$

where

θ is the angle of the scattering.
g is the anisotropy factor in the medium.

This equation can be reformulated such that given the scattering angle θ_s can be generated from a continuous random variable u (where u is uniformly distributed and bound between 0 and 1):

$$\cos\theta_S = \frac{1+g^2}{2g} - \frac{(1-g^2)^2}{2g(1-g+2gu)^2} \tag{9.6}$$

Reflection
Reflection is an optical interaction that occurs at the boundary of materials with differing refractive indices. Biological tissues contain many different structures creating such boundaries. Fresnel's equations determine the relationship between the intensities of light reflected and transmitted at the boundary of materials with different refractive indices. By definition, any light that isn't reflected is transmitted further into the tissue. For this case, Fresnel's equation can be expressed as [53]:

$$r_n = \left|\frac{n_{n+1}-n_n}{n_{n+1}+n_n}\right|^2 \tag{9.7}$$

In the case that the light incidence isn't normal, we can model this by using Snell's law together with Fresnel's. Thus, in this case the change of angle is given by [53]

$$\frac{\sin\theta_i}{\sin\theta_t} = \frac{\mathrm{n}_n}{\mathrm{n}_{n+1}} \tag{9.8}$$

Absorption and termination of the photon packet
For the purpose of modeling the OCT image, there is no need to follow all photon packets until their energy is exhausted. A number of assumptions can be made that allow for the simplification of the propagation model. If the packet travels out of the boundaries of the tissue model, it is deemed unlikely to return and the path is terminated. Another situation where the path is terminated is when the weight of the packet has been reduced below a fixed threshold level where it is deemed unlikely to have a significant effect on the light model if continued further. This may occur after a number of transmission/reflection interactions or once it has reached the end of its randomly determined free path. These conditions will allow the simulation to run significantly faster as the computation time is

not wasted on calculations that will ultimately make, at best, insignificant contributions to the final OCT image.

Class I and II paths

Light propagation paths through turbid material can be split into two classes. Class I paths are ones where the only interaction with the material is to reflect or scatter directly back directly to the source/sensor. Class II paths are ones where any number of interactions can occur with the material. Computationally, class II paths require significantly more resources than class I paths to simulate. Fig. 9.10 illustrates the difference between these classes of paths. The knowledge of such differences in requirement for resources allows the system model designer to make decisions about trades off between allocations of resources in the simulation.

9.3.1.4 Coordinate systems

A number of coordinate systems have been used for modeling the propagation of light. These include Cartesian/cylindrical [37, 39, 45, 50]-based coordinate systems. Fig. 9.11 shows the layer system used by Wang et al., Yao et al., and others [49, 54] to describe biological tissue consisting of layers with differing optical properties. As this layout can be sufficient for 1D A-scans, our proposed simulation will make use of a similar layer-based coordinate system with layers perpendicular to the incident light beam. The extents of the simulation are modeled by a bounding cuboid centered along the axis of the entering light as illustrated in Fig. 9.9.

9.3.2 Verification of simulation results

The verification of the simulation can be done in stages as components of the simulation algorithms are developed.

The interferometry component of the OCT signal simulation can be experimentally assessed. Initially FD-OCT interferometer will be set up with fully reflective and partially reflective mirrors used as targets. The results from this experiment can be compared with

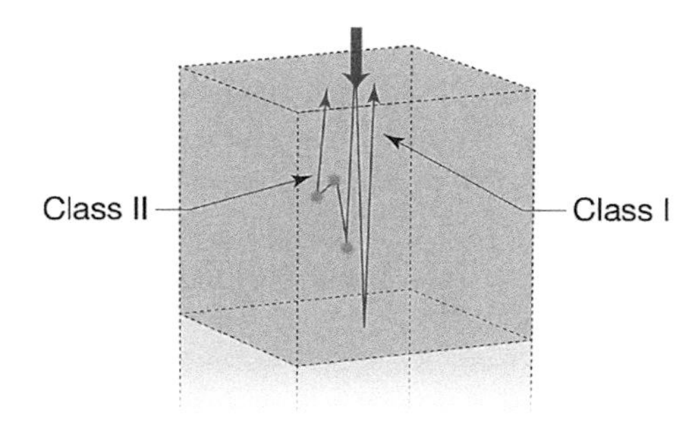

FIG. 9.10 Light propagation paths through turbid material can be split into two classes. Class I paths are the one where the only interaction with the turbid material is to reflect and head directly back to source/sensor. Class II paths are the ones where any number of interactions can occur with the material.

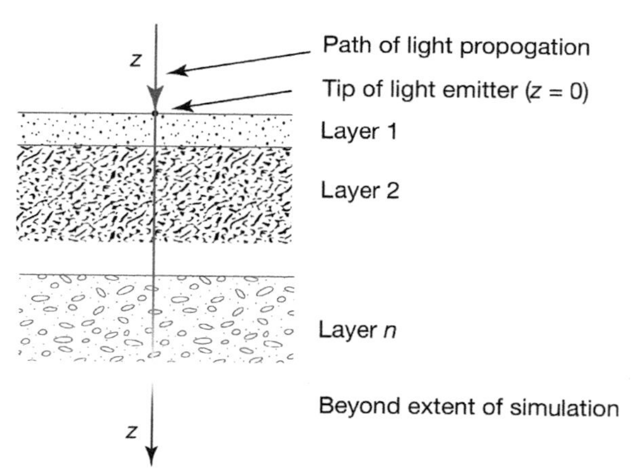

FIG. 9.11 Cartesian coordinate system layout used by Yao et al. [54] to describe a biological tissue consisting of layers with different optical properties.

the simulation results to tune the simulation to the characteristics of the particular interferometer being used.

The light propagation modeling component of the simulation can also be verified by lab experiments. Synthetic turbid material can be created by doping epoxy with varying amounts of powder that induce light scattering within the material. Again, these experimental results can be compared with simulation results to verify the suitability of initial tuning parameters of the simulation.

9.3.3 Bridging the reality gap

Given the complex nature of the light interaction process within biological tissue, system modeling of OCT imaging of heart tissues is a difficult task that relies on accurate mathematical models representing the physical interactions together with good quality empirical data about the optical parameters of the tissues. Although mathematical models of the physics of the OCT system have been well studied, empirical data about the optical parameters of tissues is limited, particularly for heart tissues over the wavelength range of the light used in a typical OCT system. To overcome these shortcomings, the physical model can be improved using a number of techniques; we will describe the main two, namely, (i) Importance sampling, which seeks to increase the accuracy of the MC simulation, and (ii) Generative Adversarial Networks, which adjust the optical parameters of system model to generate more realistic synthetic data.

9.3.3.1 Importance sampling

Because an OCT signal depends on the light returning back to the photodetector, we can focus the simulation on paths that are likely to be detected. As one of the important mechanisms for this to occur is when light is scattered one or more times and the resultant photon path ends up directed toward the photodetector, focus on this part of the simulation can yield greater efficiency and accuracy, especially as light returning to the photodetector is a very rare event. However, modeling this is critical to the simulation.

Importance sampling has been proposed [43, 55] as a way to more accurately model light-tissue interaction by reducing the number of iterations the simulation typically does by unnecessarily following paths of light that are very unlikely to return back to the photodetector. Therefore, efficient simulation should mainly focus on backscattering interactions.

A bias function is typically determined, which generates the angles of trajectories that are more likely to be of interest (backscattering), and then compensation for this bias is applied in the final distribution. A likelihood function is used as compensation for the introduced bias. This can be formulated by using a bias function f_B, defined as [55]:

$$L(\cos\theta_B) = \frac{f_{HG}(\cos\theta_B)}{f_B(\cos\theta_B)} \tag{9.9}$$

The biased H-G function is defined as:

$$f_B\cos(\theta_B) = \begin{cases} \dfrac{A}{2(1+a^2-2a\cos(\theta_B))^{3/2}} & \text{if } \cos\theta_B \in [0,1] \\ 0 & \text{otherwise} \end{cases} \tag{9.10}$$

Integrating this function across all steradians (solid angles) yields

$$f_B(\cos\theta_B) = \begin{cases} \left[1 - \dfrac{1}{\sqrt{1+a^2}}\right]^{-1} \dfrac{a(1-a)}{2(1+a^2-2a\cos(\theta_B))^{3/2}} & \text{if } \cos\theta_B \in [0,1] \\ 0 & \text{otherwise} \end{cases} \tag{9.11}$$

Substituting $f_{HG}(\cos\theta_B)$ and $f_B(\cos\theta_B)$ into Eq. (9.9), the likelihood function becomes

$$L(\cos\theta_B) = \frac{1-g^2}{2(1+g^2-2g\cos\theta_S)^{3/2}} \left[1 - \frac{1}{\sqrt{1+a^2}}\right]^{-1} \frac{a(1-a)}{2(1+a^2-2a\cos\theta_B)^{3/2}} \tag{9.12}$$

This can then be reformulated to return randomly determined biased scattering angles in terms of $\cos\theta_B$:

$$\cos\theta_{Bi} = \frac{1}{2a}\left\{1 + a^2 - \left[u_i\left(\frac{1}{1-a} - \frac{1}{(1+a^2)^{1/2}}\right) + \frac{1}{(1+a^2)^{1/2}}\right]^{-2}\right\} \tag{9.13}$$

where a is a bias coefficient in the range (0,1).

In this way, many cycles of the simulation will result in packets returning to the photodetector for image formation as opposed to calculating paths that pass through the tissue that ultimately will not contribute to the OCT image.

9.3.4 Generative adversarial network (GAN) for system identification

To more realistically synthesize a dataset from simulation, we propose a system based on GAN that performs as system model-tuning technique. As described earlier, adversarial networks are designed to create a generative model that can synthesize samples by using generator and investigator networks that are improved through a process of iteratively checking if the investigator can distinguish if a synthesized sample seems real, and if not, it adjusts the parameters of the generator until the differences between real and synthetic samples can no longer be distinguished [31].

To utilize GAN concept together with a physics model-based simulation, we introduced a generator network that generates optical parameters that are fed into the simulation. The generator takes input from the investigator, together with the optical parameters used to synthesize the image, and its ground truth (which is known due to simulation), together with labels of "real" or "not real" determined from the investigator.

The investigator takes in the real dataset and ground truth labels along with the synthetic image, and its ground truth returns whether it deems it distinguishable from the real data. Fig. 9.12 shows the components of this data generation system.

Once the networks are trained, adjusted optical parameters will result in a simulation process that produces realistic synthetic OCT images.

9.3.5 Predictive model training and test process

9.3.5.1 Data collection

Along with the real dataset gathered from the OCT imaging hardware used for training the simulation, it is anticipated that roughly 10,000 labeled OCT signal samples are required to be generated to train an accurate predictive model. The data collected from the OCT signal simulation must be separated into a training set, validation set, and test sets, and thus an additional 7000 samples for these datasets are required. The training and validations sets are initially used to train a CNN machine learning model.

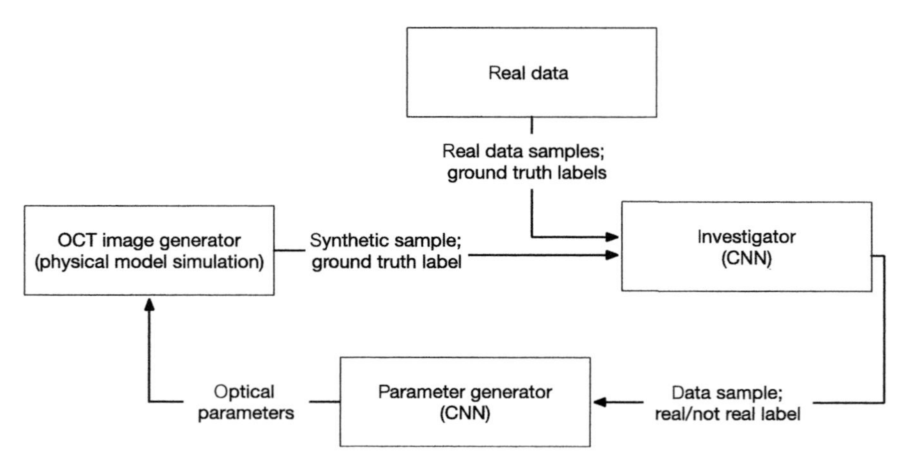

FIG. 9.12 Block diagram of proposed OCT GAN.

9.3.5.2 Verification of predictive model results

Finally, the predictive models generated are tested by classifying OCT signals obtained from in vitro trials and then measured the performance of the model with the previously mentioned metrics against the known tissue characteristics of the trialed samples.

9.3.5.3 Evaluation of predictive model(s)

To confirm improved performance once the system and predictive models have been combined, evaluation of the final models can be assessed using a number of well-established metrics. For binary classifiers, for example, if classifying a particular feature of interest, such as whether a lesion exists within an image, the model can be tested using a test set with known ground truths. If a system model simulation generates the dataset for the predictive model, then the ground truth is known upon generation of the sample, otherwise a human-assigned label will be required.

Thus, if a lesion is predicted from the data sample and this matches the ground truth label, then that would be considered a true positive (TP); in contrast, if model predicts that the lesion does not exist in that same image, then it would be considered a true negative (TN). Accordingly, for these cases, if the ground truth is that there is no lesion, then the evaluation would be false negative (TN) for the former case and false positive (FP) for the latter. Sensitivity and specificity are metrics derived using combinations of the count of these four states. These metrics are routinely reported in medical science and, as such, can be used for comparison of the generated results with other studies.

Sensitivity is defined as:

$$\text{Sensitivity} = \frac{\text{TP}}{\text{TP} + \text{FN}} \tag{9.14}$$

where TP, TN, and FN are the number of "true positives," "true negatives," and "false negatives," respectively. Specificity is defined as:

$$\text{Specificity} = \frac{\text{TN}}{\text{TN} + \text{FP}} \tag{9.15}$$

where additionally FP is the number of "false positives."

Further useful metrics for the evaluation of predictive model performance include "accuracy," "precision," "F1 scores," the "receiver operating characteristic" curve (ROC curve) [12], and area under the ROC curve (AUC) [56].

9.4 Conclusion

For supervised deep learning methods, massive data with accurate labels or ground truth is essential for high accuracy. However, in biomedical applications, exact measurement of the ground truth is often impractical or even impossible. An important avenue to generate data with ground truth is simulation. In this chapter, we have described a method based on MC simulations using a physical model of the biological and imaging systems of OCT to produce a synthetic image together with their ground truth labeling. We have also proposed Generative Adversarial Networks system to utilize the simulation as an OCT image generator and tune it such that it can produce realistic OCT images and labels ready for predictive model training.

References

[1] M. Sajjad, S. Khan, K. Muhammad, W. Wu, A. Ullah, S.W. Baik, Multi-grade brain tumor classification using deep CNN with extensive data augmentation, J. Comput. Sci. 30 (2019) 174–182.

[2] J. De Fauw, J.R. Ledsam, B. Romera-Paredes, S. Nikolov, N. Tomasev, S. Blackwell, et al., Clinically applicable deep learning for diagnosis and referral in retinal disease, Nat. Med. 24 (9) (2018) 1342–1350.

[3] H.A. Haenssle, C. Fink, R. Schneiderbauer, F. Toberer, T. Buhl, A. Blum, et al., Reader study level-I and level-II Groups, "Man against machine: diagnostic performance of a deep learning convolutional neural network for dermoscopic melanoma recognition in comparison to 58 dermatologists", Ann. Oncol. 29 (8) (2018) 1836–1842.

[4] T.J. Brinker, A. Hekler, A.H. Enk, J. Klode, A. Hauschild, C. Berking, et al., A convolutional neural network trained with dermoscopic images performed on par with 145 dermatologists in a clinical melanoma image classification task, Eur. J. Cancer 111 (2019) 148–154.

[5] M. Frid-Adar, I. Diamant, E. Klang, M. Amitai, J. Goldberger, H. Greenspan, GAN-based synthetic medical image augmentation for increased CNN performance in liver lesion classification, Neurocomputing 321 (2018) 321–331.

[6] A. Le Hou, D. Agarwal, T.M. Samaras, R.R.G. Kurc, J.H. Saltz, Unsupervised Histopathology Image Synthesis, arXiv, vol. cs.CV. 14 December, 2017.

[7] A. Halevy, P. Norvig, F. Pereira, The unreasonable effectiveness of data, IEEE Intell. Syst. 24 (2) (2009) 8–12.

[8] C. Sun, A. Shrivastava, S. Singh, A. Gupta, Revisiting unreasonable effectiveness of data in deep learning era, in: Presented at the Proceedings of the IEEE International Conference on Computer Vision, vol. 2017, 2017, pp. 843–852.

[9] K. Simonyan, A. Zisserman, "Very Deep Convolutional Networks for Large-Scale Image Recognition," (2014)arXiv.

[10] M. Mohri, A. Rostamizadeh, A. Talwalkar, Foundations of Machine Learning, MIT Press, Cambridge, MA, 2012.

[11] C.M. Bishop, Pattern Recognition and Machine Learning, Springer Science+Business Media, New York, 2006.

[12] J. Friedman, T. Hastie, R. Tibshirani, The Elements of Statistical Learning, second ed., vol. 1, Springer Science+Business Media, New York, 2001no. 10.

[13] Y. LeCun, G. Hinton, Y. Bengio, Deep learning, Nature 521 (7553) (2015) 440–444.

[14] J. Schmidhuber, Deep learning in neural networks: an overview, Neural Netw. 61 (2015) 85–117.

[15] I. Goodfellow, Y. Bengio, A. Courville, Deep Learning, MIT Press, Cambridge, MA, 2016.

[16] D. Graupe, Principles of Artificial Neural Networks, third ed., World Scientific Publishing, Singapore, 2014.

[17] G. Hu, X. Peng, Y. Yang, T.M. Hospedales, J. Verbeek, Frankenstein: learning deep face representations using small data, IEEE Trans. Image Process. 27 (1) (2018) 293–303.

[18] J. C. Chang, S. Amershi, and E. Kamar, Revolt—collaborative crowdsourcing for labeling machine learning datasets, CHI, 2334–2346, 2017.

[19] S. Robinson, Simulation: The Practice of Model Development and Use, Wiley, Chichester, England; Hoboken, NJ, 2004.

[20] J.A. Sokolowski, C.M. Banks, Modeling and Simulation in the Medical and Health Sciences, John Wiley & Sons, Hoboken, NJ, 2012.

[21] V.P. Singh, System Modeling and Simulation, New Age International, 2009.

[22] J.A. Sokolowski, C.M. Banks, Principles of Modeling and Simulation, John Wiley & Sons, Hoboken, NJ, 2011.

[23] D. Dwibedi, I. Misra, M. Hebert, Cut, paste and learn: surprisingly easy synthesis for instance detection, in: Presented at the Proceedings of the IEEE International Conference on Computer Vision, vol. 2017, 2017, pp. 1310–1319.

[24] F. Tom, D. Sheet, Simulating patho-realistic ultrasound images using deep generative networks with adversarial learning, in: Presented at the Proceedings—International Symposium on Biomedical Imaging, vol. 2018, 2018, pp. 1174–1177.

[25] P.H. Peskun, Optimum Monte-Carlo sampling using Markov chains, Biometrika 60 (3) (1973) 607–612.

[26] L. Ljung, Issues in system identification, IEEE Control. Syst. 11 (1) (1991) 25–29.

[27] R. Fong, A. Ray, J. Schneider, W. Zaremba, P. Abbeel, J. Tobin, Domain Randomization for Transferring Deep Neural Networks From Simulation to the Real World, arXiv.org, vol. cs.RO. 21 March, 2017.

[28] K. Bousmalis, A. Irpan, P. Wohlhart, Y. Bai, M. Kelcey, M. Kalakrishnan, et al., "Using simulation and domain adaptation to improve efficiency of deep robotic grasping," in: Presented at the 2018 IEEE International Conference on Robotics and Automation (ICRA), 2018, pp. 4243–4250.

[29] X.B. Peng, M. Andrychowicz, W. Zaremba, P. Abbeel, "Sim-to-real transfer of robotic control with dynamics randomization," in: Presented at the 2018 IEEE International Conference on Robotics and Automation (ICRA), 2018, pp. 1–8.

[30] M. Arjovsky, S. Chintala, L. Bottou, Wasserstein GAN, arXiv, Vol. Stat.ML, January, 2017.

[31] I.J. Goodfellow, J. Pouget-Abadie, M. Mirza, B. Xu, D. Warde-Farley, S. Ozair, et al., Generative adversarial nets, in: Presented at the Proceedings of the 27th International Conference on Neural Information Processing Systems, Cambridge, MA, USA, vol. 2, 2014, pp. 2672–2680.

[32] Z. Hussain, F. Gimenez, D. Yi, D. Rubin, Differential data augmentation techniques for medical imaging classification tasks, 2017 (2017) 979–984.

[33] L. Perez, J. Wang, The Effectiveness of Data Augmentation in Image Classification using Deep Learning, arXiv.org, vol. cs.CV. 13 December, 2017.

[34] L. Fang, D. Cunefare, C. Wang, R.H. Guymer, S. Li, S. Farsiu, Automatic segmentation of nine retinal layer boundaries in OCT images of non-exudative AMD patients using deep learning and graph search, Biomed. Opt. Express 8 (5) (2017) 2732–2744.

[35] R. Shalev, D. Nakamura, S. Nishino, A.M. Rollins, H.G. Brezerra, D.L. Wilson, R. Soumya, Automated volumetric intravascular plaque classification using optical coherence tomography (OCT), in: Presented at the Twenty-Eighth AAAI Conference on Innovative Applications, 2016, pp. 4047–4052.

[36] K. Ishida, N. Ozaki, N. Ikeda, Y. Sugimoto, Non-destructive inspection of semiconductor optical waveguide using optical coherence tomography with visible broadband light source, in: Presented at the 2017 22nd Microoptics Conference (MOC), 2017, pp. 242–243.

[37] W. Li, X. Liu, Y.-N. Wang, T.H. Chong, C.Y. Tang, A.G. Fane, Analyzing the evolution of membrane fouling via a novel method based on 3D optical coherence tomography imaging, Environ. Sci. Technol. 50 (13) (2016) 6930–6939.

[38] E. Alarousu, Low Coherence Interferometry and Optical Coherence Tomography in Paper Measurements, University of Oulu, Infotech Oulu, Oulu, 2006.

[39] J.G. Fujimoto, M.R. Hee, Handbook of Optical Coherence Tomography, CRC Press, Boca Raton, FL, 2001.

[40] A.M. Zysk, F.T. Nguyen, A.L. Oldenburg, D.L. Marks, S.A. Boppart, Optical coherence tomography: a review of clinical development from bench to bedside, J. Biomed. Opt. 12 (5) (2007) 051403.

[41] S.V. Frolov, A.Y. Potlov, D.A. Petrov, S.G. Proskurin, Monte Carlo simulation of a biological object with optical coherent tomography structural images using a voxel-based geometry of a medium, Quantum Electron. 47 (4) (2017) 347.

[42] D.C. Giancoli, Physics for scientists and engineers third edition, Phys. Educ. 35 (5) (Oct. 2000) 370–371.

[43] S. Zhao, Advanced Monte Carlo Simulation and Machine Learning for Frequency Domain Optical Coherence Tomography, California Institute of Technology, 2016.

[44] S.L. Jacques, Optical properties of biological tissues: a review, Phys. Med. Biol. 58 (11) (2013) R37.

[45] C. Zhu, Q. Liu, Review of Monte Carlo modeling of light transport in tissues, J. Biomed. Opt. 18 (May) (2013) 50902.

[46] W.F. Cheong, S.A. Prahl, A.J. Welch, A review of the optical properties of biological tissues, IEEE J. Quantum Electron. 26 (12) (Dec. 1990) 2166–2185.

[47] F. Meng, X. Ma, H. Zhao, X. Wan, F. Yin, A Study of Polarized Light Propagation in Turbid Medium by Monte Carlo Simulations and Experiments, in: Presented at the 2006 International Symposium on Biophotonics, Nanophotonics and Metamaterials, 2006, pp. 132–135.

[48] K. Wang, N.G. Horton, K. Charan, C. Xu, Advanced fiber soliton sources for nonlinear deep tissue imaging in biophotonics, IEEE J. Sel. Top. Quantum Electron. 20 (2) (2014) 50–60.

[49] S.L. Jacques, L. Zheng, L.V. Wang, MCML-Monte Carlo modeling of light transport in multi-layered tissues, Comput. Methods Prog. Biomed. 47 (2) (1995) 131–146.

[50] G. Zaccanti, A. Taddeucci, M. Barilli, P. Bruscaglioni, F. Martelli, Optical properties of biological tissues, in: Photonics West '95, vol. 2389, 1995, pp. 513–521.

[51] W. Jarosz, Efficient Monte Carlo Methods for Light Transport in Scattering Media, UC San Diego, 2008.

[52] L.G. Henyey, J.L. Greenstein, Diffuse radiation in the galaxy, Astrophys. J. 93 (January) (1941) 70–83.

[53] A.I. Lvovsky, Fresnel equations, in: Encyclopedia of Optical Engineering, Taylor and Francis, New York, 2013, pp. 1–6.

[54] G. Yao, L.V. Wang, Monte Carlo simulation of an optical coherence tomography signal in homogeneous turbid media, Phys. Med. Biol. 44 (9) (1999) 2307.

[55] A. Kalra, H.E. Hernández-Figueroa, S.S. Sherif, I.T. Lima Jr., Fast calculation of multipath diffusive reflectance in optical coherence tomography, Biomed. Opt. Express 3 (4) (2012) 692–700.

[56] A.P. Bradley, The use of the area under the ROC curve in the evaluation of machine learning algorithms, Pattern Recogn. 30 (7) (1997) 1145–1159.

[57] J. Yi, V. Backman, Imaging a full set of optical scattering properties of biological tissue by inverse spectroscopic optical coherence tomography, Opt. Lett. 37 (21) (2012) 4443–4445.

[58] G. Alexandrakis, F.R. Rannou, A.F. Chatziioannou, Tomographic bioluminescence imaging by use of a combined optical-PET (OPET) system: a computer simulation feasibility study, Phys. Med. Biol. 50 (17) (2005) 4225–4241.

[59] B.J. Tromberg, O. Coquoz, J.B. Fishkin, T. Pham, E.R. Anderson, J. Butler, et al., Non-invasive measurements of breast tissue optical properties using frequency-domain photon migration, Philos. Trans. R. Soc. Lond. Ser. B Biol. Sci. 352 (1354) (1997) 661–668.

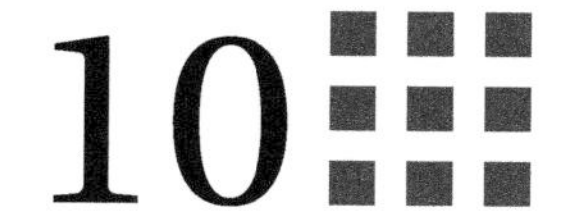

10 Deep learning-based histopathological image analysis for automated detection and staging of melanoma

Salah Alheejawi[a], Mrinal Mandal[a], Hongming Xu[b], Cheng Lu[c], Richard Berendt[d], Naresh Jha[d]

[a]DEPARTMENT OF ELECTRICAL AND COMPUTER ENGINEERING, UNIVERSITY OF ALBERTA, EDMONTON, AB, CANADA
[b]CLEVELAND CLINIC, CLEVELAND, OH, UNITED STATES
[c]CASE WESTERN RESERVE UNIVERSITY, CLEVELAND, OH, UNITED STATES
[d]CROSS CANCER INSTITUTE, EDMONTON, AB, CANADA

10.1 Introduction

With advances in high-resolution digital slide scanning, digital pathology is drawing significant attention from engineering researchers. In digital pathology, doctors can see the biopsy image on a computer monitor (instead of seeing it on a microscope), can analyze the morphological features of cell nuclei, and perform disease diagnosis of cancers such as breast, lung, and skin cancers. However, the digital whole slide images (WSI) are very large and may take significant time for analysis. Therefore, researchers are trying to develop computer-aided diagnosis (CAD) techniques using machine learning techniques to speed up diagnosis time.

As the characteristics of pathology slides vary across different organs/tissues, the CAD systems need to be tuned for each cancer type. In this chapter, we focus on skin cancer diagnosis, especially melanoma, which is an aggressive type of skin cancer with unpredictable behavior that can spread to any part of the body. As per a recent statistic, about 287,723 people have been diagnosed worldwide with invasive melanoma in 2018 [1]. Melanoma skin cancer starts in epidermis/dermis layers of the skin, when melanocytes cells begin to grow rapidly. The aggressive growth of melanocytes can create metastatic melanoma that can pass to lymph nodes through the lymph vessels. Once the melanoma invades the sentinel lymph nodes (SLNs), the melanoma may pass to any part of the body and dysfunction it. The early detection of skin cancer is very important as it helps to increase the chance of successful treatment and the survival rate. Also, some of the skin

Deep Learning Techniques for Biomedical and Health Informatics. https://doi.org/10.1016/B978-0-12-819061-6.00010-0

cancer types have similar appearances at their early stages, and it is not trivial to distinguish them. Although many techniques for melanoma detection have been developed using noninvasive devices, for example, confocal microscopy [2] and epiluminescence microscopy [3], the histopathological slide examination by the pathologist is considered the gold standard for diagnosis [4]. The digitized histopathological slide, also known as WSI, can provide the cell morphological features with a high resolution and help pathologists make precise diagnoses [5].

Fig. 10.1 shows an example of digitized images of skin and lymph node tissues stained with hematoxylin and eosin (H&E) and MART-1 stain, respectively. Fig. 10.1A shows the anatomy of skin tissue divided into three layers: epidermis, dermis, and subcutaneous layer. It is observed that the epidermis layer has a dark purple color (dark gray in print version) in H&E stain due the large density of cell nuclei. In the skin tissue, the melanocytes are typically found around the epidermis-dermis junction. The excessive exposure of the skin tissue to ultraviolet radiation may damage the DNA in the nucleus of melanocytes and cause abnormal growth and shape, resulting in melanoma. Pathologists usually use H&E-stained images to look at the morphological features of the melanocytes to diagnose the melanoma.

Fig. 10.1B shows a digitized MART-1 stained lymph biopsy image with two lymph nodes (shown with green (light gray in print version) contour). It is observed that one of the lymph nodes contains dark brown (black in print version) regions, and these brown (black in print version) regions correspond to the melanoma region. As MART-1 stain is very specific to melanoma, it is used as a confirmation of the melanoma. The Ki-67-stained images are also sometimes used to measure the growth or proliferative activity of the melanocytes.

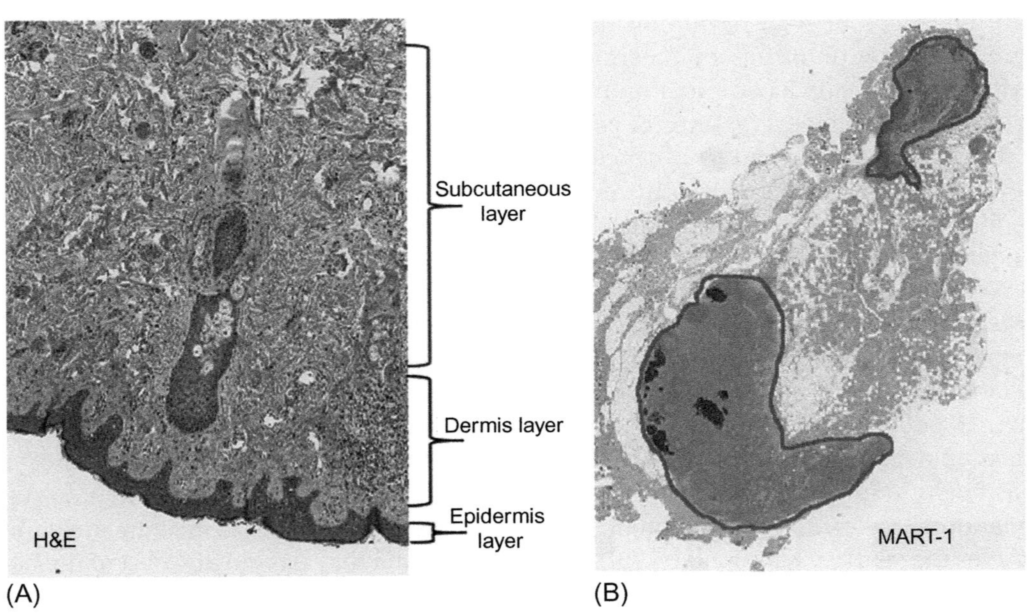

FIG. 10.1 Example of digitized biopsy images. (A) Skin tissue image with H&E stain; (B) sentinel lymph nodes (SLN) image with MART-1 stain.

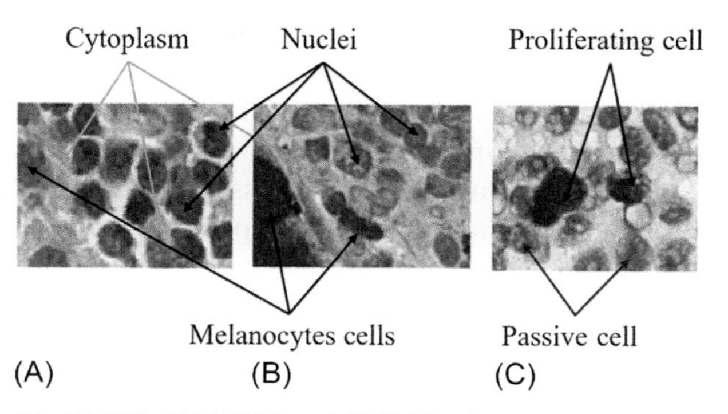

FIG. 10.2 Enlarged WSI with: (A) H&E, (B) MART-1, and (C) Ki-67 stains.

Fig. 10.2 shows examples of enlarged biopsy images obtained with H&E, MART-1, and Ki-67 stains. In the H&E-stained image, the chromatin-rich cell nuclei appear with blue shade (gray in print version), whereas the cytoplasm and other connective tissues appear with varying shades of pink (gray in print version). The melanocytes appear with heterogeneous shape and color. The MART-1 stain has high specificity to detect the melanocytes in skin histopathological images where the melanocytes appear as brown (black in print version) and other types of cell nuclei appear as blue (gray in print version). The Ki-67 stain is used to stage the melanoma by measuring the proliferative index where the proliferating cells appear in deep brown color (black in print version) and passive cells (i.e., cells that are not proliferating) appear in blue (gray in print version).

Traditionally, pathologists examine the histopathological slides using a microscope, where tissue slices are treated with different stains and mounted on glass slides. However, the pathologists' examination is based on their personal experience, and it is exposed to interobserver and intraobserver variability [6, 7]. According to one study, the variability between an average pathologist and an expert in the field (for the melanoma diagnosis) is 69% to 85% [8]. In addition, the manual examination of the WSIs with high resolution is difficult and time-consuming due to the large size of the images (typically billions of pixels) [9]. With the development of the WSI analysis systems, the computer-aided techniques can solve these problems and provide an efficient diagnosis.

This chapter is organized as follows: The dataset used for evaluating the presented CAD techniques is described in Section 10.2. The melanoma detection and staging techniques for skin and lymph node tissues are presented in Sections 10.3 and 10.4, respectively. Finally, the conclusions are presented in Section 10.5.

10.2 Data description

In this chapter, selected melanoma detection and Proliferation Index (PI) calculation techniques will be evaluated based on a WSI image datasets developed at the University of Alberta. The skin and lymph nodes biopsies have been collected at the Cross Cancer

Institute, Edmonton, Canada, in accordance with the protocol for the examination of specimens with skin melanoma. The histopathological images are obtained from formalin-fixed paraffin-embedded tissue blocks of these biopsies. The embedded tissue samples were cut into thin sections (e.g., 4 μm for light microscope) by a steel knife mounted in a microtome and then mounted to a glass slide and stained with one or more microscopical stains [10]. The digital images are captured under 40× magnification on Aperio Scanscope (CS) scanning system (0.25 μm/pixel resolution) with default calibration and illuminance settings (based on Aperio service notes) for these slides. Typical size of these images is around 30,000×50,000 (in color), and each image contains several thousand cell nuclei that need to be analyzed.

The image dataset consists of 64 WSIs for skin tissue and 39 WSIs for lymph node tissue. All skin biopsy images are obtained from H&E-stained slides only. The lymph node images are obtained as follows: nine images with H&E, nine images with MART-1, five with CD-45, seven with S-100, and nine with Ki-67 stains.

10.3 Melanoma detection

Generally, the CAD systems for melanoma detection depend on the melanocytes' distribution, population, size, and location in the skin tissue. A normal skin tissue typically contains melanocytes on the basal layer, and they occupy around 5% to 10% of the cells. The normal melanocytes appear with a regular elliptical shape. However, in the melanoma lesions, the melanocytes have irregular shapes and can be also found in the dermis or the middle layer of epidermis. The melanoma detection techniques typically have the following steps: epidermis/dermis segmentation, nuclei segmentation, and melanoma classification. Haggerty et al. [11] proposed a technique to segment the epidermis region by applying color space normalization, enhancement, and thresholding on the WSI followed by morphological operations. Xu et al. [12] proposed to use shape and color intensity features to identify the epidermis regions in H&E-stained images.

Several techniques have been proposed to segment the cell nuclei in epidermis/dermis regions. Many of these techniques use nuclei seed detection followed by seed-based-segmentation techniques. Several researchers have proposed to use thresholding techniques to identify and segment the nuclei [13–16]. But these techniques cannot segment the clustered nuclei. Cheng and Rajapakse [17] proposed a technique to segment the nuclei clusters using the inner distance map to detect nuclei seeds and adaptive H-minima transform to indicate the nuclei regions before it is merged. Parvin et al. [18] proposed to use a voting technique to detect the nuclei seeds. The technique uses an initial mask of the nuclei clusters and generates multiple voting points on the clusters boundaries, and each point will have a cone-shaped area toward the high-gradient pixels. Xu et al. [5] proposed an automated technique to segment the nuclei clusters by detecting them using an ellipse descriptor. The technique also detects the nuclei seeds using voting areas with Gaussian kernel weights. Techniques [5, 17, 18] provide a good nuclei segmentation performance, but they have high computational complexity with a large number of predefined parameters that need to be tuned. To reduce the complexity, several techniques considered the cell morphological features in addition to the thresholding-based techniques to segment the nuclei. Note that the cell nuclei tend to have circular shapes in the stained images (see Fig. 10.2), and this can be very helpful to detect and segment the cell nuclei efficiently. Xu et al. [19] proposed to perform the nuclei seed detection by applying a generalized Laplacian of Gaussian (gLoG) filter. In this technique, several of gLoG filters with different orientations and scales are convolved with WSI. The nuclei seeds are then detected based on the local maxima of the feature maps. Xu et al. [20] also proposed a nuclei segmentation technique based on the nuclei seeds detected using the gLoG technique [19]. The segmentation technique uses the gradient intensity of multiple radial lines from the nuclei seed. The technique uses high-gradient pixel locations and shape information to accurately segment the cell nuclei.

The techniques [11–20] previously mentioned generally depend on the extracted nuclei features used for nuclei detection/segmentation. The nuclei feature extraction process for these techniques usually rely on a few handcraft features, needs parameter tuning, and requires significant processing time. The deep learning algorithms using Convolutional Neural Network (CNN) have recently been used successfully in medical image analysis. It has been shown that the feature extraction process provides high performance with low computational complexity in many applications. There are different architectures of CNN models for different tasks such as classification or segmentation. Ciresan et al. [21] proposed a CNN-based technique to segment neuron membranes in electron microscopy images. The CNN contains four convolutional layers followed by two fully connected layers. The segmentation is performed by classifying each pixel individually based on a square area surrounding the pixel. Although the technique provides good segmentation results, it has a high computational complexity as each pixel is classified separately. Long et al. [22] proposed a low complexity CNN-based technique for object segmentation. Like most CNN architectures, it includes several convolutional layers, a few (max) pooling layers, and fully connected layers. However, this technique has one learnable upsampling layer instead of fully connected layers. The learnable upsampling layer helps reconstruct the class probability map (for all pixels) with same size as the input image. The class probability is then thresholded to obtain the segmented image. Although, this technique has a low complexity, some information is lost around the boundaries/edges [23]. Badrinarayanan et al. [24] proposed the SegNet architecture for efficient object segmentation. The technique uses a number of upsampling layers that are compatible with the number of max pooling layers. Furthermore, the SegNet transfers the max pooling indices into upsampling layers to reduce the blurriness around the boundaries/edges. Ronneberger et al. [25] proposed the U-Net architecture for biomedical image segmentation by using a very large number of filters. The U-Net architecture consists of the encoder and decoder sides. In the encoder side, several pooling layers and convolutional layers are applied on the input image, and coarse contextual information is then transferred to the decoder side. In the decoder side, several convolutional layers and upsampling layers along with contextual information are used to return a segmented image with the original size of the input image.

In melanoma detection for skin tissue, we have developed a technique to segment the nuclei in H&E-stained images using a CNN, and then classify the segmented nuclei into normal and melanocytes. The developed CNN used several convolutional layers with different size of filters without using pooling/upsampling layers. Fig. 10.3 shows the schematic of the melanoma detection technique, which consists of three modules: Epidermis Region Identification, CNN-based segmentation, and nuclei classification. The details of each module are presented in the following.

10.3.1 Epidermis region identification

The objective of this module is the identification the epidermis/dermis regions in an H&E-stained skin tissue image. In this work, the segmentation of epidermis region is done by applying a global threshold along with the shape feature of the epidermis region on the H&E-stained image [12]. If the estimated thickness of the epidermis mask is below a predefined threshold, the segmentation is considered accurate and the result is passed on to the next module. Otherwise, a fine segmentation module is applied, which uses the K-means algorithm on the epidermis mask to refine the epidermis segmentation results. The epidermis segmentation is a very important step for the skin cancer diagnosis based on histopathological images. The diagnosis can be done by first analyzing the segmented cell features on the epidermis/dermis or the junctional area followed by the grading and classification of the lesion [26]. The segmented epidermis and dermis regions of a typical skin tissue are shown in Fig. 10.4, where the epidermis region is contoured with blue color (black in print version).

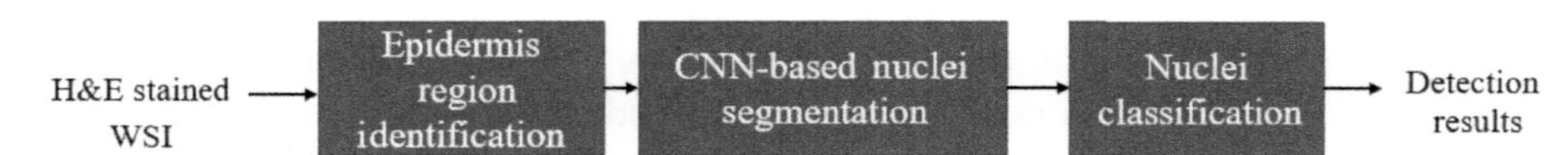

FIG. 10.3 Schematic of a melanoma detection technique.

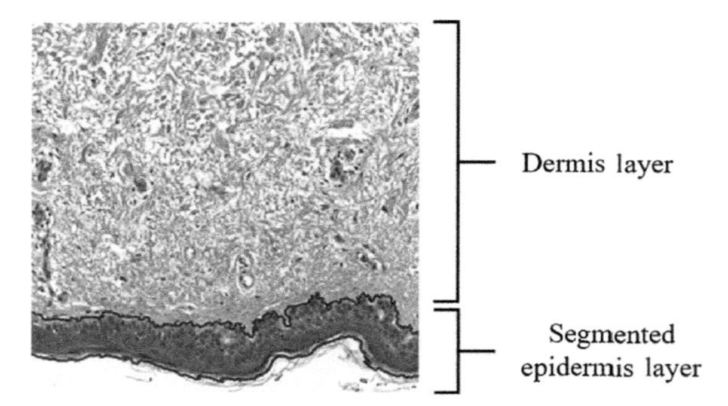

FIG. 10.4 Segmented epidermis and dermis regions in an H&E-stained skin histopathological image.

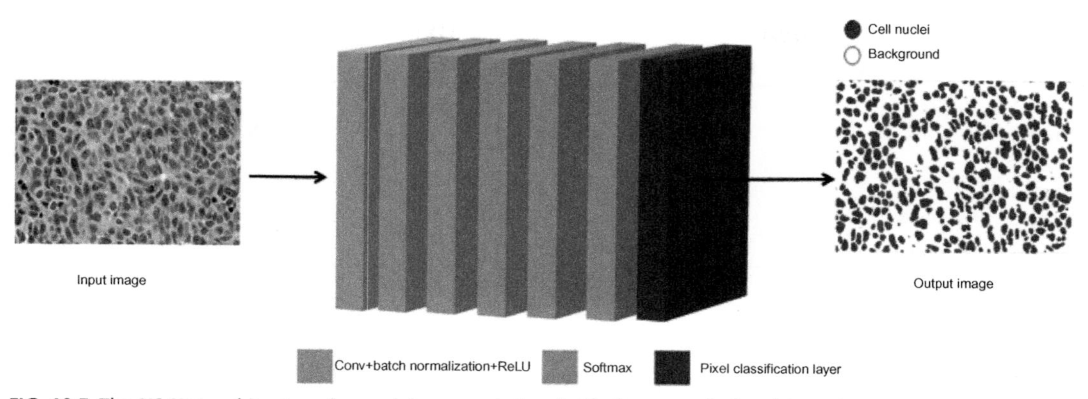

FIG. 10.5 The NS-Net architecture for nuclei segmentation (with five convolutional layers).

10.3.2 CNN-based nuclei segmentation

In this module, the input image is segmented into nuclei and background regions. A new CNN architecture, henceforth referred to as the NS-architecture (Net Nuclei Segmentation), is used to segment the H&E-stained images into two classes: nuclei and background. The NS-Net architecture provides high efficiency in terms of memory and computational time. Fig. 10.5 shows the NS-Net architecture used in this work. The CNN architecture consists of five convolutional layers (shown in gray color) and one softmax [shown in pink (gray in print version)] followed by pixel classification layer [shown in blue (dark gray in print version)]. The details of these layers are explained in the following text.

10.3.2.1 CNN architecture

(1) *Convolutional layers*: In a convolutional layer, a neuron is only connected to a local area of input neurons instead of full-connection so that the number of parameters to be learned is reduced significantly and a network can grow deeper with fewer parameters. In the NS-Net architecture, each convolutional layer consists of three operations: convolution, batch normalization, and rectified linear unit (ReLU) activation.

(a) Convolution: The output of the convolution operation is computed by convolving an input (I) with a number of filters as follows:

$$x_j = I * W_j + b_j, \quad j = 1, 2, \ldots, F \tag{10.1}$$

where F is the number of filters, x_j is the output corresponding to the jth convolution filter, W_j is the weights of the jth filter, and b_j is the jth bias. In the first convolutional layer of the network in Fig. 10.5, I, W_j, and x_j have dimensions of $K \times L \times 3$, $3 \times 3 \times 3$, and $K \times L$, respectively. In the NS-Net, the number of filters is 64 per layer with filter size varying from 3×3 to 11×11 (see Table 10.1).

Table 10.1 Details of the NS-Net architecture with five convolutional layers.

	Number of filters	Number of channels	Output image size	Filter size
Layer 1	64	3	$K \times L \times 64$	3×3
Layer 2	64	64	$K \times L \times 64$	5×5
Layer 3	64	64	$K \times L \times 64$	7×7
Layer 4	64	64	$K \times L \times 64$	9×9
Layer 5	M	64	$K \times L \times M$	11×11
Softmax layer	–	M	$K \times L \times M$	

Input image SIZE: $K \times L$ pixels (color).

(b) Batch normalization: After the convolution operation, batch normalization is used to speed up the training of NS-Net and reduce the effect of initialization [27]. It normalizes each input channel across a minibatch by subtracting the mean of the minibatch and dividing by its standard deviation. The normalized values are then scaled and shifted as follows:

$$y_i = \sigma \hat{x}_i + \beta \tag{10.2}$$

where y_i is output value, $\hat{x}_i$ is the normalized input value, and σ and β are the scale and offset factors that are learnable during the network training.

(c) ReLU: Nonlinear activation functions are applied elementwise to increase the nonlinear properties of the network. ReLU is the most commonly used nonlinear activation function [28], and it replaces all negative feature values by zeros using the following equation:

$$f = \max(0, y) \tag{10.3}$$

where y is the input value to ReLU, and f is the output. Compared with tanh and *sigmoid* activations, ReLU has the advantages of speeding up training [29] and inducing the sparsity in hidden units [30]. Therefore, the ReLU is used as the activation function in this work.

Note that steps (a)-(c) are repeated for each subsequent convolutional layer. After the ReLU module, the feature map f is used as the input (I) of the next convolutional layer (if appropriate).

(2) Softmax: The softmax layer is applied after the last convolutional layer to generate the probability distribution of the classification results for each pixel. The output of the last convolutional layer consists of C features map of size $K \times L$. Let ϕ_i denote the feature map corresponding to the ith class ($i = 1,2,\ldots,M$). Let ϕ denote the exponential summation of the features map calculated as follows:

$$\phi(k,l) = \sum_{i=1}^{M} \exp(\phi_i(k,l)) \quad 1 \le k \le K, 1 \le l \le L \tag{10.4}$$

The class probability P_i for the pixel at location (k,l) is then calculated using the following equation:

$$\mathrm{P}_i(k, l) = \frac{\exp\left(\phi_i(k, l)\right)}{\phi(k, l)} \tag{10.5}$$

In this section, all input pixels are classified into two classes (i.e., $M=2$): cell nuclei and background.

(3) *Pixel classification:* This layer performs the classification of the image pixels using the corresponding class probability. Each pixel in the image is assigned the class corresponding to the higher probability.

10.3.2.2 CNN training

There are a total of 150,336 parameters in the NS-Net architecture, which needs to be trained. In the training phase of the CNN, 24 high-resolution lymph nodes images (1920×2500 color pixels) are used. Each image is divided into overlapping blocks of 64×64 color pixels to obtain 458 block-images (see Fig. 10.6). The total number of block-images (i.e., $458 \times 24 = 10{,}992$) are divided into 70% training (7694 images), 15% validation (1099 images), and 15% testing images (1099 images). The ground truth (with segmented and classified nuclei) images were obtained by manual segmentation under doctor's supervision.

In NS-Net architecture, we use the cross-entropy as the loss function for multiclass classification. The loss function X for a pixel at (k,l) location of a training image (with size $K \times L$) is calculated as follows:

$$X(k, l) = \sum_{i=1}^{M} T_i(k, l) \ln \mathrm{P}_i(k, l) \tag{10.6}$$

where $T_i(k,l)$ is the ith-ground truth class probability and $\mathrm{P}_i(k,l)$ is ith-class probability (predicted by the softmax layer). The loss function is aggregated over a few training

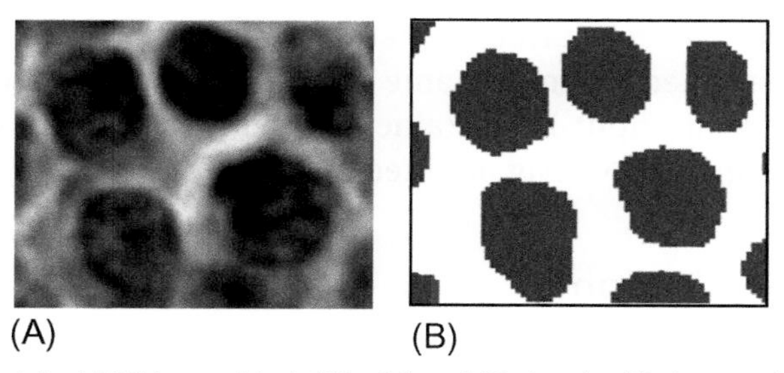

FIG. 10.6 Example of (A) original RGB image block (64×64) and (B) the classified ground truth image, where the background and cell nuclei appear in white and red color (gray in print version), respectively.

images called a minibatch. For a minibatch of S training images, the overall loss function is calculated as follows:

$$E=\sum_{s=1}^{S}\sum_{k=1}^{K}\sum_{l=1}^{L}X_s(k, l) \tag{10.7}$$

where X_s is the loss function for sth-block-image. As the two classes of pixels have different distributions for each class, the loss function is weighted for each class differently. The weights φ_i for different classes are calculated as follows:

$$\varphi_i=\frac{\text{median freq}}{\text{freq}(i)} \tag{10.8}$$

where the class frequency freq(i) is a ratio of the number of the pixels that represent the i-class on the total number of pixels in the training dataset, and median freq is the median of class frequencies.

To minimize the loss function E, the stochastic gradient descent with momentum (SGDM) optimizer [31] is used in the NS-Net architecture. The SGDM accelerates stochastic gradient descent (SGD) and reduces the oscillation problem of SGD by adding the contribution from the previous iteration to the current iteration. The network parameters at the end of nth iteration are updated as follows:

$$\theta_{n+1}=\theta_n-\alpha\nabla E(\theta_n)+\mu(\theta_n-\theta_{n-1}) \tag{10.9}$$

where θ is the parameter vector, which includes parameters such as normalization weights, filters, and biases; $\nabla E(\theta)$ is the gradient of the loss function; α is the learning rate; and μ is momentum. The training accuracy and loss are calculated on each individual minibatch.

The validation accuracy and loss are calculated on the validation dataset at the end of every 250 iteration. Fig. 10.7 shows an example of loss and accuracy during the training process. The network training stops when the validation accuracy is not increasing in the previous five validations. As observed in Fig. 10.7, the training and validation accuracy curves are generally consistent during the network training process. Note that the validation block-images are generated in the same way as the training block-images are generated. Fig. 10.8A shows an input H&E-stained image and (B) shows the masked nuclei image obtained by using the NS-Net architecture.

10.3.3 Nuclei classification

In this section, the segmented nuclei obtained using the NS-Net architecture is classified into two classes: melanocytes nuclei and other cell nuclei. The nuclei classification consists of two steps: Feature Extraction and Feature Classification. The details of each step are presented as follows:

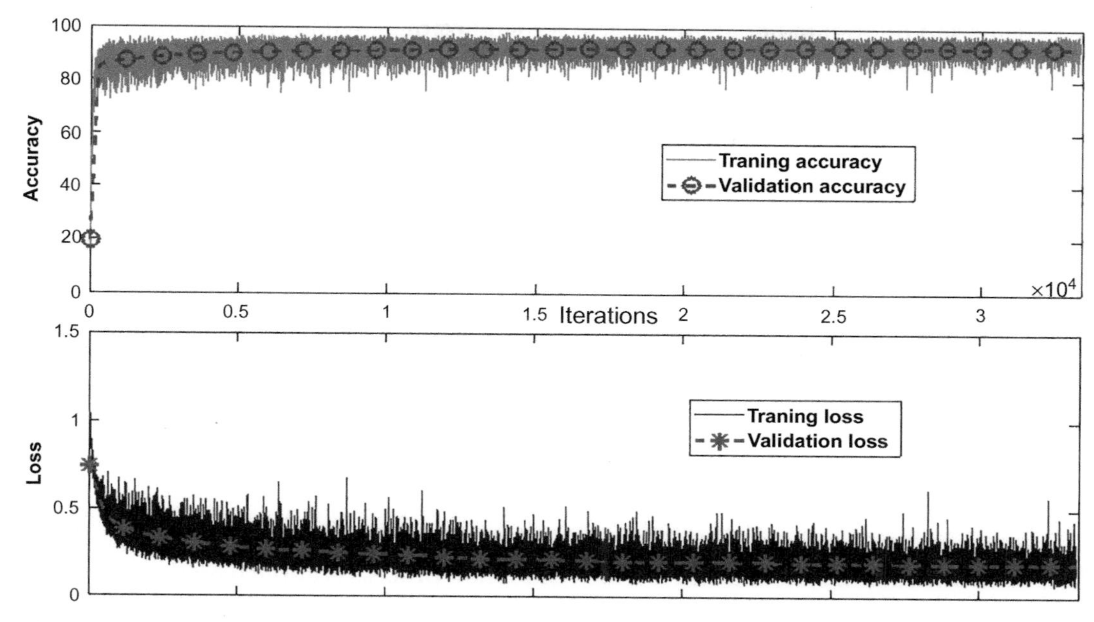

FIG. 10.7 Loss and accuracy plots during the network training process.

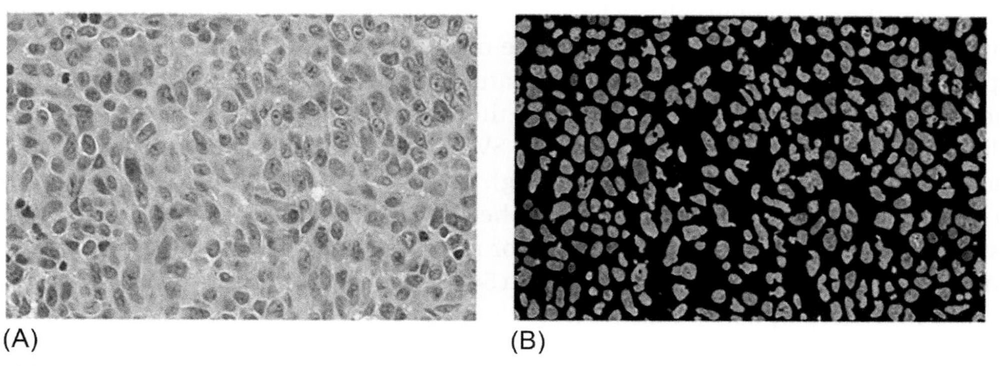

FIG. 10.8 Segmentation results. (A) Input image and (B) segmented image obtained using the NS-Net architecture.

10.3.3.1 Feature extraction

Features extraction is a very important step for analyzing an image and identifying or classifying the objects. In this work, we have used the following features to classify the segmented cell nuclei into melanocytes and other nuclei.

1. First-order texture: It includes the histogram intensity features (mean, standard deviation, third moment, smoothness, entropy, and uniformity). These features are extracted for all three-color channels (R, G, and B) of the presegmented cell nuclei (dimension is $6 \times 3 = 18$).

2. Histogram of oriented gradient (HOG): It includes the angular distribution of local intensity gradients. The gradient of each pixel is obtained by applying two filter kernels, [−1, 0, 1] and [−1, 0, 1], to the image region, and the magnitude gradient is then calculated. The magnitude gradient is then quantized into one of the nine major orientations $2(k-1)\pi/9$, $1 \leq k \leq 9$ and weighted by the gradient magnitude. This feature can distinguish shape or intensity variation of an object (dimension is 9).
3. Haralick texture features: It relates to the homogeneity of pixel intensity with neighborhood pixels using the gray-level co-occurrence matrix (GLCM) as the GLCM provides the joint frequency of intensity levels for a pair of pixels. Six Haralick features, correlation, energy, homogeneity, contrast, entropy, and dissimilarity, are selected and calculated. For calculation of the GLCM, we use a distance of one pixel in four directions (i.e., 0°, 45°, 90°, and 135°), resulting in 24 Haralick features in total (dimension is 24).
4. Morphological features: It includes three morphological features: eccentricity, solidity, and the ratio of major and minor axes of the cell nuclei (dimension is 3).

These features are combined, resulting in a feature vector of length 54, which is used for nuclei classification. Note that the features extraction process is applied on the RGB pixels of each segmented nuclei, for example, on the nuclei shown in Fig. 10.8B.

10.3.3.2 Feature classification

In this step, the segmented cell nuclei are classified into melanocytes and other cell nuclei using the features previously obtained. The nuclei classification is done by applying the support vector machine (SVM) model on feature vector corresponding to each cell nuclei. The H&E-stained image typically contains thousands of cell nuclei, and each cell nuclei will have a feature vector of length 54. The SVM model is trained and tested on 7000 cell nuclei (70% for training and 30% for testing). Let us denote a randomly selected training set of k-nuclei with $\{f_1, f_2, \ldots, f_k\}$, and their labels with $\{y_1, y_2, \ldots, y_k\}$. Note that the labels are assigned (+1) for normal nuclei and (−1) for melanoma nuclei. The SVM training model estimates weights of the best hyperplane (in 54 dimensions) to separate these two classes. The estimated hyperplane P satisfies $w^T u + b = 0$, where w is a normal vector to the hyperplane P and u is any point on the hyperplane. The value of w and b are optimized to find the best hyperplane with largest margin during training [32, 33]. The SVM classifier is a powerful tool, and it can also handle linearly nonseparable data by increasing the feature dimension with a variety of kernels (e.g., polynomial, Gaussian, radial basis function [34]). Experimental results with the test dataset show that the SVM with a Gaussian kernel provides the best classification performance. The generated nuclei mask ϕ by the trained SVM model is defined as follows:

$$\phi(N_i) = \begin{cases} 0 & \text{for Melanocytes nuclei} \\ 1 & \text{for Other cell nuclei} \end{cases}, \quad i = 1, 2, \ldots, C$$

where N_i represents the cell nuclei regions and C is the total number of cell nuclei present in the image. Note that the SVM is applied on each cell nuclei, and the training is done for around 4900 cell nuclei (2450 for each class) obtained from three H&E-stained images (labeled manually under doctors' supervision). Fig. 10.9A–C show an example of the classified cell nuclei results where melanocytes and other nuclei are contoured with red (light gray in print version) and blue color (dark gray in print version), respectively.

10.3.4 Results and discussions

The segmentation of cell nuclei using NS-Net architecture was described in Section 10.3.2, and the classification of the segmented cell nuclei using SVM was described in Section 10.3.3. In this section, the performance of cell nuclei segmentation and classification is presented.

Note that the CNN architectures do not use any handcraft features for segmentation or classification. However, the filters are trained to pick up relevant features for a given task. To demonstrate the efficiency of the deep learning algorithms using the NS-Net architecture, some selected feature maps are shown in Fig. 10.10. An input H&E image and the corresponding ground truth-segmented nuclei image are shown in Fig. 10.10A and C, respectively. Fig. 10.10C shows eight selected feature maps (out of 64) obtained at the third convolutional layer of the NS-Net architecture. The four feature maps in the upper row appear to be dedicated to detecting the background, capturing different representation of the background regions. On the other hand, the four feature maps in the bottom row appear to be dedicated to detecting the nuclei.

10.3.4.1 Segmentation performance

The segmentation performance is compared between the deep learning algorithms and a few selected classical handcraft feature-based algorithms. The handcraft feature-based algorithms extract morphological and textures features to segment the cell nuclei. These algorithms usually need a large number of parameters and feature extraction resulting in high computational complexity. On the other hand, the deep learning algorithms introduce remarkable features extraction with high efficiency and low complexity.

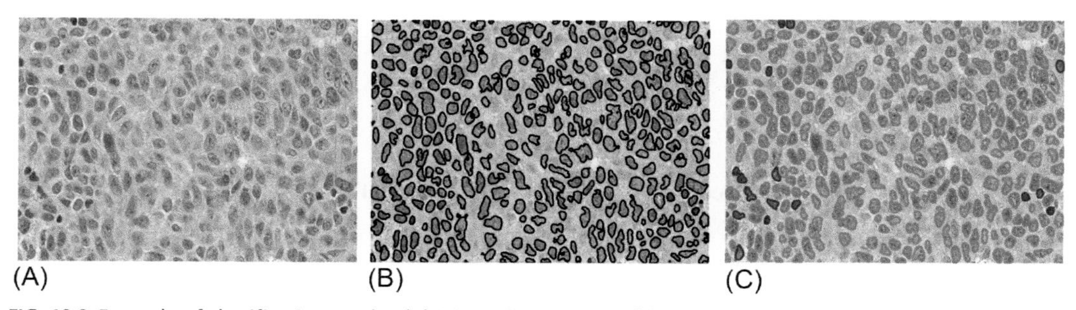

FIG. 10.9 Example of classification results. (A) NS-Net input image, (B) NS-Net output image, (C) classified image obtained using SVM, where melanocytes and other cell nuclei are contoured with red (light gray in print version) and blue color (dark gray in print version), respectively.

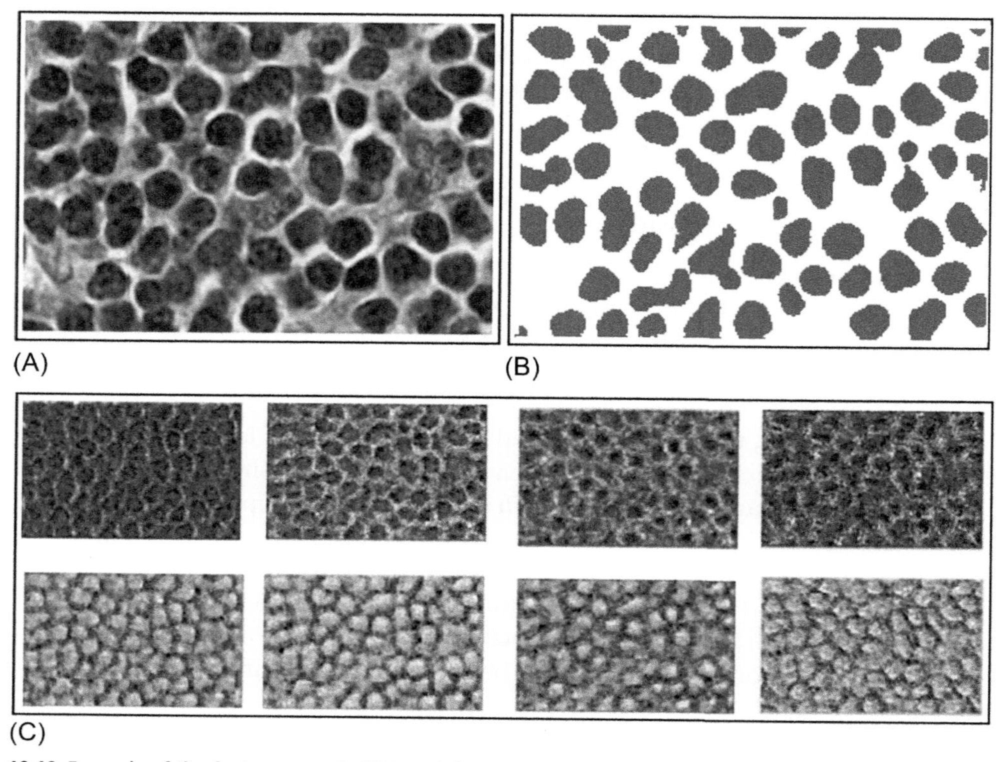

FIG. 10.10 Example of the feature maps in NS-Net. (A) An input H&E image, (B) the ground truth-segmented image, (C) eight selected features map in the NS-Net (layer 3) for segmentation of the input image into two classes: cell nuclei and background.

In this section, the segmentation performance is evaluated using Precision, Sensitivity, Specificity, Accuracy, and BF-score measures defined as follows:

$$\text{Precision} = \frac{\text{TP}}{\text{TP} + \text{FP}} \times 100\%$$

$$\text{Sensitivity} = \frac{\text{TP}}{\text{TP} + \text{FN}} \times 100\%$$

$$\text{Specificity} = \frac{\text{TN}}{\text{TN} + \text{FP}} \times 100\%$$

$$\text{Accuracy} = \frac{\text{TP} + \text{TN}}{\text{TP} + \text{FP} + \text{FN} + \text{TN}} \times 100\%$$

$$\text{BFscore} = \frac{2 \times \text{Precision} \times \text{Sensitivity}}{\text{Precision} + \text{Sensitivity}} \times 100\%$$

where TP, TN, FN, and FP denote the number of true positives, true negatives, false negatives, and false positives, respectively. Note that the segmentation techniques

Table 10.2 Segmentation performance of the deep learning algorithms and the classical feature-based algorithms.

Technique	Precision	Sensitivity	Specificity	Accuracy	BF-score	Execution time (in s)
Voting+Watershed [5]	78.24	84.64	82.91	83.64	81.31	143.71
gLoG+mRLS [20]	79.27	60.25	88.57	76.67	68.46	128.57
SegNet [24]	84.16	87.53	88.06	87.84	85.81	15.37
U-Net [25]	87.41	57.87	93.95	78.79	69.63	20.82
NS-Net	87.20	89.90	90.44	90.21	88.52	14.27

are applied on H&E-stained images that contain melanocytes and other nuclei. The nuclei segmentation is evaluated with two other well-known deep learning algorithms (SegNet [24] and U-Net [25]) and the classical feature-based algorithms gLoG+mRLS [20] and Voting+Watershed [5]. The gLoG+mRLS technique is performed to detect the nuclei seeds in H&E-stained images using a bank of gLoG filters. Based on the detected nuclei seeds, the cell nuclei are then segmented using the high gradient values of multiradial lines generated from the nuclei seeds. On other hand, the Voting+Watershed technique detects the nuclei seeds using the voting technique on nuclei boundaries and then segments the cell nuclei using watershed algorithm.

Table 10.2 shows the segmentation performance of different techniques. It is observed that the deep learning algorithms provide excellent performance compared to the classical feature-based algorithms. This is because the classical features are less sensitive to the diversity of the cell nuclei in the skin tissue. For example, the melanocytes tend to have light and inhomogeneous color (see Fig. 10.11) and that causes misdetection of the melanocytes in the gLoG+mRLS and Voting+Watershed techniques. In this work, the NS-Net, SegNet, and U-Net architectures are trained on the same number of training images (7694 images each with size 64×64 RGB pixels). The learning rate α is initialized to 0.01 and drops every 10 epochs by 20%. The momentum μ is set to 0.9, the minibatch size S is 8, and the maximum number of epochs is set to 100.

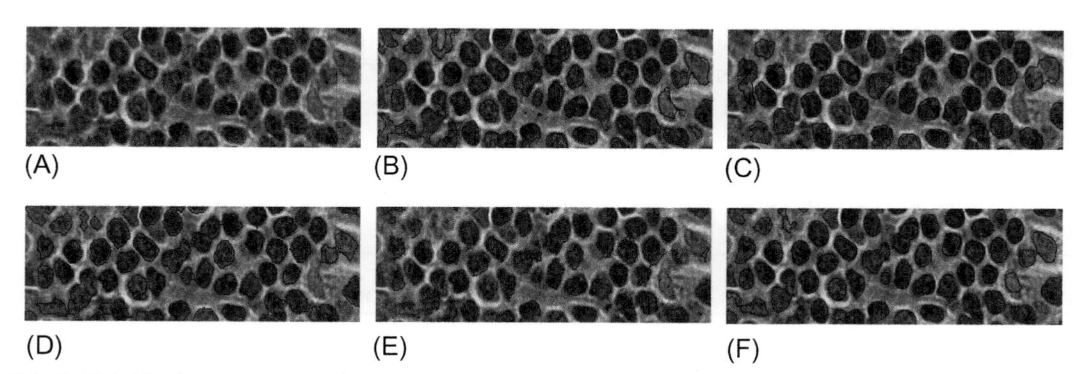

FIG. 10.11 Subjective comparison of segmentation results (A) an H&E image patch, segmentation results for (B) SegNet [24], (C) gLoG+mRLS [20], (D) Voting+Watershed [5], (E) U-Net [25], and (F) NS-Net techniques, respectively.

Table 10.3 Properties of CNN architectures used in performance evaluation.

CNN architecture	Convolutional layers	No. of trained parameters	Filter size	No. of filters
SegNet [24]	8	225,542	3×3	64
U-Net [25]	11	905,472	3×3	(64, 128, 256)
NS-Net	5	150,336	(3×3)–(11×11)	64

Table 10.3 shows the number of convolutional layers and the number of the parameters that need to be trained in each architecture. It is observed in Table 10.2 that the U-Net and SegNet have less accuracy than the NS-Net because of the overfitting in the feature extraction during the training of the SegNet and U-Net architectures.

Fig. 10.11B–F shows the subjective segmentation performance of SegNet [24], U-Net [25], gLoG+mRLS [20], Voting+Watershed, and the NS-Net architecture, respectively. It is observed that the NS-Net architecture provides excellent nuclei segmentation, whereas gLoG+mRLS [20] and Voting+Watershed techniques miss a few cell nuclei due to the inhomogeneity in the cell nuclei color. It is also observed that the U-Net architecture does not perform well compared to the other techniques because of the huge number of extracted features that cause the overfitting in the cell nuclei segmentation.

10.3.4.2 Nuclei classification performance

After the cell nuclei segmentation, the segmented nuclei are classified using different machine learning classifiers. The classifiers are trained on 4000 nuclei with 54 features (discussed in Section 10.3.3). Table 10.4 shows the nuclei classification performance of SVM and a shallow Neural Network. The SVM classifier is tested with different kernels (such as Gaussian, linear, and polynomial kernels). The Gaussian kernel provides the best performance compared to other kernels. The Neural Network in Table 10.4 used 54 neurons at the input layer, 10 hidden neurons, and 2 neurons at output layers. It is observed that the nuclei classification performance using the SVM classifier provides the best performance in terms of the Mean Absolute Error (MAE), Root Mean Square Error (RMSE), and Accuracy. Therefore, the SVM classifier with Gaussian kernel is used in this work to classify the cell nuclei into melanocytes and other cell nuclei. Table 10.4 compares

Table 10.4 Performance of the nuclei classification using different SVM Kernels and a shallow Neural Network.

	SVM Kernel			Neural Network
Evaluation measures	Gaussian	Linear	Polynomial	10 Hidden units
MAE	0.14	0.20	0.43	0.35
RMSE	0.38	0.44	0.65	0.60
Accuracy	85.72	80.52	57.28	65.44

the nuclei classification performance in terms of MAE, RMSE, and Accuracy. The MAE and RMSE are calculated as follows:

$$\mathrm{MAE}=\frac{1}{K}\sum_{k=1}^{K}|G(k)-B(k)| \tag{10.10}$$

$$\mathrm{RMSE}=\sqrt{\frac{1}{K}\sum_{k=1}^{K}[G(k)-B(k)]^2} \tag{10.11}$$

where $G(k)$ is the ground truth and $B(k)$ is the predicted value of the cell nuclei classification.

The techniques presented in this section can identify different layers of skin, and detect and classify the melanocytes and other nuclei. Based on the morphology and spatial distribution of these nuclei, melanoma can be diagnosed and classified into different types such as lentiginous, superficial spreading, and nodular [35].

10.4 Cell proliferation index calculation

Once melanoma has been diagnosed in the skin tissue, the next step is to stage the melanoma that is, how far the cancer has spread and the degree of cell proliferation activity. To determine if the melanoma has reached the subcutaneous layer in the skin and drained to the lymph nodes through lymph vessels, doctors usually excise a biopsy from the SLNs. The lymph node tissue is then stained with MART-1 stain for detecting the melanoma region and Ki-67 stain to measure the cell proliferation activity on the detected melanoma regions.

Cell proliferation is a fundamental biological process, and it is accelerated in tumors. The PI is used by doctors to assess the tumor progression and to determine the future therapy for a patient [36]. The PI value is calculated using the following equation:

$$\mathrm{PI}=\frac{\mathrm{NA}}{\mathrm{NP}+\mathrm{NA}}\times 100\% \tag{10.12}$$

where NA and NP are the number of the active and passive nuclei, respectively. The Ki-67 protein, known as MIKI-67, is found in cell nucleus during the cell division cycles (G1, S, G2, and mitosis) [37–39]. During the staining process, the tissue slice is treated with Ki-67 stain. If a cell nucleus contains Ki-67 protein (i.e., the cell is proliferating), the Ki-67 protein interacts with the Ki-67 stain and appears in brown in color (black in print version). On other hand, if the cell nucleus does not have Ki-67 protein (i.e., the cell is not proliferating), the cell nucleus appears in blue color (gray in print version) (as shown in Fig. 10.12).

Several techniques have been proposed in the literature to calculate the PI for different tumor regions such as pancreatic neuroendocrine, breast carcinoma, and melanoma on lymph nodes. Grala et al. [40] proposed a technique based on thresholding, watershed, and morphological operations for nuclei recognition in breast cancer images. The SVM model is then used to classify the RGB nuclei pixels into passive/active nuclei, and the PI value is calculated. Akakin et al. [41] used watershed segmentation and Laplacian-of-

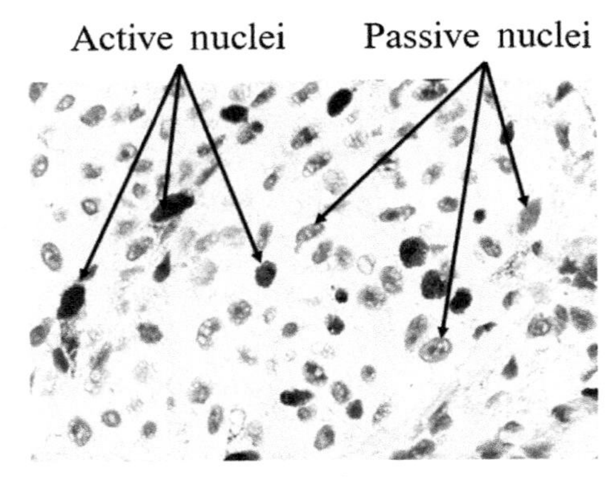

FIG. 10.12 A histopathological image stained with Ki-67 stain.

Gaussian filtering for nuclei detection and SVM as active/passive nuclei classifier. The PI value is calculated by measuring the active nuclei ratio. Al-Lahham et al. [42] proposed a technique where K-mean was used to classify the input image pixels (in $L^*a^*b^*$ color-space) into three classes: background, passive and active nuclei, followed by PI calculation. Mungle et al. [43] proposed a technique to measure the PI in a Ki-67-stained breast biopsy image using fuzzy C-means (FCM) and K-means (KM). The FCM segments the nuclei using fuzzy threshold and the KM is used to classify the masked nuclei into passive and active nuclei. Alheejawi et al. [44] proposed an automated algorithm to segment the lymph nodes on MART-1- and Ki-67-stained images and measure the PI on Ki-67-stained images of the melanoma regions. The algorithm segments the lymph nodes and melanoma regions on MART-1 and then maps the melanoma region on the segmented lymph node of Ki-67-stained image. The PI values are then calculated for the mapped melanoma regions on Ki-67-stained image of the lymph node. The PI calculation is done by segmenting the nuclei using Otsu threshold and classifying the nuclei into passive and active nuclei using SVM.

For PI calculations using deep learning, Saha et al. [45] proposed a technique to measure the PI value for breast cancer. The technique used Gamma mixture model (GMM) with Expectation-Maximization algorithm to detect the nuclei seeds, and image patches around isolated nuclei seeds are then classified into passive/active nuclei using a CNN classifier. After the CNN classifier, the PI is calculated based on the number of seeds corresponding to active and passive nuclei. The GMM based nuclei detection algorithm typically has a high computational complexity and also does not work well on the nuclei clusters.

In this section, a fully automated CNN-based technique [46] (henceforth referred to as the PI-SegNet technique) is presented to segment and classify the nuclei into active/passive types. The technique is similar to Alheejawi et al. [44], except that the classical nuclei segmentation/classification has been replaced by a CNN module. The CNN uses the SegNet architecture with two encoders and decoders, each with four convolutional layers. In this technique, three areas are identified with the highest PI values. The technique has less computational complexity and low error rate compared to the other state-of-the-art techniques [43–45].

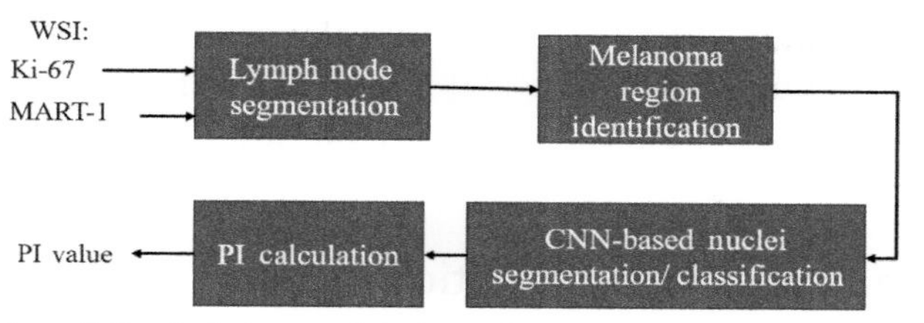

FIG. 10.13 Schematic of the cell PI calculation module for melanoma regions.

The schematic of the PI calculation is shown in Fig. 10.13, which consists of four modules: LN segmentation, Melanoma Region Identification, CNN-based segmentation, and PI calculation. The details of each module are presented as follows.

10.4.1 Lymph node segmentation

This module segments the Lymph Nodes in a biopsy image based on Histogram and High Frequency Features (LN-HHFF [44]). The lymph node segmentation consists of three steps: coarse segmentation, features extraction, and fine segmentation. The coarse segmentation is applied on MART-1-stained image to segment the lymph node image into three classes (lymph node pixels, melanoma pixels, and other tissue pixels) using two thresholds obtained from local minima in smoothed image histogram. The texture features are then extracted from the segmented lymph node and fine classification is applied using SVM classifier. Fig. 10.14 shows the lymph nodes segmentation results

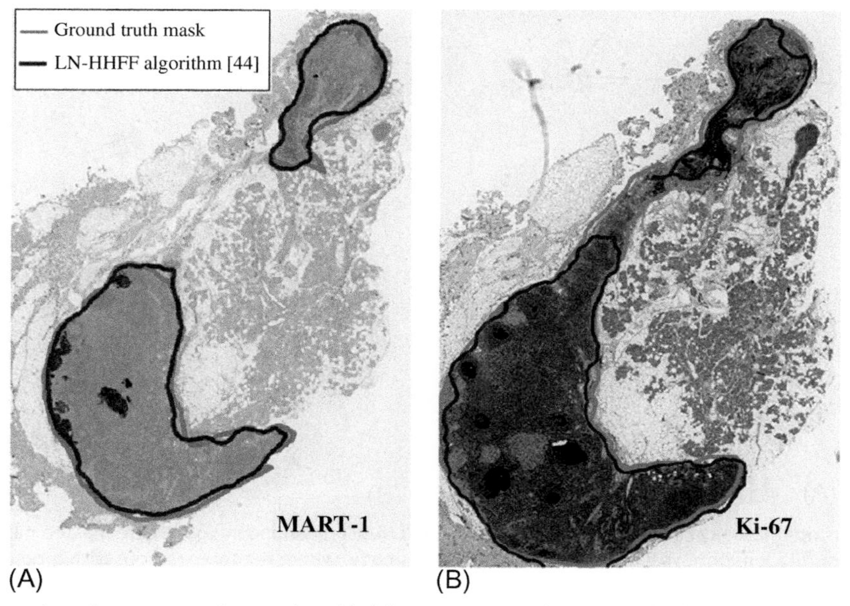

FIG. 10.14 Lymph nodes segmentation results with (A) MART-1-stained image and (B) Ki-67-stained image.

in MART-1- and Ki-67-stained images. Note that the segmented lymph node using the histogram and high-frequency features [44] are contoured with green color (light gray in print version), whereas the ground truth is contoured with blue color (dark gray in print version).

10.4.2 Melanoma region identification

The objective of this module is identifying the melanoma regions in Ki-67-stained lymph node image. The module extracts the melanoma regions in Ki-67-stained image (tumor region) with help of MART-1-stained image. MART-1 stain is first used to detect melanoma regions as it is very specific to melanoma. The detected melanoma regions are then mapped onto the corresponding Ki-67 image. The Ki-67 stain can detect, within identified melanoma regions, proliferative activity with very high accuracy. Therefore, the Ki-67-stained images are used to calculate the PI value. Fig. 10.15 shows a lymph node image stained with (A) MART-1 and (B) Ki-67. In Fig. 10.15A, the melanoma regions appear in brown color (black in print version) (in MART-1-stained image). Fig. 10.15B shows the mapped melanoma regions in green (light gray in print version) contour. Let the disconnected melanoma regions on the Ki-67-stained image

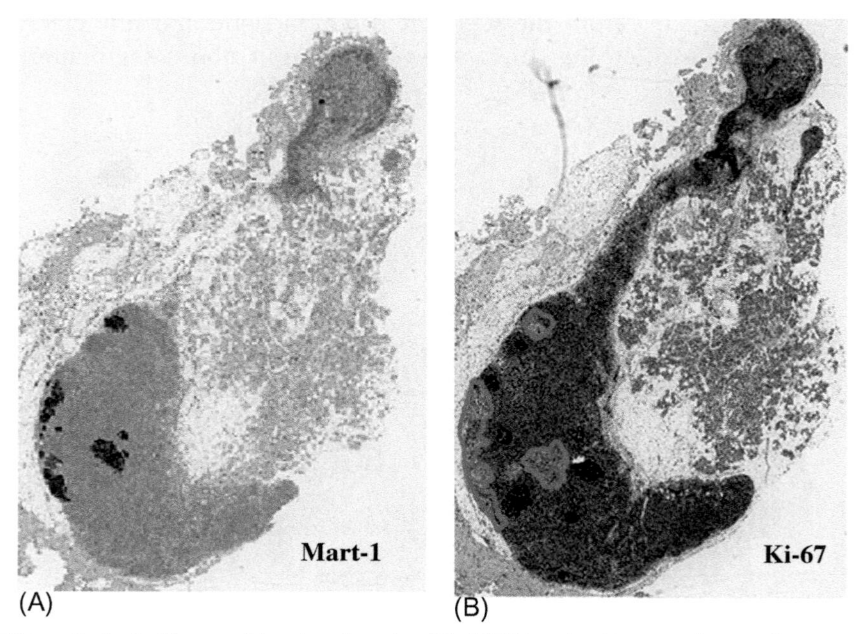

FIG. 10.15 Histopathological image of two lymph nodes: (A) MART-1-stained image, where melanoma appears in brown color (black in print version) and (B) Ki-67-stained image, where the three regions with green (light gray in print version) contour represents melanoma region. *Reproduced with permission from S. Alheejawi, M. Mandal, R. Berendt, N. Jha, Automated melanoma staging in lymph node biopsy image using deep learning, in: Proc. of the IEEE Canadian Conference on Electrical and Computer Engineering (CCECE), Edmonton, Canada, May 6-8, 2019. ©2019 IEEE.*

be denoted by R_i $(i=1,2,\ldots,N_{DR})$ where N_{DR} is the total number of disconnected melanoma regions in the lymph nodes.

10.4.3 CNN-based Nuclei Segmentation and Classification

The objective of this module is to segment an input image into active nuclei, passive nuclei, and background. The SegNet architecture is used to segment the Ki-67-stained images into three classes (background, passive, and active nuclei). The SegNet provides high efficiency in terms of memory and computational time [24]. Fig. 10.16 shows the PI-SegNet architecture used in this work, which consists of two Encoders (#1 & #2), two Decoders (#1 & #2), and one softmax followed by pixel classification layer. Each encoder contains two convolutional layers (as shown in gray color in Fig. 10.16) and one pooling layer, whereas each decoder contains one upsampling layer and two convolutional layers. In the end of the decoder side, softmax was used to classify the probability for each pixel to the assigned classes. Table 10.5 shows the number of the filters used for each layer. The details of these layers (convolution, batch-normalization, ReLU, softmax, and pixel classification layers) are similar to those presented in Section 10.3.2. The Pooling and upsampling layers are explained as follows:

(1) *Pooling*: In a pooling layer, the feature maps are downsampled using an average or a max operation to reduce the number of parameters and control overfitting in a network. In this study, a max pooling layer with filter of size 2×2 with a stride of 2 is used. After the first pooling, there are 64 feature maps of size $(K/2)\times(L/2)$, whereas the feature maps are of size $(K/4)\times(L/4)$ in the second pooling.

Table 10.5 Details of the CNN architecture with four convolutional layers at the Encoder, 3×3 filter size, and M classes.

	Layers	Number of channels	Number of filters	Output image size
Encoder-1	Conv-1	3	64	$64\times K\times L$
	Conv-2	64	64	$64\times K\times L$
	Pooling	64	64	$64\times K\times L$
Encoder-2	Conv-1	64	64	$64\times(K/2)\times(L/2)$
	Conv-2	64	64	$64\times(K/2)\times(L/2)$
	Pooling	64	64	$64\times(K/2)\times(L/2)$
Decoder-1	Upsampling	64	64	$64\times(K/2)\times(L/2)$
	Conv-2	64	64	$64\times(K/2)\times(L/2)$
	Conv-1	64	64	$64\times(K/2)\times(L/2)$
Decoder-2	Upsampling	64	64	$64\times K\times L$
	Conv-2	64	64	$64\times K\times L$
	Conv-1	64	M	$K\times L\times M$
	Softmax layer	M	–	$K\times L\times M$

Input image size: $K\times L$ pixels (color).

Adapted and reprinted with permission from S. Alheejawi, M. Mandal, R. Berendt, N. Jha, Automated melanoma staging in lymph node biopsy image using deep learning, in: Proc. of the IEEE Canadian Conference on Electrical and Computer Engineering (CCECE), Edmonton, Canada, May 6-8, 2019.

(2) *Upsampling*: In the decoder side, two upsampling processes are performed on the features maps nonlinearly using the memorized pooled indices. In the used SegNet architecture, an upsampling with size 2×2 is performed.

10.4.4 PI calculation

After the training, the trained CNN module segments the input image into three classes: background, active nuclei, and passive nuclei. The PI values are calculated on the segmented image obtained using SegNet architecture. Note that the PI values are calculated for the melanoma regions on lymph nodes using Ki-67-stained image. The melanoma mask is first generated on the MART-1-stained image, which is then mapped onto the corresponding Ki-67-stained image (see the green (light gray in print version) contour at WSI in Fig. 10.15). We assume the melanoma regions R_i $(i=1,2,\ldots,N_{DR})$ are obtained using the algorithm explained in Section 10.4.1. In each melanoma region, the PI value is calculated for a few areas with dense active nuclei. Three areas with overall highest PI values may then be selected. The area selection is done by dividing each melanoma regions into a large number of overlapping rectangular blocks corresponding to a physical size of 0.1×0.15 mm^2. The PI value corresponding to each block is calculated, and the blocks with highest PI values are selected for viewing by doctors. These blocks are likely to have dense areas of active nuclei. Fig. 10.17 shows the PI values calculated in three melanoma regions ($N_{DR}=3$) in the WSI. There are two or three nonoverlapping windows in each melanoma region, which indicates the high values of the PI. Fig. 10.17 shows the three blocks with the highest PI values using red (gray in print version) contour and the other five blocks with high values of PI using black contour. The blocks with red (gray in print version) contour may be selected for viewing by the doctors for melanoma staging.

10.4.5 Results and discussion

In this section, the performance of the deep learning algorithms and classical feature-based technique is presented. The segmentation performance is evaluated first, followed by the PI calculation performance evaluation.

The segmentation performance of the PI-SegNet technique is compared with two feature-based techniques: FCM+KM and Otsu+SVM. In FCM+KM technique, Mungle et al. [43] used FCM to segment the input image into background and nuclei. The KM algorithm is then used to classify nuclei into passive and active nuclei. In Otsu+SVM [45], the input image is segmented into background and nuclei using Otsu's threshold, and the SVM is then used as the classifier. Fig. 10.18D–F show the segmentation results of FCM+KM, Otsu+SVM, and the PI-SegNet, respectively. Fig. 10.19A–F show a comparison of the segmentation/nuclei detection results for the magnified red (gray in print version) square in Fig. 10.18A. It is observed that the PI-SegNet provides excellent nuclei segmentation. The FCM+KM and Otsu+SVM techniques show a coarse boundary in the segmentation results, and there are few pixels around the positive nuclei border, which are segmented as negative nuclei. The poor performance is mainly because these techniques

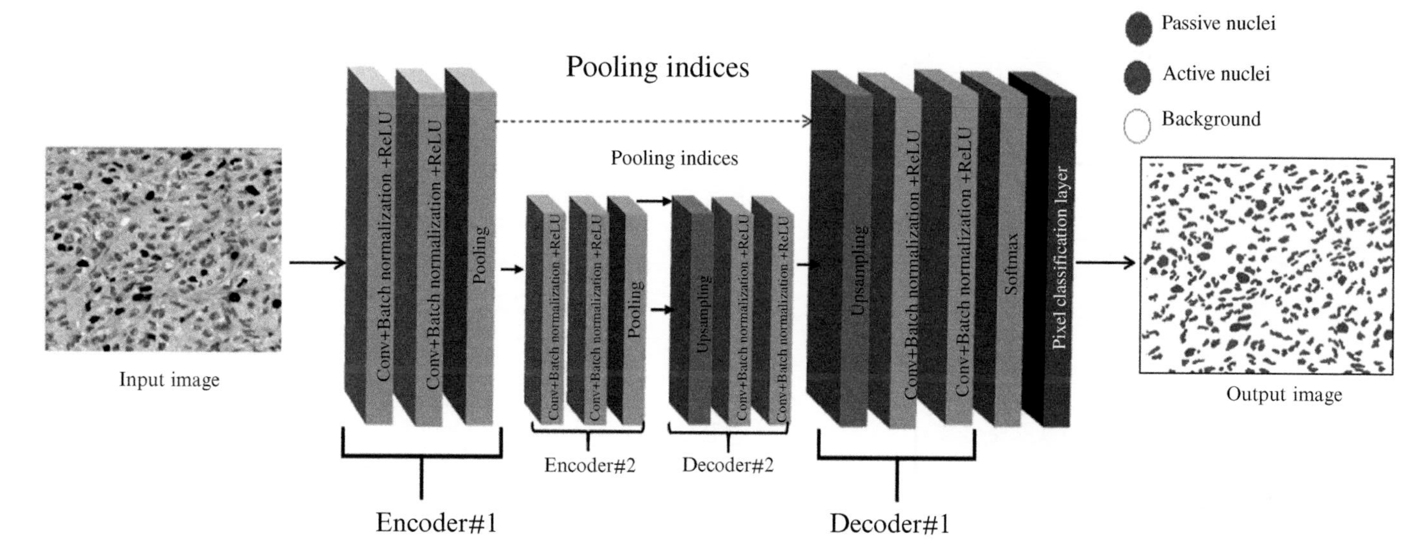

FIG. 10.16 Schematic of the PI-SegNet architecture used for nuclei segmentation and classification. *Adapted and reprinted with permission from S. Alheejawi, M. Mandal, R. Berendt, N. Jha, Automated melanoma staging in lymph node biopsy image using deep learning, in: Proc. of the IEEE Canadian Conference on Electrical and Computer Engineering (CCECE), Edmonton, Canada, May 6-8, 2019. © 2019 IEEE.*

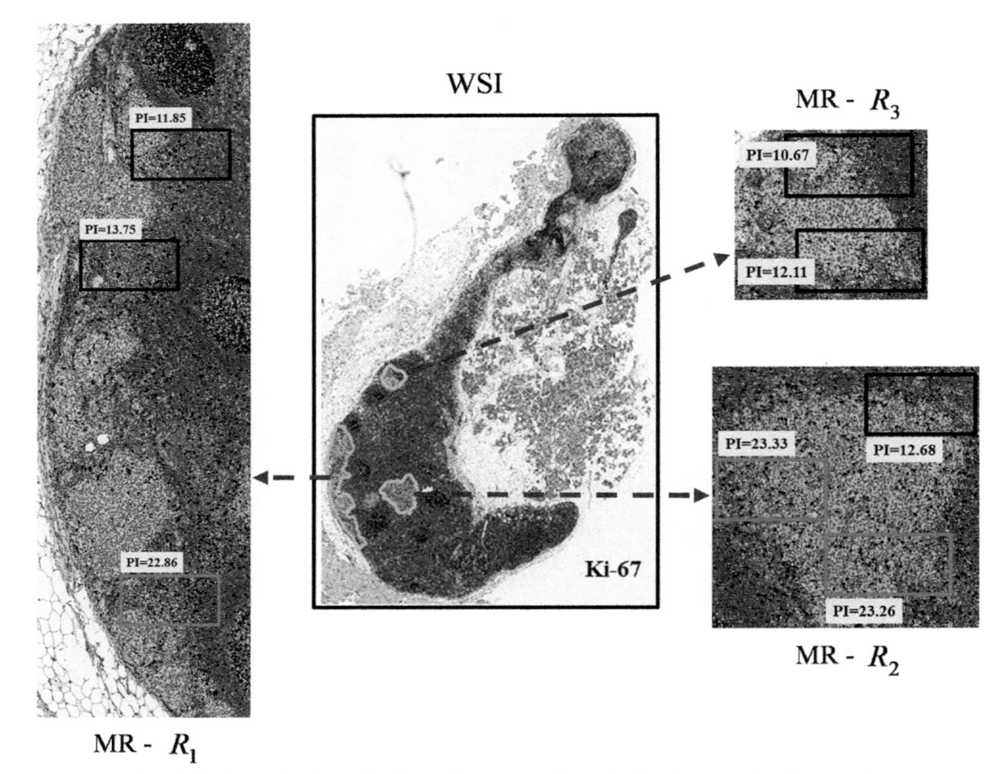

FIG. 10.17 Example of block PI calculation in the melanoma regions (MR) of a WSI. The three melanoma regions are shown with magnification. The regions R_1 and R_2 includes three blocks (with red (gray in print version) contour) with the highest PI values: 23.33, 23.26, and 22.86. The other five blocks (with black contours) are also active areas, but the PI values are smaller.

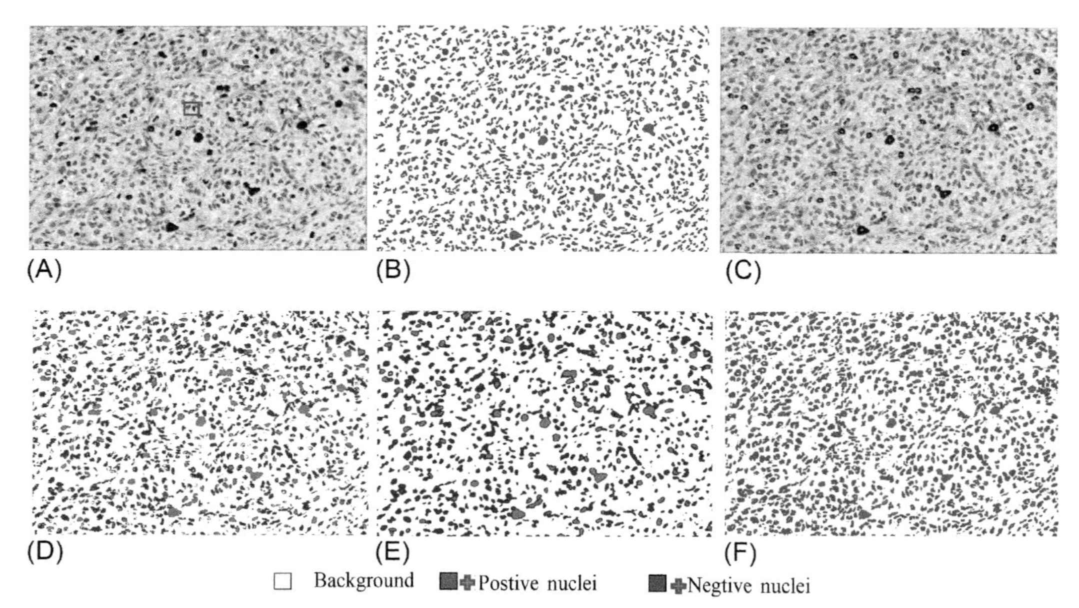

FIG. 10.18 Subjective comparison of segmentation results. (A) Original test image, (B) the ground truth image, segmentation results for (C) GMM + CNN [45], (D) FCM + KM [43], (E) Otsu + SVM [44], and (F) PI-SegNet [46] techniques, respectively.

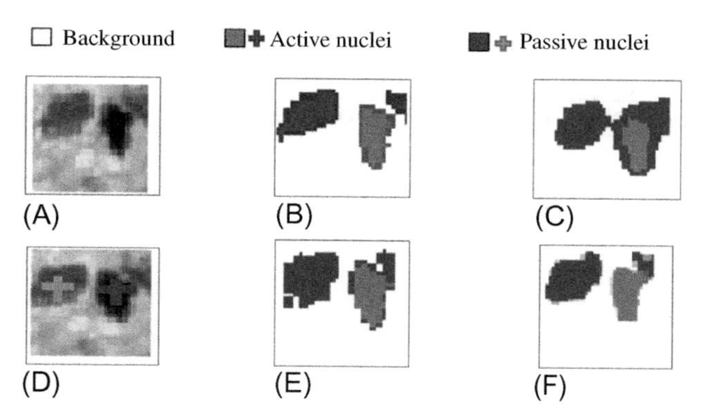

FIG. 10.19 Comparison of segmentation performance. (A) Original test image, (B) the ground truth image [red (gray in print version) square in Fig. 10.10A], segmentation results for (C) FCM+KM [43], (D) Otsu+SVM [44], (E) GMM+CNN [45], and (F) PI-SegNet [46], respectively.

are threshold-based and do not consider shape features of the nuclei during the segmentation process. Table 10.6 shows the segmentation evaluation of FCM+KM and Otsu +SVM techniques with PI-SegNet technique using accuracy measure and BF score [47]. The PI-SegNet technique provides an average 97% BF score and 88% accuracy for all classes as shown in Table 10.6.

For PI calculation performance, the PI-SegNet technique is compared with FCM+KM [43], Otsu+SVM [44], and GMM+CNN [45] techniques. Fig. 10.19E shows the nuclei detection performance of GMM+CNN. In Fig. 10.19E, the GMM+CNN technique separates these two nuclei perfectly. But the GMM+CNN technique cannot classify the patches that contain the active and passive nuclei together. Besides the complexity and the parameters that need to be provided to GMM nuclei detection, sometimes the GMM fails in detecting the nuclei that are not blob-shaped. Table 10.7 compares the PI values obtained using FCM+KM, Otsu+SVM, GMM+CNN, and the PI-SegNet techniques. By comparing the results with the ground truth, it is observed that the PI-SegNet provides the best result. The PI values of GMM+CNN technique tend to be lower than ground truth, primarily because of the inaccuracy of nuclei detection. Fig. 10.20 compares the overall performance of different techniques in terms of MAE and RMSE. It is observed that the

Table 10.6 Segmentation performance of the proposed technique.

Technique	FCM+KM [43]	Otsu+SVM [44]	PI-SegNet [46]
Accuracy	0.83 (B)	0.84 (B)	0.87 (B)
	0.97 (A)	0.95 (A)	0.97 (A)
	0.82 (P)	0.82 (P)	0.86 (P)
BF score	0.99 (B)	0.98 (B)	0.99 (B)
	0.80 (A)	0.97 (A)	0.96 (A)
	0.97 (P)	0.93 (P)	0.98 (P)

B, background mask; *A*, active nuclei mask; *P*, passive nuclei mask.

Table 10.7 PI calculation performance of THE, FCM+CNN, Otsu+SVM, GMM+CNN, and PI-SegNet techniques.

Image index	Ground truth	FCM+KM [43]	Otsu+SVM [44]	GMM+CNN [45]	PI-SegNet [46]
1	11	16.08	10.21	8.75	11.81
2	9	12.56	7.923	7.36	8.30
3	11	14.58	9.63	8.96	10.17
4	9	12.97	8.53	8.52	8.60
5	10	15.09	9.57	9.54	9.91
6	9.8	10.45	6.74	6.41	10.40
7	16	25.05	15.93	13	16.53
8	18	28.83	18.19	15.20	17.50
9	19	28.25	18.51	16.25	17.56
10	13	20.88	9.67	9.50	13.79
11	13	20.98	11.74	10.10	13.97
12	16	25.93	15.00	12.22	15.70

PI-SegNet achieves RMSE and MAE of 0.73 and 0.65, respectively, which is much smaller than those obtained by other techniques.

The run-time complexity of the PI-SegNet technique is compared with other techniques (FCM+KM, Otsu+SVM, and GMM+CNN) and is shown in Table 10.8. Note that all techniques were implemented using Matlab R2018a with the neural network toolbox. All experiments are performed on a Windows 10 computer with Intel i7-4790 CPU and 12 GB RAM. The CNN architectures are trained using an NVIDIA GeForce GTX 745 graphic card. It is observed in Table 10.8 that FCM+KM and GMM+CNN are the most compute intensive techniques. The long execution time of FCM+KM is mainly caused by the large number of iterations in the FCM and KM algorithms. The GMM+CNN technique also requires a long execution time, which is mainly due to the GMM, which takes about 90% of the total execution time for nuclei detection. The PI-SegNet technique has the lowest computational complexity among the four techniques. In other words, the PI-SegNet provides a superior performance compared to other techniques in terms of the PI accuracy as well as computational complexity.

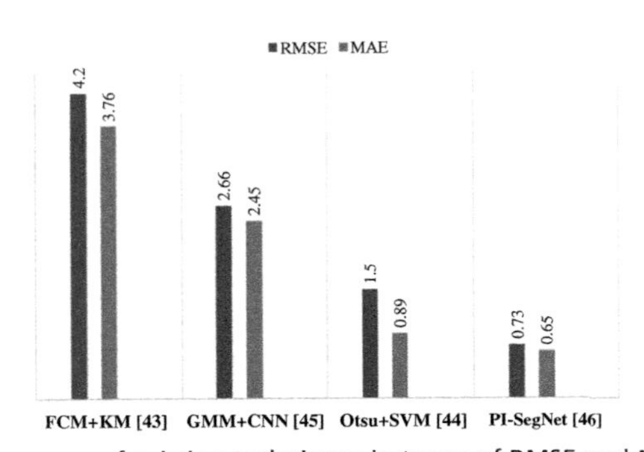

FIG. 10.20 PI calculation performance of existing techniques in terms of RMSE and MSE.

Table 10.8 Average execution time of the FCM+CNN, Otsu+SVM, GMM+CNN and PI-SegNet techniques.

Technique	Execution time (in s)
FCM+KM [43]	89.4
GMM+CNN [45]	129.6
Otsu+SVM [44]	15.7
PI-SegNet [46]	13.8

10.5 Conclusions

This chapter presented several techniques for digital pathology solutions in skin cancer applications based on histopathological image analysis. There techniques include classical feature-based as well as deep learning algorithms. The two problems considered in this chapter are melanoma detection and PI calculation. Both problems are solved based on CNN-based algorithms. The melanoma detection algorithm consists of three modules: Epidermis Region Identification, Nuclei Segmentation, and Nuclei Classification. In the Nuclei Segmentation module, the H&E histopathological images are segmented into cell nuclei and background using NS-Net architecture. The PI calculation technique uses the SegNet architecture to segment and classify the Ki-67-stained image into three classes: background, active, and passive nuclei. The PI value is then calculated based on the number of nuclei in each class.

The deep learning algorithms using CNN architectures are becoming very popular in medical image analysis compared to the classical techniques. In this chapter, the experimental results show that algorithms using CNN architectures provide an excellent performance in terms of diagnosis and staging performances.

References

[1] R.L. Siegel, K.D. Miller, A. Jemal, Cancer statistics, 2019, CA Cancer J. Clin. 69 (1) (2019) 7–34.

[2] M. Mokhtari, M. Rezaeian, S. Gharibzadeh, V. Malekian, Computer aided measurement of melanoma depth of invasion in microscopic images, Micron 61 (2014) 40–48.

[3] I. Maglogiannis, C.N. Doukas, Overview of advanced computer vision systems for skin lesions characterization, IEEE Trans. Inf. Technol. Biomed. 13 (5) (2009) 721–733.

[4] C. Lu, M. Mahmood, N. Jha, M. Mandal, Automated segmentation of the melanocytes in skin histopathological images, IEEE J. Biomed. Health Inf. 17 (2) (2013) 284–296.

[5] H. Xu, C. Lu, M. Mandal, An efficient technique for nuclei segmentation based on ellipse descriptor analysis and improved seed detection algorithm, IEEE J. Biomed. Health Inf. 18 (5) (2013) 1729–1741.

[6] S.M. Ismail, A.B. Colclough, J.S. Dinnen, D. Eakins, D. Evans, E. Gradwell, J.P. O'Sullivan, J.M. Summerell, R.G. Newcombe, Observer variation in histopathological diagnosis and grading of cervical intraepithelial neoplasia, Br. Med. J. 298 (6675) (1989) 707.

[7] S. Petushi, F.U. Garcia, M.M. Haber, C. Katsinis, A. Tozeren, Large-scale computations on histology images reveal grade-differentiating parameters for breast cancer, BMC Med. Imaging 6 (1) (2006) 14.

[8] L. Brochez, E. Verhaeghe, E. Grosshans, E. Haneke, G. Piérard, D. Ruiter, J.-M. Naeyaert, Interobserver variation in the histopathological diagnosis of clinically suspicious pigmented skin lesions, J. Pathol. 196 (4) (2002) 459–466.

[9] Y. Wang, D. Crookes, O.S. Eldin, S. Wang, P. Hamilton, J. Diamond, Assisted diagnosis of cervical intraepithelial neoplasia (cin), IEEE J. Selected Topics Signal Process. 3 (1) (2009) 112–121.

[10] O. Sertel, Image Analysis for Computer-Aided Histopathology, (Ph.D. dissertation)The Ohio State University, 2012.

[11] J.M. Haggerty, X.N. Wang, A. Dickinson, J. Chris, E.B. Martin, et al., Segmentation of epidermal tissue with histopathological damage in images of haematoxylin and eosin stained human skin, BMC Med. Imaging 14 (1) (2014) 7.

[12] H. Xu, M. Mandal, Epidermis segmentation in skin histopathological images based on thickness measurement and k-means algorithm, EURASIP J. Image. Vide. Process. 2015 (1) (2015) 1–14.

[13] H. Fatakdawala, J. Xu, A. Basavanhally, G. Bhanot, S. Ganesan, M. Feldman, J.E. Tomaszewski, A. Madabhushi, Expectation–maximization-driven geodesic active contour with overlap resolution (emagacor): application to lymphocyte segmentation on breast cancer histopathology, IEEE Trans. Biomed. Eng. 57 (7) (2010) 1676–1689.

[14] O. Schmitt, S. Reetz, On the decomposition of cell clusters, J. Math. Imaging Vision 33 (1) (2009) 85–103.

[15] B. Ehteshami Bejnordi, G. Litjens, N. Timofeeva, I. Otte-Holler, A. Homeyer, N. Karssemeijer, J. van der Laak, Stain specific standardization of whole-slide histopathological images, IEEE Trans. Biomed. Eng. 35 (2) (2016) 404–415.

[16] M.N. Gurcan, T. Pan, H. Shimada, J. Saltz, Image analysis for neuroblastoma classification: segmentation of cell nuclei, in: Proceeding of 28th Annual International Conference on Engineering in Medicine and Biology Society, IEEE, 2006, pp. 4844–4847.

[17] J. Cheng, J.C. Rajapakse, Segmentation of clustered nuclei with shape markers and marking function, IEEE Trans. Biomed. Eng. 56 (3) (2009) 741–748.

[18] B. Parvin, Q. Yang, J. Han, H. Chang, B. Rydberg, M.H. Barcellos-Hoff, Iterative voting for inference of structural saliency and characterization of subcellular events, IEEE Trans. Image Process. 16 (3) (2007) 615–623.

[19] H. Xu, C. Lu, R. Berendt, N. Jha, M. Mandal, Automatic nuclei detection based on generalized Laplacian of Gaussian filters, IEEE J. Biomed. Health Inform. 21 (3) (2017) 826–837.

[20] H. Xu, C. Lu, R. Berendt, N. Jha, M. Mandal, Automatic nuclear segmentation using multi-scale radial line scanning with dynamic programming, IEEE Trans. Biomed. Eng. 64 (10) (2017) 2475–2485 (October).

[21] D. Ciresan, A. Giusti, L. Gambardella, J. Schmidhuber, Deep neural networks segment neuronal membranes in electron microscopy images, NIPS, 2012, pp. 2852–2860.

[22] J. Long, E. Shelhamer, T. Darrell, Fully convolutional networks for semantic segmentation, in: CVPR, 2015, pp. 3431–3440.

[23] D. Eigen, R. Fergus, Predicting depth, surface normals and semantic labels with a common multi-scale convolutional architecture, in: ICCV, 2015, pp. 2650–2658.

[24] V. Badrinarayanan, A. Kendall, R. Cipolla, SegNet: A Deep Convolutional Encoder-Decoder Architecture for Image Segmentation, arXiv. Preprint arXiv: 1511.0051, 2015.

[25] O. Ronneberger, P. Fischer, T. Brox, U-Net: convolutional networks for biomedical image segmentation, Med. Image Comput. Comput. Assist. Interv. 9351 (2015) 234–241.

[26] G. Massi, P.E. LeBoit, Histological Diagnosis of Nevi and Melanoma, second ed., Springer, Berlin, 2013.

[27] S. Ioffe, C. Szegedy, Batch Normalization: Accelerating Deep Network Training by Reducing Internal Covariate Shift, arXiv preprint arXiv:1502.03167, 2015.

[28] V. Nair, G.E. Hinton, Rectified linear units improve restricted Boltzmann machines, in: Proceedings of the 27th International Conference on Machine Learning, 2010, pp. 807–814.

[29] A. Krizhevsky, I. Sutskever, G. Hinton, Image net classification with deep convolutional neural networks, in: Proc. of the 25th International Conference on Neural Information Processing Systems, 2012, pp. 1097–1105, Lake Tahoe, NV, December.

[30] X. Glorot, A. Bordes, Y. Bengio, Deep sparse rectifier neural networks, in: Proceedings of the Fourteenth International Conference on Artificial Intelligence and Statistics, 2011, pp. 315–323.

[31] C. Robert, Machine learning, a probabilistic perspective, CHANCE 27 (2) (2014) 62–63.

[32] C. Corinna, V. Vladimir, Support-vector networks, Mach. Learn. 20 (3) (1995) 273–297.

[33] A. Ben-Hur, D. Horn, H. Siegelmann, V. Vapnik, Support vector clustering, J. Mach. Learn. Res. 2 (2001) 125–137.

[34] T. Hofmann, B. Schölkopf, A.J. Smola, Kernel methods in machine learning, Ann. Statist. 36 (3) (2008) 1171–1220.

[35] H. Xu, C. Lu, R. Berendt, N. Jha, M. Mandal, Automated analysis and classification of melanocytic tumor on skin whole slide images, Comput. Med. Imaging Graph. 66 (2018) 124–134.

[36] D. Hanahan, R. Weinberg, Hallmarks of cancer: the next generation, Cell 144 (5) (2011) 646–674.

[37] S. Bruno, Z. Darzynkiewicz, Cell cycle dependent expression and stability of the nuclear protein detected by Ki-67 antibody in HL-60 cells, Cell Prolif. 25 (1) (1992) 31–40.

[38] D. Schonk, H. Kuijpers, E. Drunen, C.H. Dalen, A.H. Geurts van Kessel, R. Verheijen, F.C. Ramaekers, Assignment of the gene(s) involved in the expression of the proliferation-related Ki-67 antigen to human chromosome 10, Hum. Genet. 83 (3) (October 1989) 297–299.

[39] Bánkfalvi ", Comparative methodological analysis of erbB-2/HER-2 gene dosage, chromosomal copy number and protein overexpression in breast carcinoma tissues for diagnostic use, Histopathology 37 (5) (2000) 411–419.

[40] B. Grala, T. Markiewicz, W. Kozłowski, S. Osowski, J. Słodkowska, W. Papierz, New automated image analysis method for the assessment of ki-67 labeling index in meningiomas, Folia Histochem. Cytochem. 47 (2009) 587–592.

[41] H. Akakin, H. Kong, C. Elkins, J. Hemminger, B. Miller, J. Ming, E. Plocharczyk, R. Roth, M. Weinberg, R. Ziegler, G. Lozanski, M. Gurcan, Automated detection of cells from immunohistochemically-stained tissues: application to ki-67 nuclei staining, Proc. SPIE 8315, Medical Imaging 2012: Computer Aided Diagnosis, San Diego, California, USA, February 22(2012).

[42] H.Z. Al-Lahham, R.S. Alomari, H. Hiary, V. Chaudhary, Automating proliferation rate estimation from ki-67 histology images, Proc. SPIE 8315, Medical Imaging 2012: Computer Aided Diagnosis, San Diego, California, USA, February 23(2012).

[43] T. Mungle, S. Tewary, et al., Automated characterization and counting of Ki-67 protein for breast cancer prognosis: a quantitative immunohistochemistry approach, Comput. Methods Prog. Biomed. 139 (2017) 149–161.

[44] S. Alheejawi, H. Xu, R. Berendt, N. Jha, M. Mandal, Novel lymph node segmentation and proliferation index measurement for skin melanoma biopsy images, Comput. Med. Imaging Graph. 73 (April) (2019) 19–29.

[45] M. Saha, C. Chakraborty, I. Arun, R. Ahmed, S. Chatterjee, An advanced deep learning approach for ki-67 stained hotspot detection and proliferation rate scoring for prognostic evaluation of breast cancer, Sci. Rep. 7 (2017) 3213.

[46] S. Alheejawi, M. Mandal, R. Berendt, N. Jha, Automated melanoma staging in lymph node biopsy image using deep learning, in: Proc. of the IEEE Canadian Conference on Electrical and Computer Engineering (CCECE), Edmonton, Canada, May 6-8, 2019.

[47] G. Csurka, D. Larlus, F. Perronnin, What is a good evaluation measure for semantic segmentation? in: Proc. of the British Machine Vision Conference, 2013, pp. 32.1–32.11.

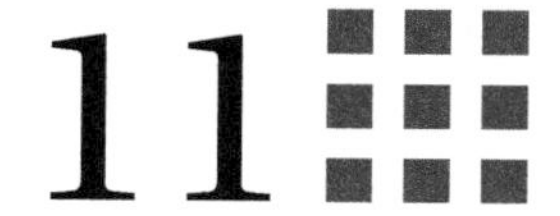

Potential proposal to improve data transmission in healthcare systems

Reinaldo Padilha França[a], Yuzo Iano[a], Ana Carolina Borges Monteiro[a], Rangel Arthur[b]

[a]SCHOOL OF ELECTRICAL ENGINEERING AND COMPUTING (FEEC), STATE UNIVERSITY OF CAMPINAS (UNICAMP), CAMPINAS, BRAZIL
[b]SCHOOL OF TECHNOLOGY (FT), STATE UNIVERSITY OF CAMPINAS (UNICAMP), LIMEIRA, BRAZIL

11.1 Introduction

Medicine is the field of study related to the maintenance of health, as well as to the prevention, treatment, and cure of diseases and injuries. Medical activity arose thousands of years ago and has been used by many people over the centuries, often mixed with religious practices. It is known that, in Ancient Egypt, complex surgeries were performed that even involved the opening of the skull. However, it was in Ancient Greece that *medicine* developed, because at that time the first techniques were developed to identify the symptoms of diseases. In this period, Hippocrates stood out and is currently considered the Father of Medicine [1].

Since ancient Greece, there have been many discoveries and advances, especially in the development of medical equipment. These instruments have a medical, dental, laboratory, or physiotherapeutic purpose and are used directly or indirectly for diagnosis, therapy, rehabilitation, or monitoring of human beings [1, 2]. One of the first highlights in the area of medical equipment was the discovery and development of X-ray equipment made in the year 1985 by German physicist Wilhelm Conrad Roentgen. This discovery paved the way for the development of new methodologies that are indispensable today for routine medicine: magnetic resonance imaging and computed tomography [3].

Nowadays, medical equipment is not restricted to just detection of pathologies but also in the transmission of obtained results. In this way, we can take notice of the emergence of a new concept: e-health. According to the Healthcare Information and Management Systems Society (HIMSS), e-health is any Internet application used in conjunction with other information technologies focused on improving clinical processes, patient treatment, and the costs of a health system, that is, improving the flow of information through electronic means to improve service delivery and coordination of health systems. This concept includes many dimensions, ranging from the delivery of information from hospital clinics to

Deep Learning Techniques for Biomedical and Health Informatics. https://doi.org/10.1016/B978-0-12-819061-6.00011-2

partners in the service chain, through the facility of interaction among all its members, arriving at the availability of this same information in difficult-to-reach and remote places [4, 5].

Modern health systems consist of a set of tools and services capable of sustaining care and improving treatment in an integrated way through the Web. These include electronic records (ePaciente); mHealth (use of mobile devices); big data; cloud computing; personalized medicine; telemedicine; Internet of things (IoT); and artificial intelligence (AI) and its aspects (machine learning, deep learning, and natural language processing) [4–6].

In this context, we can emphasize telemedicine, characterized by an advanced process for monitoring patients, exchanging medical information, and analyzing the results of different tests. These exams are evaluated and delivered digitally, providing support for traditional medicine. Telemedicine is already used around the world in a safe and legalized way, and in accordance with legislation and medical standards. With the use of information technology, which adds quality and speed in the exchange of knowledge, doctors can make decisions with greater agility and accuracy. Telemedicine allows experts to access exams from anywhere in the country using computers and mobile devices, such as smartphones and tablets connected to the Internet. Together with telemedicine is the concept of e-health [7, 8].

In developing countries such as Brazil, *telemedicine* services are mainly applied by the issuance of online reports. However, this reality is not yet consolidated throughout the country. Telemedicine was boosted in the early 1990s when Internet expansion occurred, creating a worldwide trend of medical care and generation of reports at a distance [9, 10].

In recent years, healthcare providers, medical institutions, and regulatory agencies have been making an active effort to promote, disseminate, and develop more healthcare and remote cooperation programs. It is already possible to find programs based on artificial intelligence in some reference hospitals around the world. In these places, imaging devices are able to point out possible diseases and send notifications automatically to the doctor, and other equipment will send vital signs of the patient directly to medical records, among others [7, 8]. In Brazil, the government has invested in the purchase of three supercomputers capable of increasing the capacity of data storage up to 10 times [9, 10].

Brazil presents a model of healthcare called the Unified Health System (from the Portuguese Sistema Único de Saúde—SUS). This system was created in 1988, and through a law in the constitution with the purpose of changing the situation of inequality in the healthcare of the population, it is compulsory to public service for any citizen and any charges against money are prohibited under any pretext. The SUS is destined for all citizens and is financed with resources collected through taxes and social contributions paid by the population and composite federal, state, and municipal government resources. Through the SUS, all citizens have the right to consultations, examinations, hospitalizations, and treatments in Health Units linked to SUS, whether public (at the municipal, state, and federal level) or private, contracted by the public health manager [9–11].

Despite all these benefits, this system suffers from a lack of doctors in some regions of Brazil and is affected by long waiting times, which can cause patients to wait for years for exams, medical appointments, or surgeries. However, the lack of professionals and long waiting times are not only characteristic of Brazilian hospitals, as most hospitals and health systems in underdeveloped and developing countries are subject to these situations. One of the great problems of these long queues is the aggravation of chronic diseases such as diabetes, heart diseases, renal dysfunctions, and liver dysfunctions, among others [11].

Much of the problem of long queues can be attributed to the use of bureaucratic customer service systems associated with medical data transmission equipment that use speedless technologies in the transmission of information and systems of large consumption of computational memory, which can generate crashes and even data loss [11, 12]. There are currently a wide variety of ways of managing data in telecommunication systems; however, the concept of discrete events is best suited for managing data in a queue. This concept associated with cloud technology has been widely used in industry companies and management of telemarketing centers.

11.2 Telecommunications channels

A telecommunication system follows logic that can be applied to any communication system. This logic includes the presence of three basic components: (1) a transmitter responsible for sending information and subsequent conversion in a signal, (2) a transmission channel to carry this signal, and (3) a receiver receiving the same signal, which can then convert it into useful information. The communication channel is the medium responsible for providing the physical connection between transmitters and receivers in a communication system, either as a wire or a logical connection in a multiplexed medium, for example, a radio channel in telecommunications networks and computers. Data transport typically uses two types of media: physical (twisted pair and fiber optic cable) and electromagnetic (microwave, satellite, radio, and infrared) [13].

A widely used model applicable to a large set of physical channels is the Additive White Gaussian Noise (AWGN) channel model. This model has the characteristic of introducing a statistically modeled noise, such as a white Gaussian additive process, into the transmitted signals [14]. The existence of disturbances/noise in the channel (free space/atmosphere/copper line) has multiple causes. One of them is the thermal noise by virtue of the movement of the electrons in the electronic circuit used for transmission and reception of the signal. The AWGN channel models such imperfections in a communication channel [15, 16].

In this context, there is also the wireless mobile channel. This channel refers to wireless communication and devices based on radio frequencies and where the communication path is mobile at both ends [16].

Variations in the received signal are only noticed when observed on a large scale, when the signals travel long distances or for long periods of time, so this type of variation is called large scale. In turn, the variation is determined by loss in the course and is directly related to the distance and the frequency of its propagation, presenting linear variation, expressed in decibels (dB) [17].

Both types of channels can be employed for the transmission of medical data between hospitals, hospitals and patients, and physicians and patients.

11.2.1 Discrete events

Discrete events are classified as an occurrence responsible for changing the state of the system in which they act. These actions can be classified into three types: (1) intentional, (2) spontaneously controlled occurrence, or (3) verification of a condition. All types are generally capable of producing state changes at random time intervals [18].

The universe of actions that provide events is subjective and depends on the ability of the modeler to abstract the events of the universe in which the system is modeled. A system modeled with discrete events can be defined according to its evolution, which is directly related to the occurrence of events. For this, actions are necessary, and these, in turn, generate events. Therefore, systems of this type only change state when an event occurs; if it does not, the system remains in the same state [19].

The relationship between the sequence of events with the components used in the modeling must respect the principle of operation of each component and its applicability in the modeling system. In turn, entities are defined as discrete items of interest in a discrete event simulation (DES). However, the meaning of an entity depends on what is being modeled and the type of system used. This is because the presence of certain attributes may be able to affect the way events are treated or cause changes as the entity flows through the process [20, 21]. It is important to emphasize that the concepts *entities* and *events* are different.

Events are instantaneous discrete incidents that can change a state variable, an output, and/or an occurrence from another event. In turn, an entity is somewhat dependent on what is being modeled and the type of system. Thus, the characteristic of a discrete event signal indicates something (e.g., an accident, an earthquake, a fault, a control, a person, a heartbeat, or any other desirable concept within a system) as well as the generation of a bit in a communication system. In turn, an event is a conceptual notation that denotes a state change in a system. Discrete event modeling performs discrete changes within a system [22].

Over the years, we note that the concept of discrete events presented wide applicability in several areas, generating satisfactory results for the business and entrepreneurial areas.

11.3 Scientific grounding

In 2009, a simulation of discrete events of routing protocols of wireless sensor networks was investigated. For this, a discrete event simulator called sensor networks simulator (SENSIM) was implemented for class-based routing, hierarchical routing (PEGASSIS), and location-based routing (MFR) [23].

In 2010, the framework of the wireless sensor network (WSN) was presented, which is a modeling framework in which the designer can create network components in each layer of the open systems interconnection (OSI) protocol stack. In addition, it was possible to simulate the analysis of several parameters and generate codes for different target platforms, both in hardware and in software, which was implemented with SimEvents [24].

Simulators of eight telecommunications networks were surveyed in 2011, which were classified and compared based on their type, mode of implementation, network deficiencies, and supported protocols. Thus, it was shown that the use of DES can be used as a parameter for modeling and evaluating network performance. Also in 2011, electronic collaboration over the Internet between business partners was studied. This study was performed through the exchange of messages that involve defined standards and through user-defined infrastructure patterns. At the same time, the notion of event was increasingly promoted for asynchronous communication and coordination in service-oriented architecture (SOA) systems. This communication between partners or between components was accomplished through the exchange of discrete units of data (messages or events) [25].

In 2012, a discrete event mechanism was proposed, which aimed to remove obsolete messages for delay and interrupt-tolerant networks (DTN). This study aimed to increase the probability of message delivery [26]. In 2013, the ad hoc multihop networks with chain and cross topologies were studied. This study revealed an excellent solution for low-cost computer networks. In view of this, an ad hoc multihop network model and a multihop wireless ad hoc network model with chain and cross topologies were suggested. These templates were designed using SimEvents.

In the year 2013, the processing of information flow in real time in the cloud was studied. At that time, this concept was gaining significant attention because of its ability to extract large amounts of data for a variety of applications. In cloud-based real-time streaming applications, dynamic resource management mechanisms are required to support operations. Thus, a performance-oriented approach to discrete event modeling with SimEvents was presented. This study aimed to identify the controllable properties of this communication structure.

In 2014, the performance of network coding (NC) was analyzed, which was at that time becoming an increasingly important issue in data communication systems. In this way, the behavior of an M/D/1 queuing system under an NC-based synchronous scenario was analyzed. Through SimEvents, an intermediate node represented by a first-in-first-out (FIFO) queue and a single server was modeled. The configuration was intended to receive combined and encoded packets of two streams. Also in 2014, interference was investigated in LTE-Advanced systems. This system can be attenuated using coordinate multipoint techniques (CoMP) with joint transmission of user data. Thus, there was coordination between eNodeBs through the use of X2 interface. Thus, the DES was used to evaluate the latency requirements by investigating the consequences of a contained backhaul. The results demonstrated gain in system throughput compared to the case without CoMP for low latency backhaul [27]

In 2015, the problem of event-based synchronization of discrete-time linear dynamic networks was investigated. Leader-following and leaderless synchronizations were achieved by a distributed event-triggering strategy. This way, it was shown that updating the feedback control is unnecessary until an event is triggered [28]. In 2016, the stability of the network control system (NCS) was analyzed with a discrete PID controller per even, under the condition of jitter and loss of data based on the promotion method. For this, an analogy was made to the situation of each link in the control circuit. Each link was time-driven and derived from the corresponding stability criterion. The accuracy of the stability criterion was evaluated by a practical event control system using the simulation tools of Matlab, SimEvents, and xPc-target [29].

In 2017, the problem of the existence of a controller for discrete event time network systems was addressed. For this, it was assumed that a controller (supervisor) communicates with the system to be controlled through a shared communication network. Delays and losses in communication systems and their impacts on control were investigated. The results were applied in the management of this network using a 33-node test system [30].

11.4 Proposal and objectives

Computational simulations allow for more dynamic and flexible studies, as they allow changes in several aspects, facilitating different types of evaluation without the need for a physical experimental setup. This flexibility allows for adjustment of the study's parameters, aiming to improve the performance of the system as a whole and minimize costs. Through research, it's been noted that the technique of discrete events was employed in the transport layer in telecommunications systems. Most of these systems followed a traditional data compression methodology by extracting data redundancy. The technique of discrete events can help in a greater applicability in telecommunications systems, allowing the development of the coding of bits for entities by means of discrete events (CBEDE) methodology, which applies the technique of discrete events in the stage of generation of bits, and in the physical layer, for the transmission of a signal in an AWGN channel in a simulation environment. Given this, the technique of discrete events can be applied in the treatment of the bit in its stage of its generation, which is responsible for its conversion into discrete entities. This process is the result of a methodology used at a lower level of application, acting at the physical layer rather than the typically used transport layer. As can be seen through historical records, this will contribute to the development of increasingly efficient methodologies capable of reducing the consumption of computing resources, such as memory, which is an important parameter to meet the needs of an increasingly technological world.

In this context, it is important to develop telecommunication systems based on discrete events that aim to transmit medical data with accuracy and speed. In this way, there will be positive impacts on (1) the automatic sending of data emitted by medical devices, with greater precision and speeds; (2) sending real-time data from patients with chronic diseases to the responsible physician, thus enabling better treatment and follow-up; and (3) improvement of communication between hospitals and health centers through the transmission of medical data for scheduling of medical appointments, examinations, and withdrawals of medication by patients and surgeries in a shorter time.

11.5 Methodology

The present study employs the technique of discrete events in the stage of the generation of the signal, that is, the discrete events are directly applied to the process of creation of the bits for transmission in an AWGN channel. The use of the discrete event technique is responsible for giving rise to the CBEDE methodology.

The modeling of the CBEDE methodology was carried out in the Simulink® simulation environment of Matlab software, in its version of 64 bits (2014). This tool was chosen because it is consolidated in the scientific environment and has blocks that are already tested and validated. Four libraries were used: (1) Communications System™, which is designed to design, simulate, and analyze systems, and is able to model dynamic communication systems; (2) the DSP System™, which is capable of designing and simulating systems with signal processing; (3) Simulink, which is a block diagram environment for multidomain simulation, capable of supporting system-level projects for the modeling and simulation of telecommunication systems; and (4) the SimEvents® library, which is classified as a DES mechanism and components to develop systems models oriented to specific events [31–34].

In the proposed model (Fig. 11.1), the signals corresponding to the bits 0 and 1 are generated and modulated with the advanced modulation differential quadrature phase shift keying (DQPSK) format, which uses the phase shift coming from the modulation format itself. It then proceeds to an AWGN channel according to the parameters shown in Table 11.1. The signal is then demodulated to perform the bit error rate (BER) calculation of the channel. The values obtained for BER are sent to the Matlab workspace for verification of equality and generation of the signal BER graph.

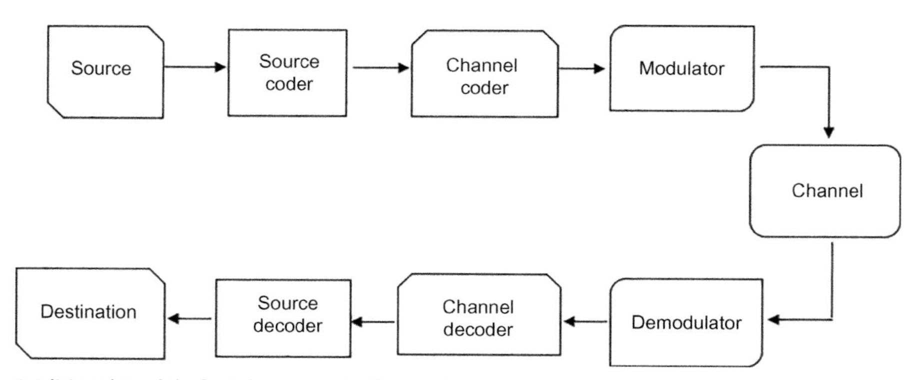

FIG. 11.1 Traditional model of a telecommunication system.

Table 11.1 Parameters channel DQPSK models.

AWGN DQPSK	
Sample time	1 s
Simulation time	1000 s
E_b/N_0	0–14 dB
Symbol period	1 s
Input signal power	1 W
Initial seed in the generator	37
Initial seed on the channel	67

11.6 Precoding bit

The modeling according to the proposal implemented with discrete events is similar to the one previously presented, differentiating that, in this model, the process of discrete events called precoding is added. The proposed bit precoding is implemented through the discrete event methodology. Bit processing is understood as the discrete event methodology in the step of generating signal bits (information) to make it more appropriate for a specific application.

The event-based signal is the signal susceptible to treatment by the SimEvents library and is converted to the specific format required for manipulation by the Simulink library. Both time-based and event-based signals are in the time domain. This treatment has the emphasis on bits 1 and 0, which are generated as a discrete entity and follow the parameters as shown in Table 11.1. Then, Entity Sink® is used to represent the end of the modeling of discrete events by the library SimEvents. This tool is responsible for marking the specific point where the Entity Sink is located, where the event-based signal conversion will be performed later for a time-based signal.

This time-based signal is converted to a specific type that will follow the desired output data parameter, an integer, the bit. By means of the real world value (RWV) function, the actual value of the input signal is preserved. Then, rounding is performed with the "floor" function. This function is responsible for rounding the values to the nearest smallest integer.

A zero-order hold (ZOH) will also be performed, which is responsible for defining sampling in a practical sense and is used for discrete samples at regular intervals. ZOH describes the effect of converting a signal to the time domain, causing its reconstruction and maintaining each sample value for a specific time interval. The treatment logic on bit 1 is presented in Fig. 11.2.

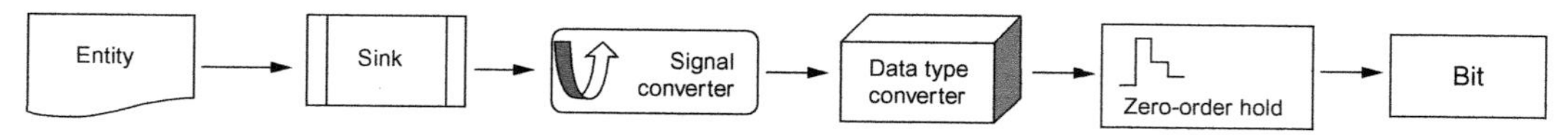

FIG. 11.2 Proposed bit precoding.

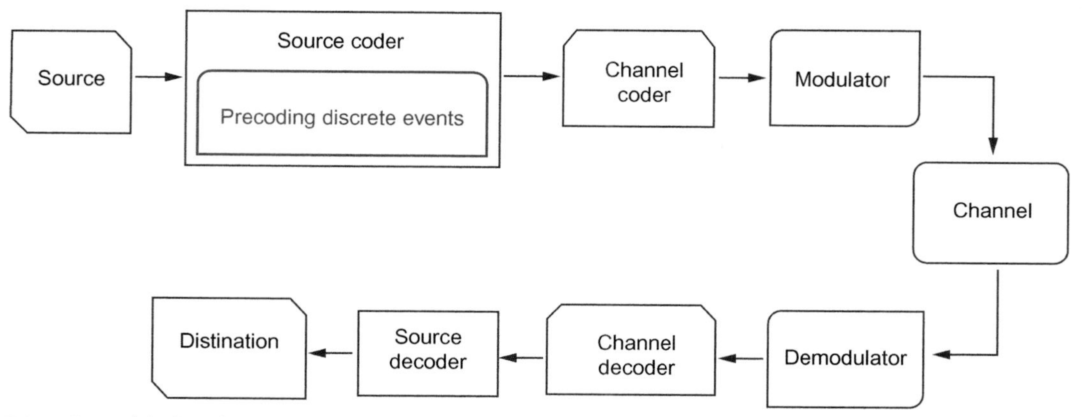

FIG. 11.3 Model of a telecommunication system with the proposal.

Subsequently, the signal is modulated with the advanced modulation DQPSK format and is inserted into the AWGN channel, and thereafter demodulated for purposes of calculating the BER of the signal. The relative values of the BER will be sent to the Matlab workspace to verify equality and generate the BER graph of the signal, as represented in Fig. 11.3.

The models shown in Figs. 11.1 and 11.3 were executed with 10,000 seconds of simulation, respecting the configuration defined according to Table 11.1.

11.7 Signal validation by DQPSK modulation

The verification of equality of the signals is performed through the "size" and "isequal" functions of the Matlab software, as well as through the BER. These functions will be responsible for the mathematical comparison proving that the signals have the same size. Together with the BER verification, it states that the same amount of information is transmitted (bits) in both the proposed methodology (CBDE) and the conventional methodology (AWGN channel). Thus, if the signals are of the same size, the logical value 1 (true) is returned and the same volume of data is transmitted, indicating that the equality of the signals is true. Otherwise, the value will be 0 (false). This check will show that the submitted proposal does not add or remove information to the original transmitted signal.

To graph the results obtained by the "size" and "isequal" functions, the DQPSK modulation is analyzed through its constellation by means of the "compass" function, which displays a compass graph with n arrows, and how the constellations will be PSKs, their representations of points are radial. Because of this feature, the graph format will be in the form of a compass, where n is the number of elements in *Z*. The location of the base of each arrow will be its origin. In turn, the location of the tip of each arrow will be determined by the real and imaginary components of *Z* relative to the constellation of the signal. In the same way and with the same intention, this will be used as the diagram of constellation.

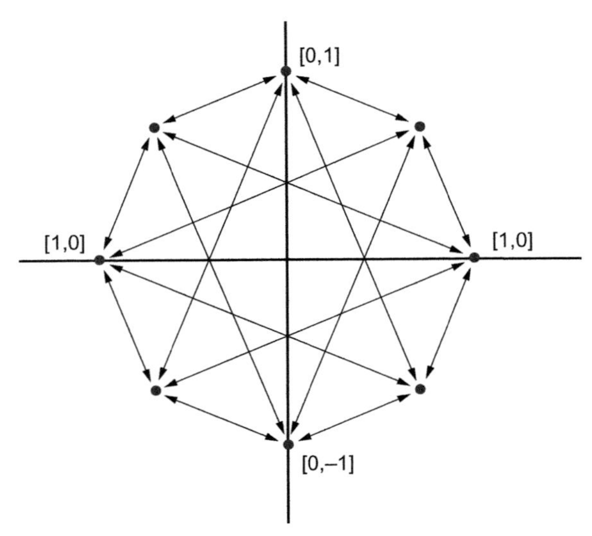

FIG. 11.4 Theoretical DQPSK constellation.

The constellation has as function to analyze both signals transmitted by the models. In the case of the DQPSK constellation, there are four possible states 0, π, $+\pi/2$, and $-\pi/2$, where each symbol represents two bits of information. The division of the binary pattern is equal to QPSK, except when a bit string is shifted to about $\pi/4$ or $\pi/2$ [14, 17]. This means that there is a total of eight status positions (compared to the four states for QPSK). Fig. 11.4 shows the constellation diagram DPQSK for the offset version $\pi/4$. This modulation is widely used in several airborne systems in association with other modulation techniques [14].

11.8 Results

The research presented in this section shows an AWGN transmission channel with DQPSK modulation, as used the Simulink simulation environment of the Matlab. The model from Fig. 11.5 incorporates the traditional method (left) and the proposed innovation of this chapter (right) is presented, showing the signal transmission flow (corresponding to bits 0 and 1) being generated and then modulated in DBPSK, passing through the channel AWGN. Fig. 11.6 shows the constellations for the proposed (left) and the traditional methods (right).

The models were investigated from the perspective of memory consumption evaluation. The first simulation of both models in each command is analyzed, because it is in the first simulation that the construction of the model in a virtual environment is performed from scratch. It is where all the variables of the model are allocated; the memory of the operating system in which the Matlab is running is reserved for the execution of the model, and the results of this model, according to the evaluation parameters are, in fact, real.

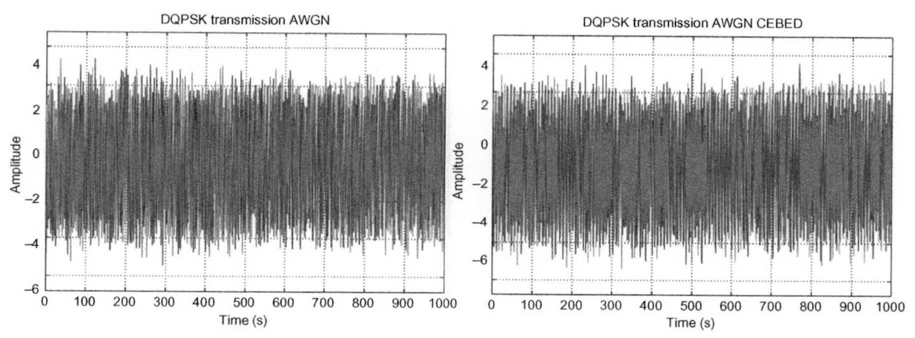

FIG. 11.5 Transmission flow for DQPSK.

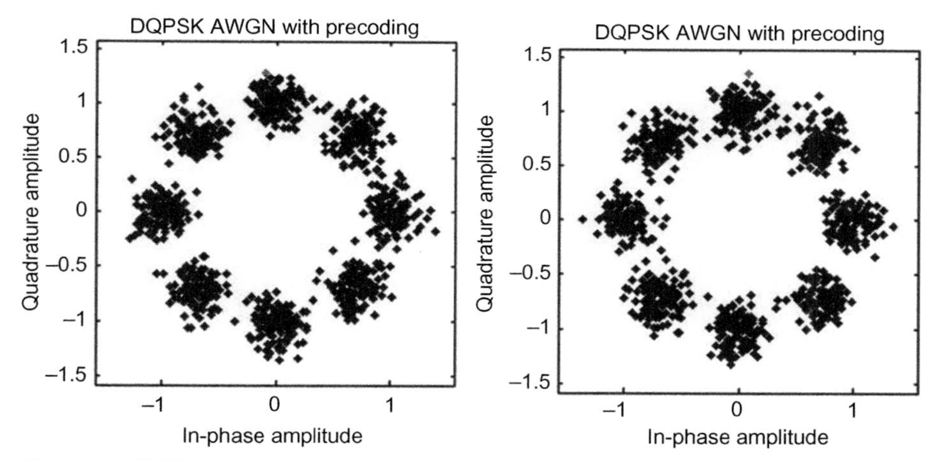

FIG. 11.6 Simulated DBPSK constellations.

Thus, the experiments considered the memory consumption. For memory calculation, the "sldiagnostics" function will be used, whereas the "TotalMemory" variable will receive the sum of all the memory consumption processes used in the model by the "ProcessMemUsage" parameter. This parameter counts the amount of memory used in each process throughout the simulation, returning the total in MB. For this, a computer was used with Intel Core i3 processor hardware configuration containing two processing cores, Intel Hyper-Threading Technology and 4 GB RAM. This machine relates the proposal to the dynamics of the real world and will affirm its efficiency and applicability.

For this, physical machines with hardware configuration were used, consisting of an Intel Core i3 processor and 4 GB RAM, as previously discussed. The experiments were carried out through five simulations of each model to develop the analysis of this chapter, as shown in Fig. 11.7.

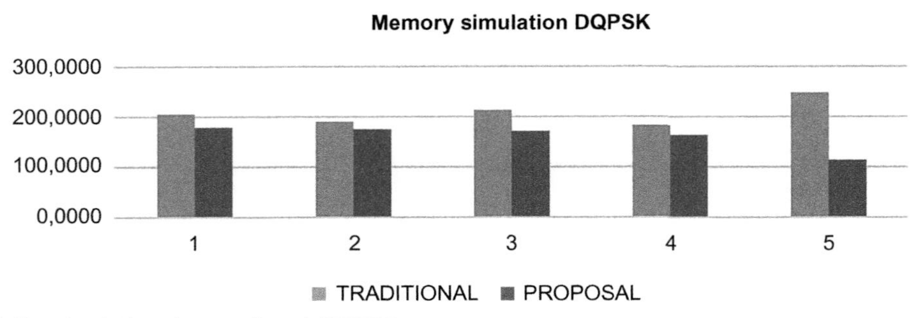

FIG. 11.7 First simulations (memory) model DQPSK.

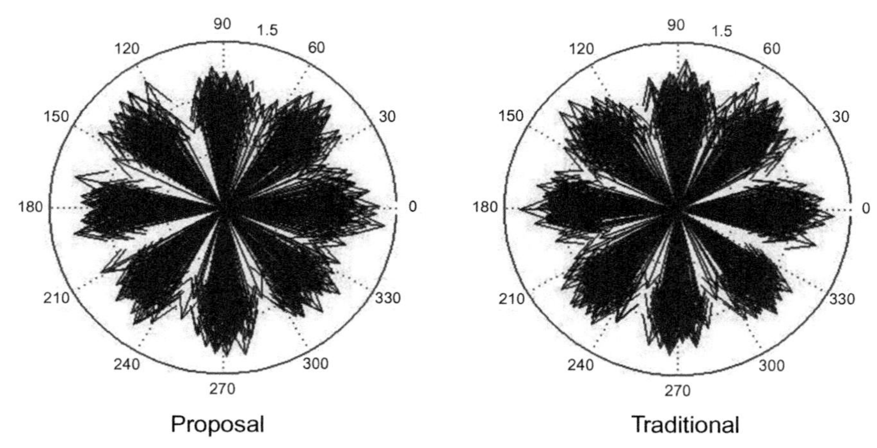

FIG. 11.8 Simulated DQPSK constellation.

The results were obtained by the "size" and "isequal" functions, the DQPSK modulation were analyzed through its constellation, by means of the "compass" function, which will display a compass graph with n arrows, and how the constellations will be PSKs, their representations of points will be radial. This feature, the graph format has a compass shape, where n is the number of elements in Z. The location of the base of each arrow will be its origin. In turn, the location of the tip of each arrow will be determined by the real and imaginary components of Z, relative to the constellation of the signal. In the same way and with the same intention, this will be used the diagram of the constellation.

Fig. 11.8 compares traditional methodology and the CBEDE methodology. In this context, the traditional methodology corresponds to a channel without discrete events.

As important as developing the methodology is to improvement of the transmission of a signal that shows better performance, as important is to make the know-how available to the academic community, as well as contributing to the area of study of the proposal as well as the theme that this chapter deals with.

Table 11.2 Amounts of memory consumption.

Simulation	Traditional	Proposal
1	207,1680	178,1797
2	191,9570	174,3555
3	215,2266	169,8281
4	183,5039	162,5273
5	247,6953	114,1602

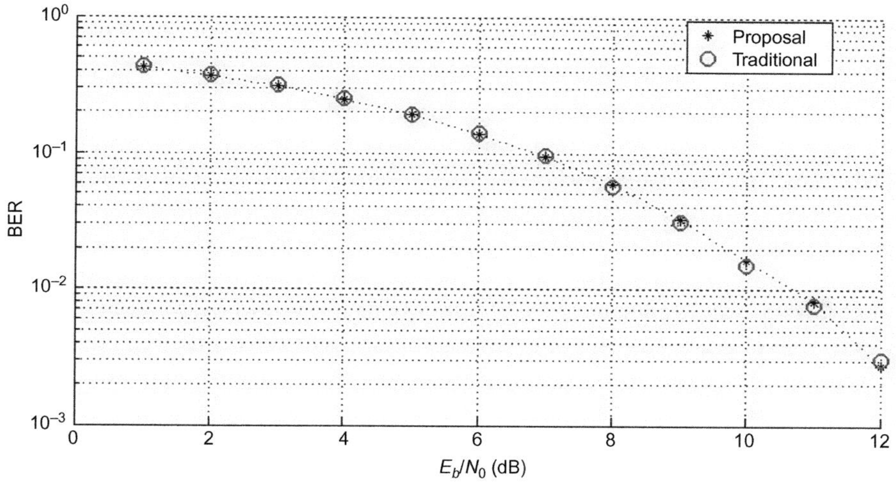

FIG. 11.9 BER between the models DQPSK.

The respective amounts of memory consumption shown in Fig. 11.7 found previously are in Table 11.2.

To analyze the relationship between the simulation methodology (proposed x traditional method) and the impact on the physical layer of the channel, scripts were made in the Matlab for processing of the graph BER. Fig. 11.9 shows the performance of the models during transmission with noise ranging from 0 to 40 dB.

This proposal brings a new approach to signal transmission. In this case, the transmission is performed in the discrete domain with the implementation of discrete entities in the bit generation process.

11.9 Discussion

The comparison between traditional methodology and CBEDE methodology shows results reaching up to 116.97% in the improvement of memory consumption through of the proposed methodology.

These results can be applied across a wide range of environments and branches of health. It is important to consider that recent research indicates the trend of growth in the population over 60 years is expected to triple in the next 40 years. Thus, in the world, there are 893 million people over the age of 60, but in the middle of the century, this number will increase to 2.4 billion [35]. On the basis of such data, it is impossible to neglect this fact.

In this way it is indispensable to create healthcare devices that help the elderly call for medical help in cases of extreme emergency such as infarction principle, stroke principle, and falls in the home, among other cases [36]. Devices with such functions are now in place, but the crucial factor for rapid care is the speed at which these data will be transmitted to help centers or relatives.

Note that the world's population, regardless of the age group, is subject to acquiring chronic diseases. Chronic diseases are those that do not present solutions in a short period of time (<3 months). Some types of chronic diseases accompany individuals throughout life, with no cure, only a stable and often controlled drug. Examples of these pathologies include diabetes, asthma, kidney diseases, liver diseases, and heart diseases, among others [37, 38]. It is important to emphasize that the number of cases of these types of chronic pathologies, mainly diabetes, have increased. It is estimated that there are around 1 million children and adolescents with diabetes in the world today [39].

In this case, with the creation of healthcare devices for the control of chronic diseases, glycemic rate results can be constantly sent to patient databases or to the doctor in charge. In this way, there is better control of these diseases, preventing them from reaching more serious complications [40].

Based on the realities of an aging population and high index of chronic diseases, the results presented by the CBEDE methodology demonstrate its great potential for employment in healthcare devices. The low memory consumption compared to traditional data transmission methods can strongly contribute to the transmission of large amounts of data streams of patients with chronic diseases, while the speed in the transmission of data by the CBEDE methodology can be seen as a crucial factor for emergency care of the elderly.

Currently, medical areas have used these telecommunications areas with the main objective of organizing scheduling queues for appointments, exams, and surgeries. However, developing countries such as Brazil suffer from large waiting lines. In some cases, patients with chronic diseases may die during the waiting time. Much of this delay is due to the use of old methodologies that consume much of the computational memory, accompanied by slowness in the transmission of medical data. As hospitals, basic health units, and specialized centers often work with large daily data flow, systems are subject to crashes and even data loss [5, 7, 11].

The CBEDE methodology demonstrates great potential to interconnect hospitals, physicians, and patients to find the necessary medical service in less time.

DES is an effective tool to approach a wide variety of healthcare issues, and this technique has been used to model concepts with a high level of abstraction in a system, such as patients, nurses, doctors; can be applied from the exchange of emails on a clinical server, even the transmission of data packets between devices connected in a hospital network; and also uses the queuing concept, which can be used to manage patient data, medical staff, or even emergency departments, intensive care units, surgical procedures, outpatient clinics, and the entire extent of a healthcare system.

11.10 Conclusion

This proposal brings a new approach for signal transmission. In this case, the transmission is performed in the discrete domain with the implementation of discrete entities in the bit generation process. This study aims to increase the information capacity for healthcare systems.

Because the differential of this research is the use of discrete events applied in the physical layer of a transmission medium, the bit itself, and this a low-level of abstraction, the results show better computational performance related to memory utilization related to the compression of the information, showing an improvement reaching up to 116.97%.

One of the applications of the CBEDE model can be seen in the following aspect: patients awaiting consultation with a medical specialist, for example, a cardiologist, in the context of a system interconnected between hospitals (which nowadays some hospitals already interconnect their data with others). However, the systems currently used are slow and consume a lot of system memory, facilitating crashes, with exchange patient data and medical consultations with each other, so such scheduling of consultations may become more effective.

This demonstrates that the CBEDE has great potential in the improvement of the hospital service's potential of improvement of already existing processes and can increase the performance of communication response between all the devices in the hospital system, because the flow of data will consume fewer resources and, therefore, improve the interactions between doctor and patient.

References

[1] M. Jaxson, The Oxford Handbook of the History of Medicine, Oxford University Press, 2011.

[2] C. Altensterter, Medical Devices: European Union Policymaking and the Implementation of Health and Patient Safety in France, Taylor & Francis, 2017.

[3] R.F. Mould, A Century of X-Rays and Radioactivity in Medicine With Emphasis on Photographic Records of the Early Years, Taylor & Francis, 2018.

[4] S. Barello, S. Triberti, G. Graffigna, C. Libreri, S. Serino, J. Hibbard, G. Riva, eHealth for patient engagement: a systematic review, Front. Psychol. 6 (2016).

[5] R.J. Jacobs, Q.L. Jennie, L.O. Raymond, C. Joshua, A systematic review of ehealth interventions to improve health literacy. Health Inform. J. 22 (2) (2016) 81–98, https://doi.org/10.1177/1460458214534092.

[6] G. Ruggeri, O. Briante, A framework for IoT and E-Health systems integration based on the social Internet of Things paradigma, in: 2017 International Symposium on Wireless Communication Systems (ISWCS), Bologna, 2017, pp. 426–431.

[7] L.S. Wilson, A.J. Maeder, Recent directions in telemedicine: review of trends in research and practice, Healthc. Inform. Res. 21 (4) (2015) 213–222.

[8] M.M. Ward, M. Jaana, N. Natafgi, Systematic review of telemedicine applications in emergency rooms, Int. J. Med. Inform. (2015).

[9] J.M.S.V.M. Maldonado, B.C. Alexandre, Telemedicine: challenges to dissemination in Brazil, Cadernos Saúd. Públ. 32 (14) (2016) 1–12.

[10] D.C.M.T.n.B. Saade, Nova Regulamentação Incentiva Pesquisa e Inovação em Soluções Seguras para Saúde Digital, J. Health Inform. 1 (2019) 11.

[11] G.W.D.S. Campos, R. Bedrikow, J.A. Santos, L.S.V. Terra, J.A. Fernandes, F.T. Borges, Right to health: is the Brazilian National Health System (SUS) at risk? Interface-Comun. Saúd. Educ. 20 (2016) 261–266.

[12] T.M.G. Menicucc, L.A. Costa, J.A. Machado, Pacto pela saúde: aproximações e colisões na arena federativa, Ciência Saúd. Colet. 23 (2018) 29–40.

[13] E.P. Tozer, Broadcast Engineer's Reference Book, first ed., FOCAL PRESS, 2012.

[14] J.G. Proakis, Digital Communications, in: McGraw-Hill, fifth ed., 2008.

[15] A. Rama-Krishna, A.S.N. Chakravarthy, A.S.C.S. Sastry, Variable modulation schemes for AWGN channel based device to device communication. Indian J. Sci. Technol. 9 (20) (2016) https://doi.org/10.17485/ijst/2016/v9i20/89973.

[16] R.L. Freeman, Telecommunication System Engineering, fourth ed., John Wiley & Sons, 2004.

[17] L.W. Couch, Digital and Analog Communication Systems, eighth ed., Prentice Hall, 2013.

[18] M.A. Helal, Hybrid System Dynamics-Discrete Event Simulation Approach to Simulating the Manufacturing Enterprise, PhD Thesis, Department of Industrial Engineering and Management Systems, College of Engineering and Computer Science, University of Central Florida, 2008.

[19] T.B. Brito, E.F.C. Trevisan, R.C. Botter, A conceptual comparison between discrete and continuous simulation to motivate the hybrid simulation technology, in: S. Jain, R.R. Creasey, J. Himmelspach, K.P. White, M. Fu (Eds.), Proceedings of the 2011 Winter Simulation Conference, 2011, pp. 3915–3927.

[20] R. Padilha, Proposta de Um Método Complementar de Compressão de Dados Por Meio da Metodologia de Eventos Discretos Aplicada Em Um Baixo Nível de Abstração, Dissertação (Mestrado em Engenharia Elétrica)—Faculdade de Engenharia Elétrica e de Computação, Universidade Estadual de Campinas, Campinas, SP, Brazil, 2018.

[21] R. Padilha, B.I. Martins, E. Moschim, Discrete event simulation and dynamical systems: a study of art, in: BTSym'16, Campinas, SP, Brazil, 2016.

[22] F. Semchedine, L. Bouallouche-Medjkoune, S. Moad, R. Makhloufi, D. Aïssan, Discrete events simulator for wireless sensor networks, in: 3rd International Workshop on Verification and Evaluation of Computer and Communication Systems, VECoS'2009, 2009.

[23] L. Pomante, A. Spinosi, M. Mostafizur, R. Mozumdar, S.O.L. Lavagno, An Extended Framework for the Development of WSN Applications, IEEE UltraModern Telecommunications and Control Systems and Workshops (ICUMT), 2010.

[24] Y. Lee, Event-centric test case scripting method for SOA execution environment, in: 2011 6th International Conference on IEEE Computer Sciences and Convergence Information Technology (ICCIT), 2011.

[25] E.N. Gomes, M.S.R. Fernandes, C.A.V. Campos, A.C. Viana, Um Mecanismo de Remoção de Mensagens Obsoletas para as Redes Tolerantes a Atrasos e Interrupções, CSBC, 2012.

[26] K. An, A. Gokhale, Model-driven performance analysis and deployment planning for real-time stream processing, in: 19th IEEE Real-Time and Embedded Technology and Applications Symposium, 2013.

[27] M. Artuso, H.L. Christiansen, Modeling and event-driven simulation of coordinated multi-point in LTE-advanced with constrained Backhaul. in: A. Tolk, S.Y. Diallo, I.O. Ryzhov, L. Yilmaz, S. Buckley, J.A. Miller (Eds.), Proceedings of the 2014 Winter Simulation Conference: Exploring Big Data Through Simulation. IEEE, 2014, pp. 3131–3142, https://doi.org/10.1109/WSC.2014.7020150.

[28] Mediouni N, Abid S.B., Kallel O., Hasnaoui S. High level NoC modeling using discrete event simulation. In: 2015 IEEE 10th International Design & Test Symposium (IDT).

[29] Y. Zhang, Y. Yang, H. Gu, Y. Lu, Stability analysis of NCS under PID control, in: IEEE Control and Decision Conference (CCDC), 2016.

[30] B. Zhao, F. Lin, C. Wang, X. Zhang, P.M. Polis, Y.L. Wang, Supervisory control of networked timed discrete event systems and its applications to power distribution networks, IEEE Trans. Control Netw. Syst. 4 (2) (2017).

[31] MATLAB and Communications System™ Release, The MathWorks, Inc, Natick, MA, United States, 2014.

[32] MATLAB and DSP System™ Release, The MathWorks, Inc, Natick, MA, United States, 2014.

[33] MATLAB and SimEvents® Release, The MathWorks, Inc, Natick, MA, United States, 2014.

[34] MATLAB and Simulink® Release, The MathWorks, Inc, Natick, MA, United States, 2014.

[35] B. Hon, R. John, D.E. Bloom, Towards a comprehensive public health response to population ageing, Lancet (Lond., Engl.) 385 (9968) (2015) 658.

[36] M.S. Bernardes, C.S. Santana, O uso de aparelhos de monitoramento à saúde por idosos com condições crônicas no domicílio, Rev. Facul. Med. Ribeirão Preto Hosp. Clín. 49 (Suppl. 2) (2016) 15.

[37] P.R. Alcalde, G.M. Kirsztajn, Expenses of the Brazilian Public Healthcare System with chronic kidney disease, J. Bras. Nefrol. São Paulo 40 (2) (2019) 122–129.

[38] S. Hamine, E. Gerth-Guyette, D. Faulx, B.B. Green, A.S. Ginsburg, Impact of mHealth chronic disease management on treatment adherence and patient outcomes: a systematic review, J. Med. Internet Res. 17 (2) (2015).

[39] S. Saydah, G. Imperatore, Y. Cheng, L.S. Geiss, A. Albright, Disparities in diabetes deaths among children and adolescents—United States, 2000–2014, MMWR—Morb. Mort. Wk. Rep. 66 (19) (2017) 502.

[40] R. Iljaž, A. Brodnik, T. Zrimec, I. Cukjati, E-healthcare for diabetes mellitus type 2 patients—a randomised controlled trial in Slovenia. Zdravst. Varst. 56 (3) (2017) 150–157, https://doi.org/10.1515/sjph-2017-0020.

Further reading

A.E. Bone, et al., What is the impact of population ageing on the future provision of end-of-life care? Population-based projections of place of death, Palliat. Med. 32 (2) (2018) 329–336.

12 Transferable approach for cardiac disease classification using deep learning

P. Gopika, V. Sowmya, E.A. Gopalakrishnan, K.P. Soman
CENTER FOR COMPUTATIONAL ENGINEERING AND NETWORKING (CEN), AMRITA SCHOOL OF ENGINEERING, AMRITA VISHWA VIDYAPEETHAM, COIMBATORE, INDIA

12.1 Introduction

Cardiovascular disease is caused due to irregular rhythms and affects heart muscles, valves, and blood vessels. According to a survey of cardiovascular diseases in India [1], myocardial infarction and arrhythmia are the predominant cardiac diseases affecting India. Atrial fibrillation is another type of cardiac disease affecting 12 million North Americans and Europeans [2]. Due to this, we considered cardiac diseases like arrhythmia, myocardial infarction, and atrial fibrillation for this study. Cardiac arrhythmia is a group of conditions in which the heartbeat is irregular, accelerated, or slow. Arrhythmia is classified according to duration of heartbeat. Myocardial infarction is a cardiac condition where blood flow to the heart muscles is disrupted, which leads to full or partial blockage of the coronary arteries. Atrial fibrillation is another cardiac condition, which arises due to disorder of the electrical system of the heart. It is a kind of arrhythmia, which is characterized by irregular heartbeat.

Globally, millions of people are affected by cardiac diseases. Nearly one-third of mortality is due to cardiovascular disease [1]. In the case of cardiac diseases, the correct diagnosis at an early stage is very important. Therefore, paramount importance must be given for a fast and efficient diagnosis technique. Disease diagnosis is considered as an effort to classify a person's health condition into specific grades, which can assist the physicians for treatment. In the literature for the automatic classification of cardiac diseases, many studies explored using a conventional machine learning algorithm.

One of the pathological measures to diagnose cardiovascular disease is the electrocardiogram (ECG). ECG is the process of recording the electrical activity of the heart over a period of time. An ECG signal consists of P, QRS, T, and U components, which are known as features [3]. These features together define one ECG beat, which is known as one cardiac cycle. The P wave in the ECG signal denotes the atrial contraction, QRS wave

Deep Learning Techniques for Biomedical and Health Informatics. https://doi.org/10.1016/B978-0-12-819061-6.00012-4

denotes depolarization of the ventricles, and T wave indicates repolarization of the ventricles. The characteristic of the ECG signal depends on the functionality of the heart [3].

Most of the conventional machine learning approaches involves three different phases like preprocessing, feature extraction, and feature normalization. Preprocessing phase involves cleaning the signal (denoising). Denoising techniques used in the literature includes low-pass linear phase filter, linear phase high-pass filter, bandpass filter, fast Fourier transform (FFT), adaptive filter, and notch filter [3]. In the second phase, disease-specific handcrafted features are extracted by applying many statistical approaches [3]. The handcrafted features extraction includes computation of RR interval, ST interval, PR interval, QT interval, detection of QRS wave, R wave, P wave, and T wave. Some of the feature extraction techniques used in the literature are continuous wavelet transform (CWT), discrete wavelet transform (DWT), discrete Fourier transform (DFT), discrete cosine transform (DCT), principal component analysis (PCA), Pan-Tompkins algorithm, and Daubechies wavelet [3]. In the literature, the extracted features are normalized using standard deviation and z-score techniques [3]. These handcrafted features are fed to conventional machine learning algorithms like support vector machine (SVM), multilayer perceptron, K-nearest neighbor, quantum neural network, radial basis functional neural network, fuzzy clustering neural network, and decision trees for the final classification task [3].

In recent years, deep learning is a subset of machine learning, which gained popularity due to its high performance and effectiveness. Deep learning techniques outperformed the conventional machine learning techniques by extracting the required features itself. Deep learning has less processing overhead and has high accuracy, when compared to the traditional methods. CNN has been used in beat classification, coronary artery disease detection, arrhythmia detection, and myocardial infarction classification [4]. In one work [5], one-dimensional ECG signal is beat segmented and converted to two-dimensional image. They have used CNN, VGG net, and AlexNet for the classification. CNN was able to achieve a better result with data augmentation and VGG net shows the unfavorable result. In another work [6], they used recurrent networks: RNN, LSTM, and GRU for the classification of atrial fibrillation. The previously mentioned work avoids trivial denoising of the data and uses raw data, which reduces complexity and enables real-time detection. It motivated us to use the recurrent networks for the study. However, Sujadevi et al. [6] tested the robustness of the model with only 10 samples. In our study, we considered 19,430 instances to ensure the robustness of the model. Recurrent networks are also used in ECG-based biometrics, sleep apnea detection, and beat classification. Deep learning requires a large amount of data to train the network, which increases the computational time. To reduce the computational time and burden of training the network from scratch, a transfer learning approach is introduced.

Transfer learning is the methodology where the model trained for one particular task is used for another task. In Salem et al. [7], the one-dimensional ECG signal is converted to a two-dimensional image using spectrogram, and the standard dense net architecture is used for the classification. In Kachuee et al. [8], ECG signal is directly used, and the residual CNN model trained for the arrhythmia is used for the classification of myocardial infarction. It was able to provide benchmark accuracy of 95% in the myocardial infarction classification. In the case of arrhythmia classification, benchmark accuracy of 94% was obtained. This motivated us to consider the residual skip CNN architecture for our study. A paper by Fazeli [9] shows that there is a performance improvement in the classification of myocardial infarction, when the model is trained from scratch using the same residual CNN used in Kachuee et al. [8]. All these existing works are disease-specific approaches.

The proposed work aims for the single best architecture for the classification of the diseases arrhythmia, myocardial infarction, and atrial fibrillation by transferable approach. The network parameters considered for the classification of a particular disease are considered the same for the classification of other cardiac diseases, which we denote as transferable approach. To achieve the single best architecture, we consider the previously mentioned disease-specific benchmark architectures like RNN, LSTM, GRU, and residual CNN and retrained the model by transferable approach. We analyzed the performance of all the models for all the three diseases based on precision, recall, and F1 score.

This chapter is organized as follows: Section 12.2 describes the proposed work, Section 12.3 describes the background information of the network architecture, Section 12.4 describes the architectural details, Section 12.5 describes experimental analysis, and the chapter is concluded in Section 12.6.

12.2 Proposed work

12.2.1 Dataset description

12.2.1.1 Arrhythmia

In this work, we used Kaggle arrhythmia ECG heartbeat categorization database as a data source [9]. This database is derived from the famous MIT-BIH Arrhythmia dataset. The database is resampled with the sampling frequency of 125 Hz. It is a preprocessed and beat-segmented database that has 109,446 ECG beats of 47 different subjects. Each instance is annotated by the cardiologist in accordance to the standard of Association for the Advancement of Medical Instrumentation (AAMI) EC57 [10]. It has five different categories, and the corresponding mapping of each beat according to the AAMI standard is shown in Table 12.1.

12.2.1.2 Myocardial infarction

In this work, we used the publically available Kaggle myocardial infarction ECG heartbeat categorization database [9]. This database is derived from PTB Diagnostics database of 200 subjects. Among 200 subjects, 148 subjects are diagnosed as myocardial infarction and 52 subjects are diagnosed as healthy. The database contains 14,552 ECG beats of

Table 12.1 Summary of beat mappings in accordance with AAMI standard [10]

Category	Annotation
N	Normal
	Left/right bundle branch block
	Atrial escape
	Nodal escape
S	Supra-ventricular premature
	Nodal premature
	Aberrant atrial premature
	Atrial premature
V	Premature ventricular contraction
	Ventricular escape
F	Fusion of ventricular and normal
Q	Paced
	Fusion of paced and normal
	Unclassifiable

two categories. The two categories are normal and myocardial infarction. The sampling frequency of this database is 125 Hz.

12.2.1.3 Atrial fibrillation

In this work, we used the data available in the PhysioNet Challenge 2017 [11]. We considered 5154 subjects of normal and 771 atrial fibrillation signals. Each ECG signal is 60 seconds long and sampled at 250 Hz.

12.2.2 Methodology

The input to the deep learning architecture is the ECG signal of the three cardiovascular diseases, namely arrhythmia, myocardial infarction, and atrial fibrillation. We considered four benchmark architectures for the study. Residual skip convolutional neural network (RCNN) is the benchmark architecture that exists in the literature for the arrhythmia and myocardial infarction classification. RNN, LSTM, and GRU are the benchmark architectures that exist in the literature for the atrial fibrillation classification.

The hyperparameter determines the performance of the model. The batch size, number of epoch, number of hidden layers, number of neurons, and the number of learning rate are known as hyperparameters. All these hyperparameters tuned for the particular disease classification are kept the same, and the model is trained to classify cardiac diseases. This approach is defined as transferable approach.

We evaluated the performance of the model using the standard evaluation metrics like precision, recall, and F1 score. The performance comparison was performed among the CNN, RNN, LSTM, and GRU for all the considered cardiac diseases to find the best single

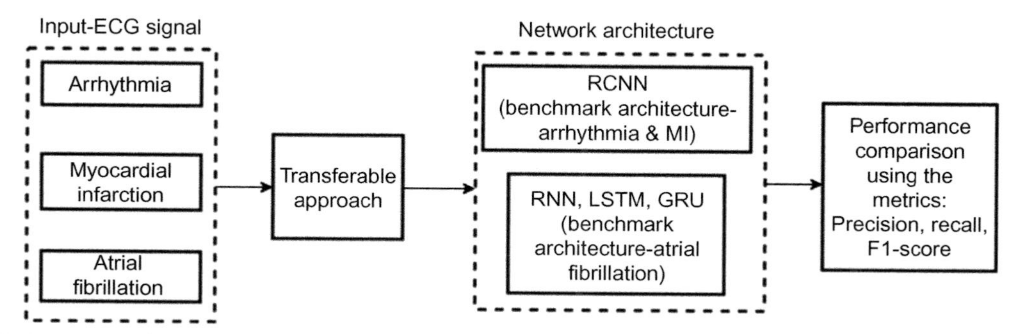

FIG. 12.1 Proposed methodology for cardiac disease classification.

architecture and to verify the concept of a transferable approach in deep learning. Fig. 12.1 demonstrates the proposed methodology for the cardiac disease classification.

Classical machine learning algorithms are disease-specific approaches where disease-specific features are extracted and the performance comparison is done to validate that they are not transferable. We have used different classical machine learning algorithms like Naive Bayes, K-nearest neighbor, SVMs with linear and RBF kernels, AdaBoost, decision tree, and logistic regression.

12.3 Background

12.3.1 Recurrent neural network

RNNs are a powerful neural network, which is popular because of its robustness. It was initially used for time-series data modeling but has wide variety of applications like machine translation, speech recognition, generating sequences, and text prediction.

RNN is similar to a feed forward network (FFN) with an additional cyclic loop. This cyclic loop carries the information from one time-step to another. Cyclic loops are short-term memory which stores and retrieves the past information over time scales. RNN learns the temporal patterns and calculates the current time-step from its previous state and the present state.

To train the RNN, it is converted to FFN by unfolding or unrolling. This unfolded RNN will not have cyclic connections, and it known as deep feed forward neural network (DFFN). It consists of input sequences of length N and N hidden layers as shown in Fig. 12.2. The network parameters are shared across time-steps. It calculates the gradient of the current time-step from the present and past time-steps. This is known as back propagation through time (BPTT). The vanishing gradient problem arises when RNN has to learn long-term dependencies in time-steps. It is due to the fact that the gradient vector either grows or decays exponentially when propagating through many layers of RNN to learn long-term dependencies in time-steps. This vanishing gradient problem is addressed in LSTM.

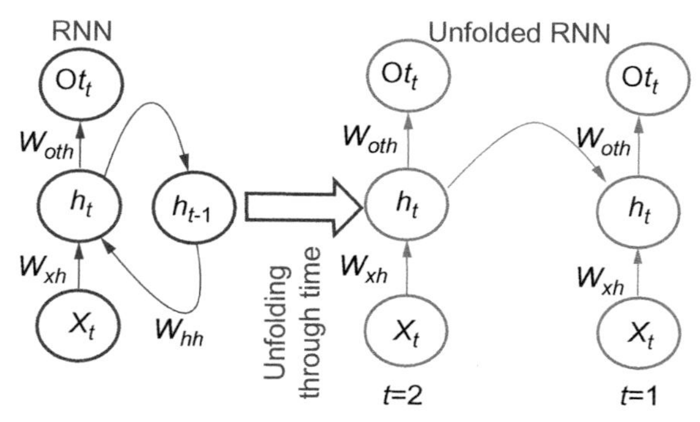

FIG. 12.2 RNN and unfolded RNN across time-steps [6].

12.3.2 Long short-term memory

LSTM was introduced by the researchers to solve the vanishing gradient problem addressed by traditional RNN. Unlike traditional RNN, it has an additional memory block that is an subnet of LSTM architecture. A memory block consists of one or more memory cells with a pair of adaptive multiplicative gates as input and output gates, as shown in Fig. 12.3. A memory block stores the information. The input and the output flow of information to the memory cell are controlled by the input and output gates. Based on the input and output gates, it updates the information across time-steps. In addition to the input gate and the output gate, it has the forget gate, which forgets the past value at a specific time-step. All gates have peephole connections to the memory cell to learn precise timing of the output.

In general, LSTM accepts sequences of arbitrary length $x = (x_1, x_2, \ldots x_{T-1}, x_T)$ to LSTM architecture as input that estimates an output sequence $o = (o_1, o_2, o_3, \ldots o_T - 1, o_T)$ with

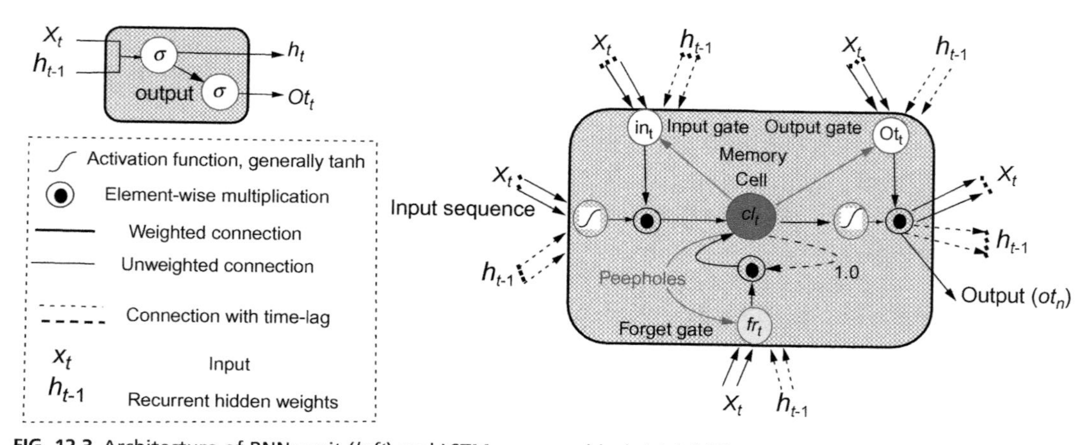

FIG. 12.3 Architecture of RNNs unit (*left*) and LSTMs memory block (*right*) [6].

continuous write, read, and reset operations by three multiplicative units (input, output, and forget gates) on a memory cell in an iterative manner.

12.3.3 Gated recurrent unit

An alternative to LSTM networks to reduce computations is GRU. GRU looks simpler as it combines input gate (*in*) and forget gate (*fr*) to form a new gate called update gate (*in_fr*). The update gate balances the state between the previous activation and the candidate activation (*h*) without any output activations. In addition to this, it does not have peephole connections. The forget gate resets the previous state (*cl*). Fig. 12.4 represents the architecture of GRU where information propagates through the dashed line, and the solid line represents modifying the flows of information.

12.3.4 Residual convolutional neural network

CNN has revolutionized the field of computer vision [4]. Due to its inexplicable efficacy, it is also applied in the biomedical field. The main block of CNN is convolution. Convolution is the mathematical operation that performs element-wise multiplication on input data and the filter value to then produce a feature map. The convolution is performed by striding the filter across the entire input; each time it performs element-wise multiplication and sums the result onto the feature map. The number of neurons determines the size of the filter, which is known as receptive field. Stride determines the number of steps the filter moves each time. The result of the convolution operation is based on the mathematical equation,

$$O = (I - F + 2P)/s + 1 \tag{12.1}$$

where O denotes the output length, I denotes the input length, F denotes filter size, P denotes padding size, and s denotes number of stride. To match the input size and the output size, padding is used. In pads, zeros are added around the input border. After convolution layer, the results are applied to the activation layer for the nonlinear transformation. The most commonly used activation function is rectified linear units (ReLU). ReLU

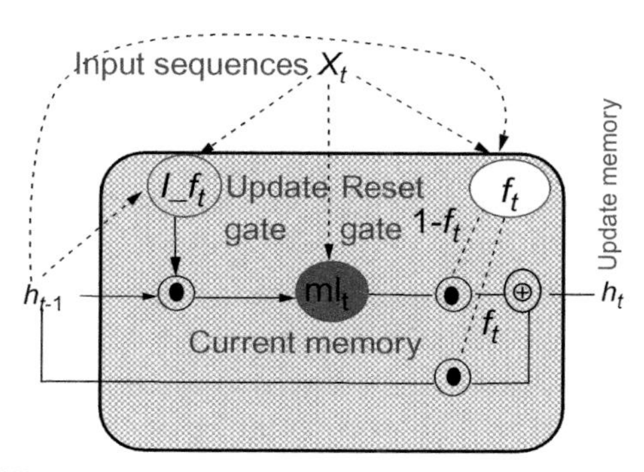

FIG. 12.4 A memory block of GRU.

works based on mathematical function, $F(X) = \max(0, X)$. This function is applied across the entire convolution output. To make computation easier, pooling layers are used. Max pooling function is most commonly used. It basically takes stride of length K. The stride is applied to the input and outputs the maximum value in that subregion. There are also some other pooling operations like average pooling and min pooling. The pooling layer has two main advantages: First, the computation is reduced by dimension reduction; and the second advantage is that it controls overfitting. Fully connected layer is the next layer, where each neuron in the fully connected layer is connected to all activations in the previous layer. Output layer has a set of neurons same as the number of classes. The residual connection means skipping one or more layers and connecting to the next layer. It trains the network deeper when compared to the traditional network with the similar number of network parameters. The deeper network faces a degradation problem during training, whereas the residual block overcomes this problem. It trains the networks deeper with the same number of parameters.

12.3.5 Classical machine learning algorithms

12.3.5.1 Feature extraction

Arrhythmia and myocardial infarction is preprocessed, and the corresponding beat-segmented data is available in the Kaggle. In the case of atrial fibrillation, we used feature extraction technique proposed by Andreotti et al. [12]. They used 10th-order bandpass Butterworth filters with cut-off frequencies 5–45 Hz (narrow band) and 1–100 Hz (wide band) for filtering. They used four well-known QRS detectors such as gqrs (WFDB toolkit), Pan-Tompkins, maxima search, and matched filtering to detect QRS wave. A consensus based on kernel density estimation is output as final decision. They have extracted 8 time-domain features, 8 frequency domain features, 22 morphological features, 95 nonlinear features, and 36 signal quality features; in total, 169 features are extracted in a signal.

12.3.5.2 Classification algorithms

We use seven different classification algorithms for the classification of diseases. Decision tree approaches the problem in a structured way including the chance outcomes to obtain a logical conclusion. Logistic regression is the statistical way of modeling the outcomes by estimating the probabilities using logistic function. AdaBoost is an ensemble algorithm that has many weaker models independently trained, and the probabilities are combined to make an overall decision. K-nearest neighbor (KNN) is an instance-based learning algorithm used for classification. The class with the highest frequency from the K-most similar instances will be the output. SVM is another popular algorithm used for classification. It is a discriminate classifier that classifies by finding a hyperplane. It works by tuning parameters like kernel, regularization, etc. Nave Bayes is a simple and powerful classifier that works based on Bayes algorithm. The performance of the classifier varies based on its application [13].

12.4 Network architecture

12.4.1 Recurrent networks

The layers in recurrent network architectures are as follows:

- The first layer is the recurrent layers: RNN, LSTM, and GRU, which are discussed in Section 12.3.
- The second layer is the dense layer. Dense layer is a classical fully connected layer that connects each input node to each output node. It uses sigmoid activation function. The sigmoid activation function is used where we have to predict the output probability of the models. The role of activation function is to map it to nonlinear operation.
- The last layer is the output layer where the number of neurons in the output layer denotes the number of classes. The output layer uses softmax activation function. Softmax activation function is a logistic activation function mainly used for multiclass classification.

The network architecture for the proposed methodology is shown in Fig. 12.5. The network parameters used in the proposed work are given in Table 12.2.

The input layer shape of arrhythmia and myocardial infarction is 187 and for atrial fibrillation is 169. The network architecture has 100 neurons in the dense layer. The output layer neurons are according to the number of classes. It contains two neurons for the diseases myocardial infarction and atrial fibrillation, and in the case of the arrhythmia, it has five neurons.

12.4.2 Residual convolution neural network

The residual convolution neural network (RCNN) has 13 weight layers. Each convolution filter size is 32. This network architecture has five residual blocks. The residual blocks are shown using dotted lines in Fig. 12.6. It has five pooling layers to reduce the dimension. ReLU is activation function used at all layers except the output layer. ReLU has an advantage

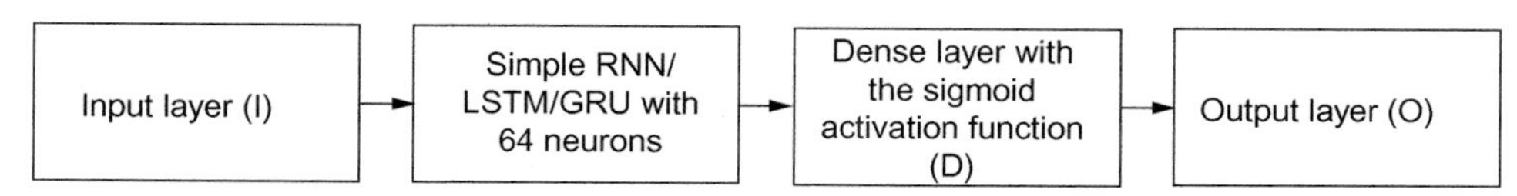

FIG. 12.5 Flow diagram of RNN, LSTM, and GRU architectures for cardiac disease classification.

Table 12.2 The network parameters used in the recurrent architectures

Parameters	Arrhythmia	Myocardial infarction	Atrial fibrillation
I	(None, 187)	(None, 187)	(None, 169)
D	100	100	100
O	5	2	2

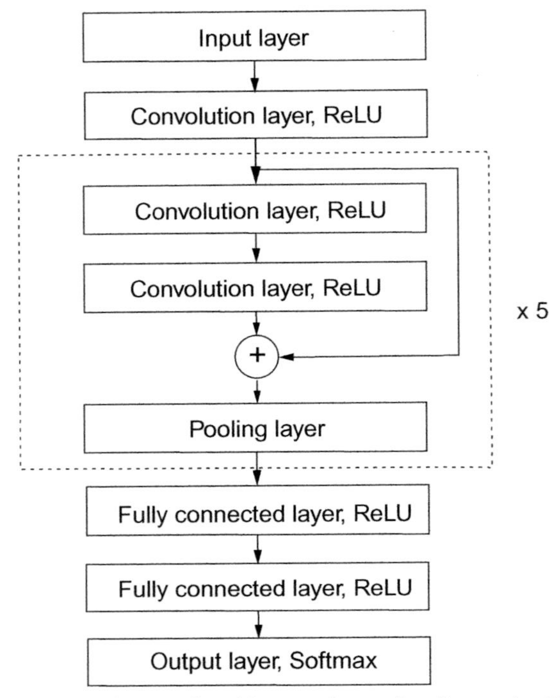

FIG. 12.6 Residual convolutional neural network architecture for cardiac disease classification.

that it avoids vanishing gradient problem [8]. The final output layer has number of neurons same as that of the number of class. It has five neurons in the case of arrhythmia, and two in the case of myocardial infarction and atrial fibrillation. Softmax activation function is used at the output layer. The softmax activation function returns the output probabilities for each class, and maximum probability value is considered as target class. The summary of architectural details for all the three considered diseases are shown in Table 12.3.

Table 12.3 The architectural details of residual skip convolution neural network for cardiac disease classification using ECG signal

	Arrhythmia	Myocardial infarction	Atrial fibrillation
Layers		Layer size	
Input	187, 1	187, 1	169, 1
Convolution	183, 32	183, 32	165, 32
Convolution	183, 32	183, 32	165, 32
Convolution	183, 32	183, 32	165, 32
Fully connected	None, 32	None, 32	None,32
Fully connected	None, 32	None, 32	None, 32
Output	5	2	2

12.5 Experimental results

12.5.1 Train/test split

We consider the train/split as in the literature [4, 5]. Table 12.4 shows the summary of train/test split considered for specific architectures for all three diseases. The ratio of train/test split in the case of:

- RCNN considers 800 samples from each class for testing the RCNN model. In total, 4000 samples are considered for testing and the remaining ECG samples are considered for training the RCNN model cardiac disease classification.
- Recurrent architectures like RNN, LSTM, gated recurrent neural network (GRU) consider a ratio of 20% from each class for testing and 80% from each class is considered for training the recurrent models for the cardiac disease classification.

Table 12.4 The details of number of signals used to train and test the classification of three different cardiac diseases

Architecture	Disease	Class	Category	Train	Test
CNN	Arrhythmia	0	N	89,789	800
		1	S	1979	800
		2	V	6436	800
		3	F	2412	800
		4	Q	7239	800
		Total		107,855	4000
	Myocardial infarction	0	N	9706	800
		1	A	3246	800
		Total		12,952	1600
	Atrial fibrillation	0	N	55,442	800
		1	A	7725	800
		Total		63,167	1600
Recurrent networks RNN/LSTM/GRU	Arrhythmia	0	N	72,471	18,118
		1	S	2223	556
		2	V	5788	1448
		3	F	641	162
		4	Q	6431	1608
		Total		87,554	21,892
	Myocardial infarction	0	N	8405	2101
		1	A	3237	809
		Total		11,642	2910
	Atrial fibrillation	0	N	39,369	16,873
		1	A	5968	2557
		Total		45,337	19,430

12.5.2 Hyperparameters

Hyperparameters have high impact in determining the model's performance. The benchmark architectures have the most optimum parameters for the specific disease. To validate our proposed transferable approach, we considered the same hyperparameters trained for the particular disease. The model is retrained for the remaining diseases with the same optimum hyperparameters.

Batch size, learning rate, number of epochs, and the hidden layers are the network or hyperparameters considered for the analysis. Table 12.5 denotes the summary of hyperparameters considered in training the recurrent networks (RNN, LSTM, GRU) [6] and RCNN [9].

12.5.3 Evaluation metrics

We considered certain evaluation metrics such as classification accuracy, loss, precision, recall, F1 score, area under curve, and confusion matrix to evaluate the performance of the model. Each metric has its own efficacy, as described in the following:

- Confusion matrix describes the complete performance of the model.
- F1 score describes the preciseness and robustness of the classification model.
- Precision and recall measures the relevance of the model.
- ROC-AUC curve describes the specificity and sensitivity of the model.
- Loss is a continuous value that describes the model's behavior after every single iteration of optimization. The better the classification model, the lower the loss value.

12.5.3.1 Transferable approach for arrhythmia classification

We compared the performance of arrhythmia using the recurrent networks and the RCNN. After the data augmentation for class 4, the residual skip CNN was able to give precision, recall, and F1 score of 0.94 and 0.95. The recurrent network could perform better than the existing residual skip CNN without any data augmentation and less training samples. The detail comparison of the scores for the arrhythmia classification is shown in Table 12.6. It is evident from the analysis that the performance of LSTM and GRU is high. GRU is computationally efficient. Table 12.7 shows the confusion matrix of the deep learning architectures for the arrhythmia classification. Confusion matrix scores are based on macro average method, which is the reason for the high class-wise scores in RCNN. But it is

Table 12.5 Hyperparameters considered in the architectures for the cardiac disease classification

Hyperparameters	Residual networks	RCNN
Epochs	1000	75
Learning rate	0.01	0.001
Batch size	4	500
Number of hidden layers	2	21

Table 12.6 Performance comparison of the deep learning architectures for the arrhythmia disease classification

Models		**RNN**	**LSTM**	**GRU**	**RCNN**
Accuracy (%)		96.88	97.86	97.94	94.25
Loss		0.11	0.13	0.12	0.087
ROC: AUC (macro average)		0.92	0.90	0.90	0.92
Number of parameters		16,389	64,581	48,517	50,853
Precision: Average		0.97	0.98	0.98	0.95
Precision	0	0.97	0.99	0.98	0.83
	1	0.91	0.85	0.88	0.99
	2	0.92	0.95	0.95	0.93
	3	0.82	0.78	0.82	0.99
	4	0.99	0.99	0.99	1.00
Recall: Average		0.97	0.98	0.98	0.94
Recall	0	0.99	0.99	0.99	0.99
	1	0.60	0.72	0.72	0.87
	2	0.90	0.94	0.94	0.96
	3	0.54	0.75	0.69	0.90
	4	0.92	0.97	0.97	0.99
F1 score: Average		0.97	0.98	0.98	0.94
F1 score	0	0.98	0.99	0.99	0.90
	1	0.71	0.77	0.79	0.93
	2	0.91	0.94	0.94	0.95
	3	0.65	0.79	0.75	0.94
	4	0.95	0.98	0.98	0.99

Table 12.7 Confusion matrix of deep learning architectures for the arrhythmia disease classification

Architecture		**Confusion matrix**					
RNN	Actual class	Predicted class					
			0	1	2	3	4
		0	**0.99**	0	0	0	0
		1	0.39	**0.58**	0.02	0	0
		2	0.08	0	**0.9**	0.01	0
		3	0.28	0	0.19	**0.54**	0
		4	0.07	0	0.01	0	**0.92**
LSTM		0	**0.99**	0	0	0	0
		1	0.26	**0.72**	0.02	0	0
		2	0.05	0	**0.93**	0.01	0
		3	0.15	0	0.08	**0.77**	0
		4	0.02	0	0	0	**0.97**
GRU		0	**0.99**	0	0	0	0
		1	0.33	**0.66**	0.01	0	0
		2	0.07	0	**0.91**	0.01	0
		3	0.2	0	0.1	**0.7**	0
		4	0.03	0	0	0	**0.9**

Continued

Table 12.7 Confusion matrix of deep learning architectures for the arrhythmia disease classification—cont'd

Architecture	Confusion matrix					
RCNN	0	**0.99**	0	0	0	0
	1	0.18	**0.81**	0.01	0	0
	2	0.02	0	**0.95**	0.02	0.01
	3	0.03	0	0.01	**0.96**	0
	4	0.01	0	0	0	**0.99**

Bold denotes the corrected predicted values.

Table 12.8 Performance comparison between different machine learning algorithms for arrhythmia disease classification

Algorithms	Accuracy (%)	Precision	Recall	F1 score
Decision tree	0.95	0.95	0.95	0.95
SVM (linear)	0.92	0.92	0.88	0.9
SVM (rbf)	0.88	0.79	0.88	0.88
AdaBoost	0.88	0.86	0.88	0.87
Logistic regression	0.9	0.87	0.9	0.87
KNN	0.96	0.96	0.96	0.96
Nanve Bayes	0.28	0.81	0.28	0.38

evident that recurrent networks could give comparable class-wise results although with class imbalance data.

Arrhythmia data is beat-segmented, and the preprocessed data is classified using eight different classical machine learning algorithms. We observed that machine learning algorithms like KNN and decision tree could give comparable performance to deep learning. Table 12.8 shows the comparison between machine learning algorithms, although the performance of the deep learning is high compared to the conventional machine learning algorithms.

12.5.4 Transferable approach for myocardial infarction classification

The RCNN performs better in the case of myocardial disease classification with more training samples. But recurrent networks could give comparable performance with residual skip CNNs with the less training samples. Table 12.9 shows the performance comparison of the deep learning architectures for the myocardial disease classification. It is evident from Table 12.9 that all the deep learning architectures performance are almost comparable. The data of myocardial infarction is beat-segmented, and the data is preprocessed.

Table 12.10 shows the confusion matrix of the deep learning architectures for the myocardial disease classification. It is evident from the confusion matrix that all the deep

Table 12.9 Performance comparison of the deep learning architectures for the myocardial disease classification

Models		**RNN**	**LSTM**	**GRU**	**RCNN**
Accuracy (%)		97.32	98.04	97.73	99.31
Loss		0.10	0.12	0.15	0.0407
ROC: AUC (macro average)		1.00	1.00	0.99	0.99
Number of parameters		16,194	64,386	48,322	50,574
Precision: Average		0.97	0.98	0.98	0.99
Precision	0	0.96	0.96	0.96	1.00
	1	0.98	0.99	0.98	0.99
Recall: Average		0.97	0.98	0.98	0.99
Recall	0	0.95	0.97	0.96	0.99
	1	0.98	0.99	0.98	1.00
F1 score: Average		0.97	0.98	0.98	0.99
F1 score	0	0.95	0.96	0.96	0.99
	1	0.98	0.99	0.98	0.99

Table 12.10 Confusion matrix of deep learning architectures for the arrhythmia disease classification

Architecture		**Confusion matrix**		
RNN	Actual class	Predicted class		
			0	1
		0	**0.94**	0.06
		1	0.02	**0.98**
LSTM		0	**0.95**	0.05
		1	0.02	**0.98**
GRU		0	**0.96**	0.04
		1	0.02	**0.98**
RCNN		0	**0.99**	0.01
		1	0.02	**0.98**

Bold denotes the corrected predicted values.

learning architectures (RNN, LSTM, GRU, and RCNN) classify the abnormality (myocardial infarction) with the classification accuracy of 98%. The misclassification rate in classifying normal ECG signal is less in GRU compared to RNN and LSTM, and is similar to existing RCNN. GRU gives almost best classification performance similar to the existing RCNN.

Table 12.11 shows the performance comparison of the different classical machine learning algorithms for the myocardial disease classification. Among the considered machine learning algorithms, decision tree gives comparable performance because the data is feature extracted. We observed from the experimental analysis that deep learning algorithms perform better compared to machine learning algorithms.

Table 12.11 Performance comparison between different machine learning algorithms for the myocardial disease classification

Algorithms	Accuracy (%)	Precision	Recall	F1 score
Decision tree	0.95	0.95	0.95	0.95
SVM (linear)	0.84	0.84	0.84	0.84
SVM (rbf)	0.73	0.53	0.73	0.61
AdaBoost	0.89	0.89	0.89	0.89
Logistic regression	0.82	0.82	0.82	0.82
KNN	0.92	0.92	0.92	0.92
Nanve Bayes	0.76	0.79	0.76	0.77

12.5.5 Transferable approach for atrial fibrillation classification

The recurrent networks are the existing benchmark architectures available in the literature for the atrial fibrillation classification using raw ECG data. The present study compares the performance of AF classification with feature-extracted ECG data using deep learning and machine learning algorithms.

The recurrent networks also perform well in the case of feature-extracted data shown in Table 12.12. We observed from Table 12.12 that recurrent networks have maximum accuracy. The residual skip CNN's performance is very low in the atrial fibrillation (AF) classification although with the more training samples. It may be due to the model overfitting. ROC:AUC range is higher for LSTM and GRU (0.95) compared to RNN and RCNN. GRU is computationally efficient compared to the LSTM. The F1 score of GRU is high (0.98) when compared to the other deep learning algorithms. Table 12.12 shows the performance comparison of the deep learning architectures for atrial fibrillation classification.

Table 12.12 Performance comparison of the deep learning architectures for the atrial fibrillation disease classification

Models		RNN	LSTM	GRU	RCNN
Accuracy (%)		94.84	94.76	96.47	50.00
Loss		0.13	0.22	0.13	8.0591
ROC: AUC (macro average)		0.94	0.95	0.95	0.32
Number of parameters		15,170	60,290	45,250	49,730
Precision: Average		0.95	0.95	0.96	0.25
Precision	0	0.95	0.98	0.98	0.50
	1	0.93	0.76	0.86	0.00
Recall: Average		0.95	0.95	0.98	0.50
Recall	0	0.99	0.96	0.87	1.00
	1	0.66	0.88	0.96	0.00
F1 score: Average		0.94	0.95	0.98	0.67
F1 score	0	0.97	0.97	0.87	0.00
	1	0.77	0.82	0.96	0.03

Table 12.13 Confusion matrix of deep learning architectures for the myocardial disease classification

Architecture		Confusion matrix		
RNN	Actual class	Predicted class		
			0	1
		0	**0.99**	0.01
		1	0.34	**0.66**
LSTM		0	**0.96**	0.04
		1	0.12	**0.88**
GRU		0	**0.98**	0.02
		1	0.13	**0.82**
RCNN		0	**1.00**	0.00
		1	1.00	**0.00**

Bold denotes the corrected predicted values.

Table 12.13 shows the confusion matrix of the deep learning architectures for atrial fibrillation classification. It is evident from the confusion matrix that LSTM and GRU could give better classification performance than RNN. It is observed from Table 12.13 that GRU accurately distinguishes the normal ECG signal (0.98). The abnormal (atrial fibrillation) signal classification performance is also high in the GRU (0.88) when compared to RNN and LSTM. Therefore, GRU is considered as the best model in the case of atrial fibrillation classification.

Table 12.14 shows the comparison of the classical machine learning algorithms for AF classification. It is evident from the table that KNN (0.95) and decision tree (0.96) gives comparable performance. We observed from the experimental analysis that the performance of the deep learning algorithms is high when compared to all the conventional machine learning algorithms.

12.5.6 Comparison of the performance for the proposed method against the existing benchmark results

The performance comparison of the proposed work against the existing system is shown in Table 12.15. It is evident from Table 12.15 that the proposed system has better precision, recall, and F1 score (0.98) compared to the existing system. In the case of atrial fibrillation,

Table 12.14 Performance comparison between different machine learning algorithms for atrial fibrillation disease classification

Algorithms	Accuracy (%)	Precision	Recall	F1 score
Decision tree	95	0.95	0.95	0.95
SVM (linear)	91.9	0.91	0.91	0.91
SVM (rbf)	87.8	0.79	0.87	0.83
AdaBoost	86.4	0.86	0.86	0.86
Logistic regression	89.6	0.88	0.89	0.88
KNN	97.4	0.96	0.96	0.96
Naive Bayes	76.5	0.90	0.76	0.80

Table 12.15 Performance comparison of the proposed work against the existing system

Disease	Work	Precision	Recall	F1 score
Arrhythmia	**Proposed**	**0.98**	**0.98**	**0.98**
	Existing [8]	0.95	0.94	0.94
Myocardial infarction	**Proposed**	**0.98**	**0.98**	**0.98**
	Existing [8]	0.95	0.95	0.95
Atrial fibrillation	**Proposed**	**0.96**	**0.98**	**0.98**
	Existing [6]	1	1	1

The proposed results are highlighted in bold.

the proposed system was able to achieve almost the score of the existing system. The existing system score may be high because they considered only 10 samples for testing the model.

12.6 Conclusion

The existing works for cardiac disease classification are based on the disease-specific approach. Unlike the existing works, the proposed work determines the single best architecture for the classification of the following cardiac diseases: arrhythmia, myocardial infarction, and atrial fibrillation. We achieved this analysis by comparing the benchmark architectures (CNN, RNN, LSTM, and GRU) available in the literature. In our analysis, the transferable approach denotes that hyperparameters fixed for particular disease classification is considered the same for the other diseases.

We observed that recurrent network, which is the benchmark architecture for atrial fibrillation classification, also performs better in the case of arrhythmia and MI. It is evident from the analysis that LSTM and GRU performs better when compared to RNN. GRU is computationally efficient compared to LSTM. Therefore, GRU is the single architecture for all three cardiac diseases considered in the study. Also, the recurrent networks are trained with fewer training samples compared to the existing residual CNN.

The main conclusions derived from the proposed work are as follows:

- A single architecture for different cardiac disease classification such as arrhythmia, myocardial infarction, and atrial fibrillation by transferable approach.
- Unlike the conventional machine learning algorithms, the deep learning algorithms are transferrable for different diseases classified using the same source of the signal.

As a future work, these architectures can be extended to the cardiac diseases that are diagnosed using different pathological measures such as phonocardiogram.

References

[1] D. Prabhakaran, P. Jeemon, A. Roy, Cardiovascular diseases in India: current epidemiology and future directions, Circulation 133 (16) (2016) 1605–1620.

[2] I. Savelieva, J. Camm, Update on atrial fibrillation: part I, Clin. Cardiol. 31 (2) (2008) 55–62.

[3] S.H. Jambukia, V.K. Dabhi, H.B. Prajapati, Classification of ECG signals using machine learning techniques: a survey, in: 2015 International Conference on Advances in Computer Engineering and ApplicationsIEEE, 2015, pp. 714–721.

[4] A. Krizhevsky, I. Sutskever, G.E. Hinton, Imagenet classification with deep convolutional neural networks, in: Advances in Neural Information Processing Systems2012, pp. 1097–1105.

[5] T.J. Jun, H.M. Nguyen, D. Kang, D. Kim, D. Kim, Y.H. Kim, ECG arrhythmia classification using a 2-D convolutional neural network, arXiv preprint arXiv:1804.06812 (2018).

[6] V.G. Sujadevi, K.P. Soman, R. Vinayakumar, Real-time detection of atrial fibrillation from short time single lead ECG traces using recurrent neural networks, in: The International Symposium on Intelligent Systems Technologies and ApplicationsSpringer, 2017, pp. 212–221.

[7] M. Salem, S. Taheri, J.-S. Yuan, ECG arrhythmia classification using transfer learning from 2-dimensional deep CNN features, in: 2018 IEEE Biomedical Circuits and Systems Conference (BioCAS)IEEE, 2018, pp. 1–4.

[8] M. Kachuee, S. Fazeli, M. Sarrafzadeh, ECG heartbeat classification: a deep transferable representation, in: 2018 IEEE International Conference on Healthcare InformaticsIEEE, 2018, pp. 443–444.

[9] S. Fazeli, ECG heartbeat categorization data set, Kaggle, (2018) http://www.kaggle.com/shayanfazeli/heartbeat.

[10] ANSI-AAMI EC57, Testing and Reporting Performance Results of Cardiac Rhythm and ST Segment Measurement Algorithms, Association for the Advancement of Medical Instrumentation, Arlington, VA, 1998.

[11] G.D. Clifford, C. Liu, B. Moody, H.L. Li-Wei, I. Silva, Q. Li, A.E. Johnson, R.G. Mark, AF classification from a short single lead ECG recording: the PhysioNet/computing in cardiology challenge 2017, in: 2017 Computing in Cardiology (CinC)IEEE, 2017, pp. 1–4.

[12] F. Andreotti, O. Carr, M.A.F. Pimentel, A. Mahdi, M. De Vos, Comparing feature-based classifiers and convolutional neural networks to detect arrhythmia from short segments of ECG, in: 2017 Computing in Cardiology (CinC)IEEE, 2017, pp. 1–4.

[13] K.P. Soman, S. Diwakar, V. Ajay, Data Mining: Theory and Practice [With CD], PHI Learning Pvt. Ltd., New Delhi, Delhi, 2006.

Automated neuroscience decision support framework

I.D. Rubasinghe, D.A. Meedeniya
UNIVERSITY OF MORATUWA, MORATUWA, SRI LANKA

13.1 Introduction

Psychophysiological chronic disorders are one of the main challenges in today's medical science with the growth of the number of patients with neurological disorders. The common psychophysiological disorders can be listed as attention-deficit/hyperactivity disorder (ADHD), autism spectrum disorder (ASD), and cerebral palsy (CP) [1–3]. Some of the psychophysiological disorders are mostly encountered among children and continuing into adulthood, resulting in a life-long health problem. Moreover, additional conditions can occur due to a comorbidity of disorders. Thus, the identification of these psychophysiological conditions is crucial in medical science. However, reliable solutions are still in research-based experimental levels. Thus, the early identification of these disorders is important to start treatments with well-defined behavioral therapies, medications, and psychological activities vital to reducing the negative consequences.

At present, health informatics is a rapidly evolving area that acquires, analyzes, and manages biomedical and healthcare data in conjunction with computer engineering to provide better healthcare services. The development of the automated computational learning models has shown advancements over traditional biomedical models with the interaction of domain experts. Their capability of learning from raw data to automatic decision making and applicability on large-scale heterogeneous data has shown impressive results in biomedical data analysis [4, 5].

Supervised and unsupervised machine learning classification models such as random forest (RF), support vector machine (SVM), deep belief network (DBN), generative adversarial network (GAN), and convolutional neural network (CNN) in deep learning have been widely applied to neuroimaging computational solutions [6, 7]. However, existing studies have only considered single disorders, although there are commonalities among many of the medical disorders. Thus, there is a lack of support for multiple disorders and comorbidities due to the computational complexity in neuroimaging data preprocessing, feature extraction, and classification [2, 8]. These limitations have motivated a need for a generic platform that can integrate and support the identification of multiple psychophysiological disorders at early stages.

Deep Learning Techniques for Biomedical and Health Informatics. https://doi.org/10.1016/B978-0-12-819061-6.00013-6

The main objective of this chapter is to present a generic model to concatenate the computational solutions of psychophysiological chronic disorders together into a single platform. Because most of these disorders have some common symptoms, there can be instances when a patient may suffer from more than one psychophysiological disorder [9]. Hence, it is important to have an interconnected computational diagnosis methodology rather than disease-specific methods. The proposed model in this chapter is focused on the interconnectivity among computational diagnosis solutions in the context of psychophysiological disorders, such as ADHD and ASD, allowing for extendibility to other disorder types, such as Down syndrome, cerebral palsy, sickle cell disease, Alzheimer disease, depression, and anxiety.

We propose a novel computationally feasible neuroscience decision support system for psychophysiological disorder identification using neuroimaging data, such that the practitioners can experience confidence in efficiently diagnosing these disorders. This study aims to provide a computational framework that can process heterogeneous neuroimaging data relevant to multiple medical disorders that would output the identification of disorders based on different analytical measures and on different classifiers. Thus, multiple disorders can be diagnosed simultaneously, ensuring reliability and usability in the clinical practice. The development of the framework is based on different learning techniques that give high performances on large datasets.

This chapter is structured as follows: Section 13.2 describes the existing psychophysiological measures. Section 13.3 explains the data preprocessing techniques with their related work. Section 13.4 summarizes the significant related work. Section 13.5 presents the proposed neuroscience decision support framework. Methodology and evaluation are described in Sections 13.6 and 13.7, respectively. Finally, Section 13.8 provides a discussion of this work, and Section 13.9 concludes the chapter with possible future research directions.

13.2 Psychophysiological measures

Several psychophysiological measures are available in the literature and each has unique features. Electroencephalography (EEG) is a method used to measure the electrical activity of the brain by using electrodes placed on the scalp [10]. This surface measurement captures the activity of the brain, which is helpful to determine variations in response to a given stimulus. EEG data is focused on temporal resolution and does not capture spatial resolution. An average level of human expertise involvement is required in obtaining EEG data of a patient. Many researchers often use EEG data to identify disorders such as ASD, ADHD, and depression [4, 11, 12].

By contrast, magnetic resonance imaging (MRI) is a technique that provides structural information in the form of a map of the brain at a given time. MRI is used to determine the sizes of brain regions to detect abnormalities, such as tumors. MRI does not target temporal resolution such as EEG and instead focuses on spatial resolution. The MRI process requires a high level of expertise and is expensive compared to EEG. However, researchers are extensively applying classification and analysis methods on MRI data with the help of various public datasets and clinical data. For instance, MRI data has been used to identify disorders such as Alzheimer disease [5].

Similarly, the primary technology behind functional MRI (fMRI) is same as MRI. Specifically, fMRI is calculated by determining the number of changes in oxygenated blood, which is known as the blood-oxygenation level-dependent response known as BOLD [10]. The fMRI data is also expensive and requires expert involvement similar to MRI. fMRI is a main data type used for computational analysis in psychophysiological disorder identification. For instance, there are many ADHD classification [8, 13], ASD discrimination [7], and anxiety identification [2] studies performed with fMRI data.

There are other neuroimaging data types such as computed tomography (CT) scans, electroconvulsive therapy (ECT), single photon emission computed tomography (SPECT), quantitative electroencephalography (qEEG), and magnetoencephalography (MEG). Moreover, there are biometric data such as eye movement tracking [14], demographics, blood tests, and cognitive ability scores that are mainly used for psychophysiological disorder identification. Additionally, there are various other clinical diagnosis measures in practice such as child behavior analysis and anatomical analysis. The SNAP-IV rating scale is a popular instance of a child behavior analysis that consists of a questionnaire to gather patient data [11, 15].

13.3 Neurological data preprocessing

13.3.1 Importance of data preprocessing

Neuroimaging data preprocessing is a complex process compared to general digital image preprocessing. The reason is the higher dimensionality of most neuroimaging data such as EEG, MRI, and fMRI, which varies from three to four dimensions. Because the medical images contain more combined data, preprocessing needs to be carried out carefully with expert knowledge, such that data are not affected. There are many software applications and development platforms that support different steps in neuroimaging preprocessing. Primarily, Python-based platforms are currently in practice due to their higher capability in mathematical computations. The Python libraries Nilearn, scikit-learn, NumPy, NiBabel, and Pandas are just a few among many. Also, popular tools such as Matlab, OpenCV, ImageJ, and Weka provide various functions to preprocess medical images to a certain extent. These applications are described in Section 3.3.

13.3.2 Data preprocessing techniques

The features required for extraction from a medical image may vary depending on the type of image, type of diagnosis, and the addressed problem domain. Often, the texture, voxels, grayscale, and shape are some common brain image features extracted from neuroimaging data [16]. The proposed approach in this chapter is evaluated using fMRI data.

Table 13.1 summarizes the related preprocessing techniques applied to fMRI data.

Because medical images are in higher dimensions such as 3D and 4D, dimensionality reduction is important before further analysis. Among the many procedures to minimize dimension in neuroimaging data, principle component analysis (PCA), generalized linear model (GLM), mean, and covariance-related computations are some techniques applied in related works [39].

Noise removal refers to the elimination of unwanted data in an image [10]. In complex neuroimaging, it is helpful to ease into further analysis when the irrelevant data is removed. This is known as denoising, which involves different denoising algorithms. Theoretically, there are various types of noise such as Gaussian noise, Salt and Pepper, Brownian noise, Speckle, and many more. Some of the widely used denoising algorithms in the context of medical images are wavelet transform, independent component analysis (ICA), and Histogram equalization.

Table 13.1 fMRI preprocessing techniques used in related studies.

Technique		Related study
Dimension reduction	Principal component analysis (PCA)	[3, 17, 18]
	Generalized linear model (GLM)	[19]
Noise removal	Wavelet transform	[20]
	Adaptive filters	[21]
	Independent component analysis	[18, 22]
	Histogram equalization	[23]
	Density estimation filters	[3]
Image filtering	Motion correction	[5, 24]
	Slice timing correction	[25]
	Head motion correction	[26]
Image correction	Distortion correction	[3]
	Statistical (mean, standard deviation)	[17, 24, 27–29]
Normalization	Volume-based	[16, 30]
	Homomorphic filtering	[31]
	Laplacian	[32]
	Spatial smoothing	[5, 18, 26, 33]
Smoothing	Atlas-based (brain coordinates)	[24, 29]
Segmentation	Interactive segmentation	[34]
	Singular value decomposition (SVD)	[35]
	Discrete wavelet transforms (DWT)	[36]
	Fast Fourier transform (FFT)	[37]
Feature extraction	Max pooling	[3]
	Contrast stretching	[38]
	Spatial domain methods	[18]
	Frequency domain methods	[27]

The recovery process of a degraded image is known as restoration. This involves mathematical modeling of the degradation and the application of filters based on the image dimension [40]. Further, image registration can be performed by adjusting the alignment of the image, if necessary.

In neuroimaging, it is crucial to determine the regional sections for accurate medical diagnosis. Normalization helps to align regions in an image with the creation of reference frames as templates. There are different normalization techniques such as surface-based, statistical, histogram-based, landmark-based, and volume-based approaches mainly applied to neuroimaging modalities [40].

Smoothing is another important preprocessing step in neuroimaging to suppress noise, which may exist as a result of sampling, radiation, and transmission. Because the noise in medical images is mostly in high frequencies, low-pass filtering and high-frequency suppression techniques are commonly in practice for smoothing [41]. The amount of smoothing needs to be taken care of because excessive smoothing can blur the edges and degrade useful information in images if the intensities are reduced unnecessarily. Otherwise, sharpening is required as an enhancement step to reduce any blurring effects with the use of filters such as Laplacian and Differencing filter, depending on the image dimension [40].

Partitioning an image into certain meaningful portions is referred to as segmentation in image preprocessing, which is the same for medical images. In neuroimaging modalities, the segments can be often brain regions, tissues, and any other body organs. Accordingly, the often applied medical image segmentation methods include shape-based, atlas-based, subjective-surface, image-based, and interactive segmentation [42].

13.3.3 Software application support for neuroimage processing

Several neuroimage preprocessing, analysis, and state-of-the art tools are available in the literature. Salford Predictive Modeler (SPM) is a MiniTab-based software platform, which can be used in ultrafast predictability [22, 24]. It comprises of regression, classification, and clustering related machine learning models to analyze different datasets regardless of their size and complexity. This tool enables efficient model creation, exploration, and refinement. Advanced normalization tools (ANTs) provide several image preprocessing features to extract data from complex neuroimaging datasets [43]. Those include image registration, correction, segmentation, and 4D transformations in biomedical images. ANTs are an open source platform applicable to any complex dataset with multi-dimensional data visualization abilities.

Another widely used application is EEGLAB that is included in the Matlab toolbox. It supports interactive, high-performance computing in any operating system environment [44]. This is mainly used for EEG and MEG data processing activities such as time-frequency analysis, and ICA with visualization. This supports multiformat, high-density data types. As an open source platform, EEGLAB provides a structured programming facility to analyze features with more flexibility. Similarly, FreeSurfer is a software package that has multisubject statistical features to analyze brain images [45]. This open source tool is widely used for human MRI and fMRI image preprocessing and analysis. Image registration, segmentation, skull-stripping, reconstruction, and vitalization are mainly supported preprocessing activities in structural and functional neuroimaging data.

Analysis of functional neuroimaging (AFNI; [46]) and Cartool Community [47] are particularly used on EEG, fMRI, and MRI data. AFNI supports batch processing, slice timing, motion correction, masking, smoothing, and visualization on fMRI data in Unix-based operating system environments. Being an open source platform built using C language, AFNI is popular among researchers for fMRI analysis. Cartool Community is another tool used in EEG image analysis including interpolation, segmentation, fitting, and maps visualization.

Additionally, there are several open source libraries and packages that programmatically ease the medical image data accessibility. For instance, Table 13.2 shows the functions provided by some of the popular Python libraries in neuroimaging processing.

Further, popular tools such as OpenCV [48] and ImageJ [49] provide several functions to preprocess medical images to a certain extent. For instance, OpenCV provides a programming platform in C++ for generic computer vision features such as facial recognition, motion correction, segmentation, and augmented reality with machine learning models included in a separate library. On the other hand, ImageJ is a Java-based tool executable on any operating system environment. This supports processing various raw types of images including medical image formats such as DICOM. Contrasting, convolution, smoothing, sharpening, and Fourier analysis are some main preprocessing activities supported by ImageJ.

Table 13.2 Python library support for neuroimaging.

Python library	Function	Supported data types
MNE	Preprocessing, time-frequency analysis, visualization, and connectivity analysis	EEG, MEG
NiBabel	Access image meta information and image data	Any saved in GIFTI, NIfTI, DICOM, MINC, AFNI, CIFTI, ECAT, MGH
NIPY	Analyze functional and structured neuroimaging data	MRI, fMRI, PET, EEG
Nilearn	Statistical learning and analysis of neuroimage data	MRI, fMRI
scikit-learn	Preprocessing, dimensionality reduction, regression, classification, and clustering	Any
NumPy	Mathematical functions on arrays and matrices	Arrays, matrices
Pandas	Mathematical functions on numerical tables and time series functionality	Numerical tables, time series

13.4 Related studies

Several learning models have been used in the literature to classify a given image according to a set of possible categories. Machine learning (ML) and deep learning (DL) algorithms are widely used in related psychophysiological disorder identification studies with image classification [3, 7]. Tables 13.3 and 13.4 summarize the benefits and limitations of different supervised and unsupervised learning algorithm categories used in the related literature, respectively.

The SVM classifier is majorly applied on EEG, MRI, and fMRI data to identify ADHD, ASD, and other neurological disorders [27]. Some have used multiple SVM classifiers together with a radial basis function (RBF) kernel [33, 50, 58], linear SVM [2, 3, 17, 51, 55, 59], and SVM along with other classifiers such as random forest and Naïve Bayes for comparative analysis.

The ML classifiers SVM, random forest, and Naïve Bayes have shown higher accuracy and performance due to minimal overfitting and robustness to noise [54, 55]. However,

Table 13.3 Supervised machine learning techniques.

Machine learning technique		Benefits	Limitations
Classification	Support vector machine (SVM) (linear and nonlinear) [3, 45, 50–52]	Better classification performance and accuracy on training data, efficient, not overfitting	Depend on the kernel type, speed versus time is a trade-off, can be slow in testing, extensive memory consumptions
	K-nearest neighbor (KNN) [40]	Robust to noisy data due to averaging the neighbors	Memory consumption can be high
	Naïve Bayes [17]	Easy to build and understand; useful to scale on large datasets due to linear time	Zero probability problem, need to apply Laplacian estimator to overcome this
	Random forest [27, 45, 53, 54]	Corrects the DTs' overfitting in training set, saves time in prepreparation, parallel built	Can be slow, hard to understand predictions, not suitable for categorical variables
Regression	Decision tree (classification trees and regression trees) [27, 55]	Robust to noise and errors, efficient due to simplicity, readable, easy to implement, make optimal decisions	Needs tuning, not powerful for complex data handling, considers only one attribute at a time, not fit for continuous variables
	Linear regression [56]	Easy to understand, minimum tuning is required, fast	Overfitting tendency is high
	Logistic regression [33, 37, 45]	The output is informative, uncover hidden relationships in data	Difficulty in implementing
	Lasso [33]	Minimizes prediction error; helps to identify the most important predictors	Selects only single feature from correlated feature groups
	Support vector regression (SVR) (linear and nonlinear)	Computational complexity is independent of input space dimensionality, high capability in generalization, high accuracy	Complexity depends on the selection of kernel

Table 13.4 Unsupervised machine learning techniques.

	Machine learning technique	Benefits	Limitations
Clustering	*k*-Means [37]	Produce tighter clusters in globular clusters, faster for a larger number of variables	Difficulty in predicting k value, less performs with clusters in different sizes and density
	Gaussian mixture model (GMM) [57]	A better sum of square compared to *k*-Means clustering; flexible for cluster covariance	Accuracy may depend on the input data characteristics
Dimensionality reduction	Principal component analysis (PCA) [16, 18]	Provides a lower-dimensional viewpoint	Results based on relative scaling, have high-order dependences
	Independent component analysis (ICA) [22, 37]	Removes both correlations and high order dependences	Require data preprocessing before applying ICA to remove correlations

ML methodologies lack the accuracy over DL methods in neurological disorder classifications, as neuroimaging data are complex with higher dimensions [7]. Moreover, the ML techniques have limitations such as speed versus time trade-off, high memory consumption, and zero probability problem [3, 17].

Other than supervised and unsupervised learning models, several reinforcement learning algorithms such as Actor-Critic Algorithm [60], Deep Deterministic Policy Gradient (DDPG) [61], Deep Q Network (DQN) [30], State-Action-Reward-State-Action (SARSA) [60], and Q-Learning [61] have been used in related studies. These algorithms perform efficiently and do not require manually defined rules such as for expert systems. Additionally, they have shown high performance. However, reinforcement algorithms consume high memory and require the support of optimization techniques.

The work by Kuang et al. [8] has focused on a region of fMRI brain images from an ADHD-200 raw dataset for ADHD identification. They have primarily applied DBN architecture in DL with Gaussian-Bernoulli RBM that has three hidden layers. Additionally, this work has been performed for a neural network with a single hidden layer as well. For some cases, they have shown that the neural network-based results are better over DBN.

The discrimination of ASD patients has been the focus of the work [7]. They have applied denoising Autoencoder (AE) architecture in DL for the brain activation patterns of resting state fMRI data obtained from autism brain imaging data exchange (ABIDE) dataset. In addition, they have also compared the work with SVM and random forest classifiers in ML. The cross-validation results have shown higher accuracy for the DL approach and higher performance for SVM classifier solution in ML. The WB voxel-level features of resting-state fMRI data obtained from ADNI dataset has been used for the identification of Alzheimer disease (AD) in the work [5]. Moreover, the use of LeNet-5 in CNN architecture has shown higher accuracy over SVM in ML. Although there are more related

Table 13.5 Summary of features in related studies with fMRI data.

Related work	Whole brain	Partial brain regions	Functional connectivity	DL models	ML models	Larger dataset	Single disorder
[3]	X		X		X	X	
[8]		X		X	X		X
[7]	X			X		X	X
[58]		X			X		X
[37]	X		X		X		X
[5]	X			X	X		X
[2]	X		X		X		X
[28]	X						X
[24]	X				X		X
[59]	X		X		X		

studies on ADHD, ASD, and AD, the other disorders such as anxiety, Down syndrome, cerebral palsy, and SCD have research hindrances in terms of classification models. The literature [2] has addressed anxiety disorder based on whole brain functional connectivity in fMRI data. The prediction results for SVM and RBF classifiers in this work have shown significant accuracy though the used data count is smaller. As summarized in Table 13.5, most of the related studies have only applied ML classifiers and have addressed only one disorder in their studies. They have been mostly limited for a smaller dataset because of the performance limitations in ML models.

13.5 Neuroscience decision support framework

This section presents a model to concatenate the computational solutions on psychophysiological disorders together into a single platform. Because psychophysiological chronic disorders can have similarities in symptoms, there can be instances when a patient may suffer from more than one such disorder. Hence, it is important to have an interconnected computational diagnosis methodology rather than disease-specific methods. This model proposes a common platform for computational diagnosis solutions in the context of psychophysiological chronic diseases.

Thus, given a set of neuroimaging test reports such as EEG, MRI, and fMRI scan images, the model identifies the existence of any psychophysiological disorder. Different ML and DL techniques are used in this work. Accuracy and performance are crucial to ensure when it is required to diagnose a given input over multiple diseases than checking for a single disease. Fig. 13.1 illustrates the high-level architecture of the proposed neuroscience decision support system platform.

The proposed neuroscience decision support system is supposed to fulfill the excessive amount of manual expert involvement required during the diagnosis process. The inputs

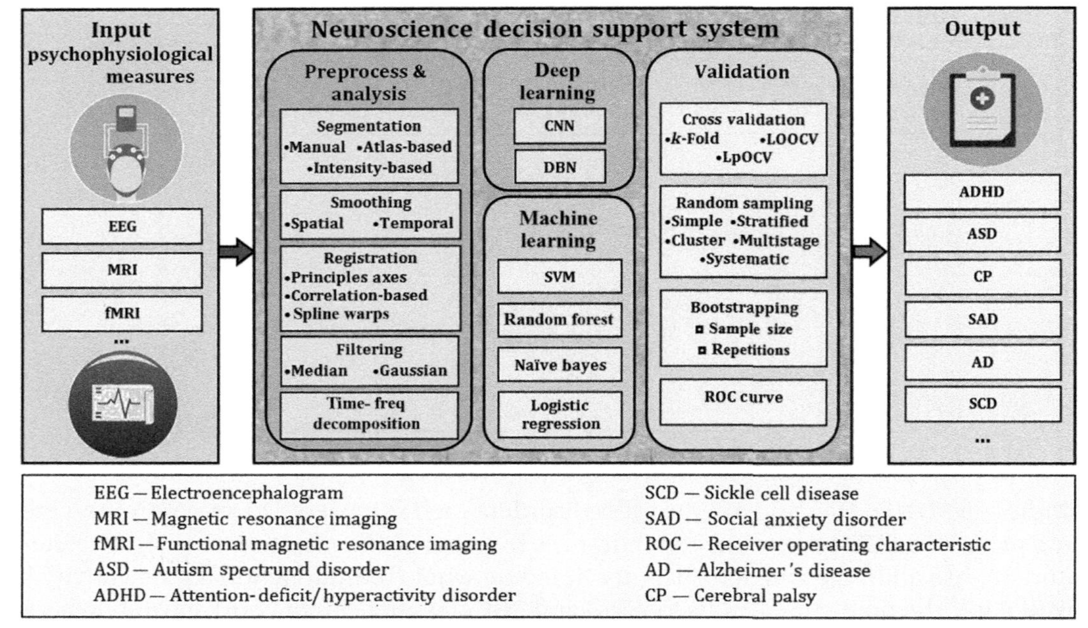

FIG. 13.1 Neuroscience decision support system model architecture.

of the system are the patients' testing results using different neuroimaging. In addition, a patient's database containing medical history and nonimaging measures such as demographics can be integrated for further analysis [51]. Then, the heterogeneous neuroimaging types are preprocessed and analyzed using feature extraction. This can be supported by several techniques such as image registration, segmentation, smoothing, filtering, and time-frequency decomposition particularly for EEG and missing value analysis.

Feature extraction is important to avoid the overhead of the excessive number of features during learning mode classifications. In Fig. 13.1, DL and ML components represent the applicable techniques such as CNN, DBN, SVM, random forest, Naïve Bayes, and linear regression. Result validation is also significant to ensure the acceptability of the outcomes based on accuracy. Cross-validation methods such as *k*-Fold, LOOCV, random sampling, bootstrapping, and ROC curve can be used to assess the proposed methodology. Finally, the outcome of this system is used as a support tool for ease and to speed up expert reviewing but not to replace human expert involvement. Thus, expert knowledge can be shared among patients.

This framework can be extended for other data types and related disorders as well. Accordingly, this proposed neuroscience decision support system model adds another layer to mitigate human error, encourage awareness, quicken the diagnosis process, and reduce expenses in consulting, which tends to discourage the public to attend check-ups.

13.6 System design and methodology

13.6.1 Datasets

This prototype solution is tested with two datasets, namely a ADHD-200 sample that consists of 776 biased fMRI and anatomical datasets across eight imaging sites, and ABIDE dataset that consist of biased 1112 functional and structural neuroimaging datasets regarding ASD collected from 16 imaging sites.

The ADHD-200 preprocessed dataset [62] is a collection of preprocessed resting state fMRI and structural MRI images of healthy and ADHD patients. This has been presented as a solution to reduce the extra effort required to process complex medical images before continuing with their characteristics toward solid solutions. Thus, computer specialists can use direct data computationally. There are several versions that have been processed through different analytical pipelines to support various data analysis requirements. Those preprocessing strategies include Athena, Burner, NIAK, and CIVET. This study considered Athena, which is a fully released version that has used slice time correction, reorientation, motion corrections, masking to exclude nonbrain parts, averaging volumes to obtain a mean image, coregistration to anatomic image, writing down fMRI data and mean image into template space, downsampling, band-pass filtering to exclude the frequencies not implicated in functional connectivity, and blurring and filtering using Gaussian filter. The Athena data in volumetric NifTi (.nii) format is mainly used, although data processed through any strategy is usable. Fig. 13.2 shows a view of a data instance from preprocessed a ADHD-200 dataset.

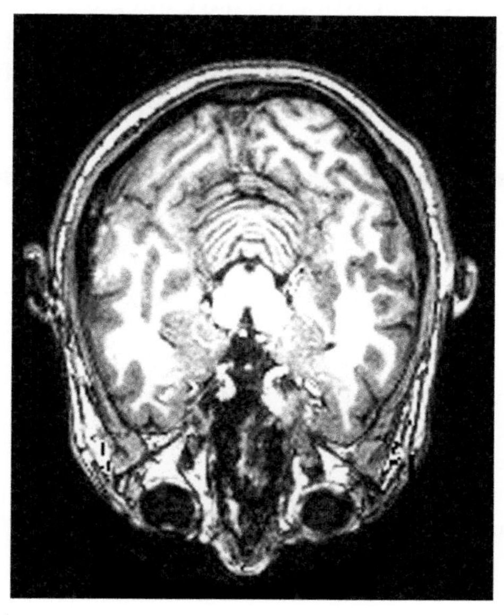

FIG. 13.2 ADHD fMRI subject instance.

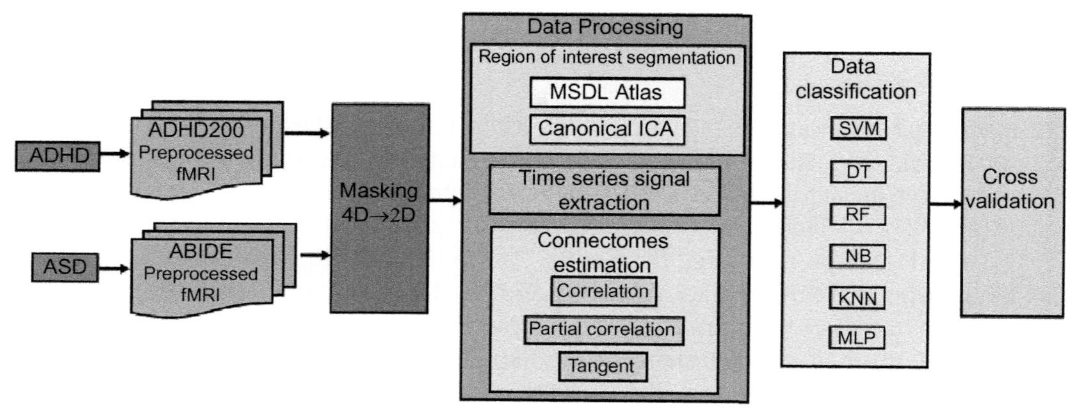

FIG. 13.3 Implementation process flow.

Similarly, the ABIDE Preprocessed Connectomes Project (PCP) also provides preprocessed versions of neuroimaging data involved in this study [63]. This has passed through a similar set of stages such as slice time correction, motion realignment, intensity normalization, nuisance signal removal, and registration.

13.6.2 Solution implementation

The proposed framework is implemented as a prototype tool for the purpose of multiple disorder identification based on different neuroimaging data. The focus of this solution is to provide a common platform to classify, analyze, and view the results for a disorder-related dataset. Thus, as a proof-of-work, we have integrated two different datasets, each addressing ADHD and ASD, respectively. Fig. 13.3 illustrates the abstract process flow of discriminating disorders based on functional connectivity coefficients in fMRI data, and Algorithm 1 shows the followed sequence of abstract steps.

Algorithm 1

Neuroimaging data analysis

```
 Input: preprocessed neuroimaging data (d), number of subjects (n),
classifier (c)
 If (d = ADHD) {
   Obtain probabilistic ROIs from MSDL atlas
    Region signals extraction
    Compute region signals
   Extract phenotypic information
   For (each connectivity type: Correlation, Partial-correlation,
Tangent){
      Compute ROI to ROI Connectomes
      Classify (c)
```

```
        Cross-validate
      }
    }
    Else if (d = ABIDE) {
      Extract time series signals from brain parcellations
      Extract time series signals from atlas
      Region extraction from atlas
    For (each connectivity types: Correlation, Partial-correlation,
   Tangent) {
          Connectomes estimation
          Classify (c)
          Cross-validate
      }
    }
    Output: Cross validation results for classification, the number of
   patients among n, analytical plots for results
```

The user is given the option to select a dataset, the number of subjects from each, and the type of classification. The fMRI data are 4D images in coordinate space as *nii* files. They are required to be transformed from 4D to 2D arrays to proceed with analysis and computations. Correspondingly, we have performed four-dimensional to two-dimensional masking to convert volumetric data into 2D voxels over time to extract regions of time series.

Data processing module prepares the data features required for classification. We have applied probabilistic Region of Interest (ROI) extraction based on MSDL atlas for ADHD data and an estimated atlas using Canonical ICA for ABIDE data, because it is ideal to follow probabilistic ROIs [29]. Canonical ICA is a group-level analysis approach for fMRI data supported in Python packages. This extracts the brain regions from the computed atlas. The tool can be extended for other atlas types as well as toward a more generalized platform. Next, we have extracted region time series or signals with phenotypic information that maps all the neural connections. We have used the default bandwidth options such as low pass and high pass as 0.1 that can be extended to allow user input data for deep analysis purposes.

In fMRI data processing, connectivity is an important factor estimated using correlations among time series. However, different connectivity types lead to different results. Thus, identifying the most useful type of connectivity can be crucial for a better classification result. The correlations among the extracted probabilistic ROIs is computed according to three functional connectivity measure types: correlation, partial correlation, and tangent using correlation matrices of data. The correlation matrix is the simplest connectivity that identifies brain regions with similar activities. Partial correlation is also widely addressed and measures the direct-linear functional connectivity induced by neural activity. Tangent represents a combination of both types that focus on building group connectivity matrices, which is not widely addressed. The purpose of this study is to create an analytical space. Similarly, a generic computation platform can be supported by incorporating different analytical specifications.

After that, the classification is performed for each connectivity type at a time using the predefined learning models to output the patient's count for a given disorder. Classifiers include the existing learning algorithms, implemented in the Nilearn Python module designed for neuroimaging data. The incorporated classifiers include SVM, decision trees, random forest, Naïve Bayes, KNN, and multilayer perception classifier representing insight of neural networks.

The cross-validation is performed to test the classification results. We have used a test set size as a 0.2 proportion from the given number of subjects in the dataset. Straight k-Fold cross-validation variations are used to generate the random folds in a way that the percentage of samples for every class is preserved. Accordingly, multiple validation methods can be incorporated into the tool to allow deep result comparisons.

13.7 Solution evaluation

The solution classifies a given dataset with a given number of subjects on a selected type of classifier over three types of correlation matrix. The results for three connectivity types are plotted for graphical analysis as shown in Fig. 13.4. It shows an instance of SVM classifier results for 12 data items, where 9 are ADHD subjects and 3 are healthy.

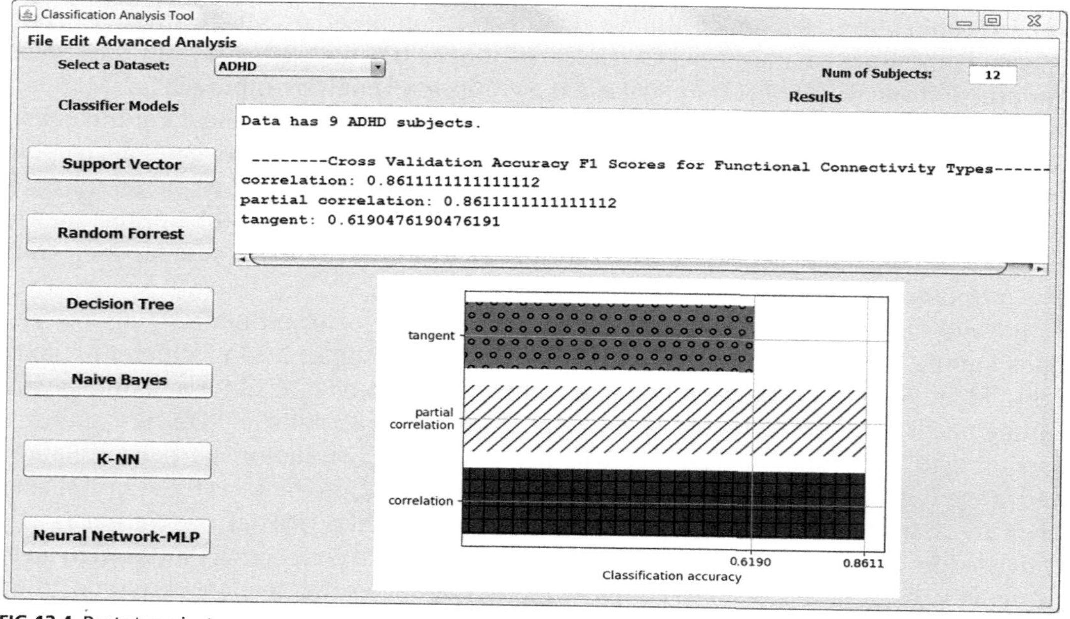

FIG 13.4 Prototype instance.

According to the results, the highest classification accuracy in terms of F1 score is achieved when using correlation and partial correlation functional connectivity analysis types. However, the tangent, which is a combination of both types, has led to lesser accuracy. Considerably, correlation connectivity analysis method provides higher accuracy due to its large number of connectomes. This may subject to change when new datasets are integrated. Hence, it is observable that the classifier accuracy highly depends even on the functional connectivity analysis method used for classification and the type of applied classifier. Thus, this solution can be used by researchers and practitioners to differentiate the results by integrating more analysis methods and datasets types.

Table 13.6 presents the obtained classifier F1 score results for 12 ADHD subjects and 100 ABIDE subjects. Among the used functional connectivity matrix types, the correlation has helped to obtain the maximum classification F1 score for most of the cases. In comparing ADHD results versus ABIDE data results, both results have a uniform behavior for three types of functional connectivity matrices. For instance, in ADHD, a decision tree classifier has given maximum accuracy when used correlation, second highest accuracy when used tangent, and least when used partial correlation. It has been in the same behavior in ABIDE dataset decision tree classifier results. Similarly, it is observable that the use of different functional connectivity matrices affects the results independent of disorder type in data. Moreover, according to the distribution of results, decision tree has been the weakest classifier for ADHD data but the best in ABIDE data. It is observable that the results vary with the number of data and the type of dataset on different classifiers. Thus, this type of results distribution via a single platform can be useful to select the optimum classifier for different datasets as the same classifier is not ideal for any dataset. Further, a deeper comparative analysis with the involvement of medical experts can help to identify commonalities and characteristics in results.

Figs. 13.5–13.7 show an instance of the differentiation of these three connectome types: correlation, partial correlation, and tangent, respectively. There are 9 ADHD subjects over a total of 12 subjects. The mean correlation of identified ADHD subjects is shown in this visualization as functional networks by cords on top of brain glass schematics. The left and right directions of hemispheric projections are denoted by L and R. The maximum connectivity can be seen in the correlation category, which showed higher classification

Table 13.6 An instance of classifier results.

Dataset	Functional connectivity matrix kind	Support vector machine	Random forest	Decision tree	Naïve Bayes	*K*-nearest neighbor	Neural network-MLP
ADHD	Correlation	0.8611	0.7403	0.6493	0.8611	0.8611	0.8611
12 subjects	Partial correlation	0.8611	0.8611	0.5555	0.8611	0.8611	0.8611
	Tangent	0.6190	0.8611	0.6166	0.8611	0.8611	0.3333
ABIDE	Correlation	0.8683	0.8524	0.8756	0.6218	0.8682	0.8545
(Autism) 100	Partial correlation	0.8571	0.8546	0.7439	0.8571	0.8571	0.8571
subjects	Tangent	0.6249	0.8588	0.7608	0.8571	0.8624	0.4131

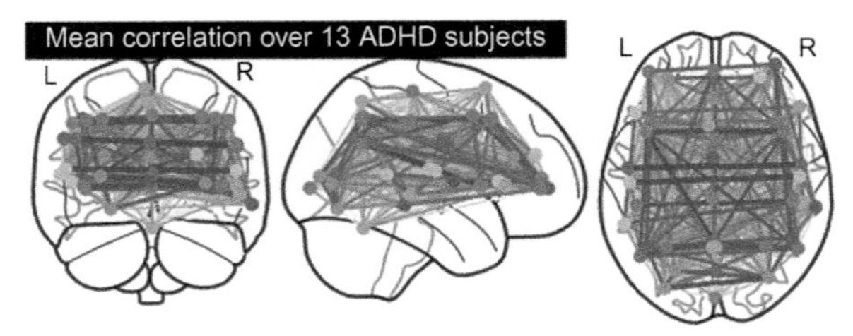

FIG. 13.5 Differentiation of correlation results for ADHD subjects.

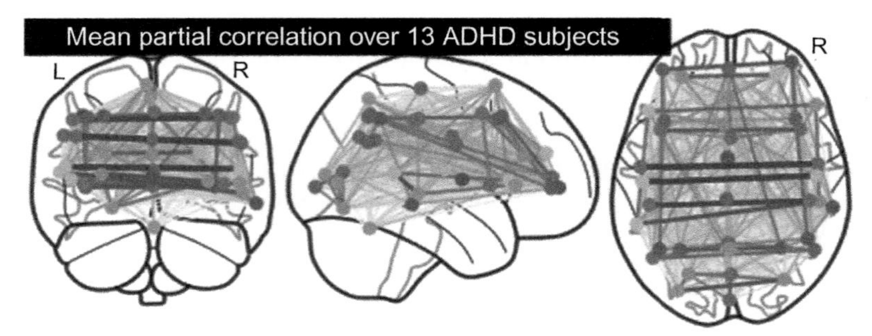

FIG. 13.6 Differentiation of partial correlation results for ADHD subjects.

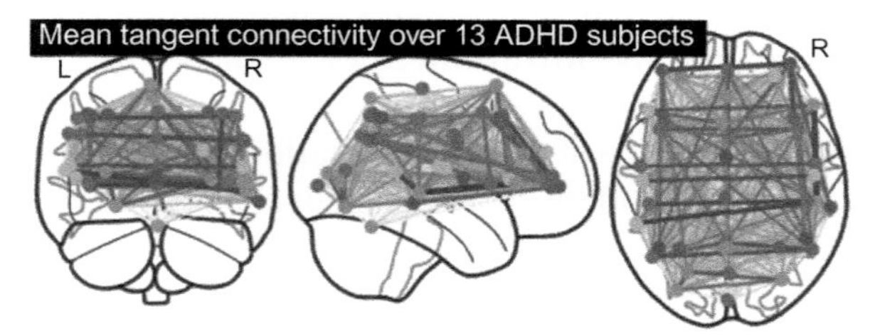

FIG. 13.7 Differentiation of tangent functional network results for ADHD subjects.

accuracy in both ADHD and ASD discrimination. Further, these behaviors remain true when varying the number of subjects. Hence, it is observable that the correlation functional connectivity type that creates the maximal functional network is useful for classification over partial correlation and tangent types regarding ADHD and ASD subjects.

Fig. 13.8 shows an instance of a statistical map of the computed brain atlas for an ABIDE dataset using Canonical ICA. This can be used as an input to deep learning models. In this work, we rather focus on the differences in results when different methods are applied and facilitates the ability for researchers and experts to analyze them scientifically with respect to disorder types.

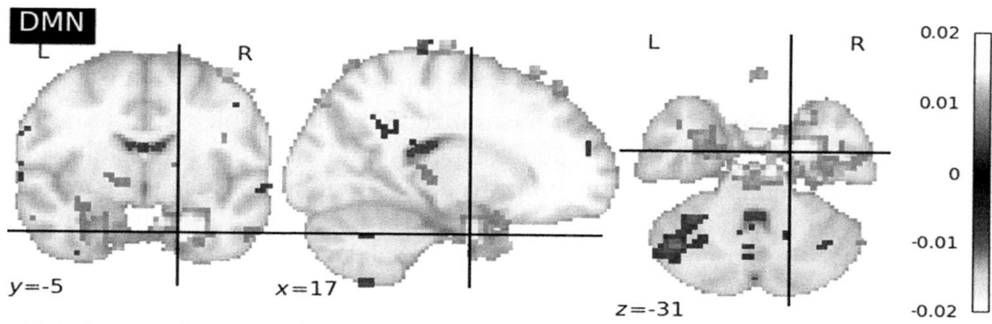

FIG. 13.8 An instance of a computed brain atlas for ABIDE data.

13.8 Discussion

Considering the related studies that have used the ADHD-200 and ABIDE datasets, there are several features addressed. With the preprocessed ADHD-200 dataset, Riaz et al. [64] have focused on the fusion in fMRI images, and Riaz et al. [65] have introduced a CNN named FCNet with a maximum accuracy of 68.6% to calculate functional connectivity. In contrast, we have used three functional connectivity analysis types: correlation, partial correlation, and tangent with multiple classifiers in an analyzable perspective. These techniques test the correlations against a hypothesis independent of how brain regions are coupled. Further, they are used to identify the brain region patterns that are useful in classification. In existing works, the consideration has been narrowed into a single functional connectivity analysis method. Because we focus on bringing the current research-level solutions into a common platform, our approach acts as a middle layer to plug any analytical feature identified in these related studies.

Moreover, several studies have used an ABIDE-preprocessed dataset to address ASD. For instance, Thyreau et al. [32] have addressed an aspect in segmentation, Dvornek et al. [66] have used LSTM networks for ASD classification with an accuracy level of 68.5%, and Guo et al. [26] have diagnosed functional connectivity patterns with a deep neural network with an accuracy of 86.36%. In our work, we use ABIDE dataset for classification based on three functional connectivity types via several classifiers. We have achieved a maximum accuracy up to 0.87 F score with minimal classifier customizations. Our aim is to provide flexibility to get the comparative results for several disorder types. Accordingly, our work differs from the existing works that have involved ADHD-200 preprocessed and ABIDE preprocessed datasets in terms of flexibility to multiple disorder types, openness to different datasets, and the integrability of various analytical measures. Although the existing classification supportive tools such as Weka and Matlab allow classification on different classifiers, they cannot be fully customized with different analytical measures according to requirements. Hence, the need for having a separate neuroimaging analysis platform for psychophysiological disorders is beneficial to broaden the focus on analysis.

Accordingly, our focus is to introduce a common platform rather than to propose a single strong classifier that addresses a specific disorder as in a majority of related work. This was identified as a limitation in existing related literature [11, 13]. Thus, a limitation of this work is the lack of customized classifier implementations. Therefore, the accuracy levels of the conducted experimental classification results are not completely strong compared to the improved classifier model accuracies in some related studies. However, the long-term goal of this is to achieve the ability to integrate a strong classification or clustering model on any selected dataset and to analyze any measure. The proposed solution enables multiple classifiers and better dataset integrations and broadens decision making because most of the psychophysiological disorders contain commonalities. Further, this can be extended further with more input types, analytical measures, and output variations for rigorous analytical purposes.

13.9 Conclusion

A neuroscience decision support system is proposed in this chapter with the primary focus to integrate multiple psychophysiological chronic disorder identification via a single platform to make it more presentable toward a clinical usage in practice. Another aim is to propose a generalized platform with customizable and extensible features, in terms of datasets, learning models, and analysis approaches. It is essential to ensure the learning models can identify each disorder with higher accuracy. Heterogeneous neuroimaging preprocessing, analysis, testing, validation techniques, and tools need to be incorporated together with further extendibility. As an initial step for this generalized solution, this chapter has proposed a prototype tool developed to analyze multiple psychophysiological chronic disorder types with learning classifiers based on different analytical measures and datasets. As proof of work, we have used two dataset types as sets of learning classifiers. Mainly, we have shown the use of three functional connectivity measures to the computer revealing the correlation among the regions of interests, in a way that increases the classifier accuracy. The analytical measure used is three types of functional connectivity: correlation, partial correlation, and tangent. Correspondingly, this can be used for the analytical purpose of different datasets of multiple disorders simultaneously.

The proposed neuroscience decision support system can be extended in many directions. For instance, the psychophysiological chronic disorders that are hardly addressed in related studies such as depression, anxiety, and Down syndrome can be incorporated into this learning model. Moreover, performance and accuracy improvements are crucial when integrating into a single platform. Thus, strengthening the feature extraction process to cope with multiple disorder types simultaneously and enhancing the accuracy and performance to cope with larger datasets would be a significant future direction to generalize the outcomes.

References

[1] R.N. Baumgartner, S.B. Heymsfield, A.F. Roche, Human body composition and the epidemiology of chronic disease, Obes. Res. 3 (1) (1995) 73–95.

[2] F. Liu, W. Guo, J.-P. Fouche, Y. Wang, W. Wang, J. Ding, L. Zeng, et al., Multivariate classification of social anxiety disorder using whole brain functional connectivity, Brain Struct. Funct. 220 (1) (2015) 101–115.

[3] B. Sen, N.C. Borle, R. Greiner, M.R.G. Brown, A general prediction model for the detection of ADHD and Autism using structural and functional MRI, in: B.C. Bernhardt (Ed.), PLoS One, vol. 13(4), Public Library of Science, 2018, p. e0194856.

[4] U.R. Acharya, S.L. Oh, Y. Hagiwara, J.H. Tan, H. Adeli, D.P. Subha, Automated EEG-based screening of depression using deep convolutional neural network, Comput. Methods Progr. Biomed. 161 (2018) 103–113.

[5] S. Sarraf, G. Tofighi, Classification of Alzheimer's Disease using fMRI Data and Deep Learning Convolutional Neural Networks, 2016, arXiv:1603.0863.

[6] A. Gibson, J. Patterson, Major architectures of deep networks, in: Deep Learning: A Practitioner's Approach, O'Reilly, 2017.

[7] A.S. Heinsfeld, A.R. Franco, R.C. Craddock, A. Buchweitz, F. Meneguzzi, Identification of autism spectrum disorder using deep learning and the ABIDE dataset, NeuroImage: Clin. 17 (2018) 16–23.

[8] D. Kuang, X. Guo, X. An, Y. Zhao, L. He, Discrimination of ADHD Based on fMRI Data With Deep Belief Network, Springer, Cham, 2014, pp. 225–232.

[9] F.H. Wilhelm, S. Schneider, B.H. Friedman, Psychophysiological assessment, in: Clinician's Handbook of Child Behavioral Assessment, Academic Press, 2006, pp. 201–231.

[10] R. Abreu, A. Leal, P. Figueiredo, EEG-informed fMRI: a review of data analysis methods, Front. Hum. Neurosc. 12 (2018) 29.

[11] G. Brihadiswaran, D. Haputhanthri, S. Gunathilaka, D. Meedeniya, S. Jayarathna, A Review of EEG-based Classification for Autism Spectrum Disorder. J. Comput. Sci. 15 (08) (2019) 1161–1183, https://doi.org/10.3844/jcssp.2019.1161.1183.

[12] D. Haputhanthri, G. Brihadiswaran, S. Gunathilaka, D. Meedeniya, S. Jayarathna, M. Jaime, Y. Jayawardena, An EEG based channel optimized classification approach for autism spectrum disorder. in: Moratuwa Engineering Research Conference (MERCon), IEEE, 2019, pp. 123–128, https://doi.org/10.1109/MERCon.2019.8818814.

[13] S. De Silva, S. Dayarathna, G. Ariyarathne, D. Meedeniya, S. Jayarathna, A survey of attention deficit hyperactivity disorder identification using psychophysiological data, Int. J. Online Biomed. Eng. (iJOE) 15 (13) (2019), pp. 61–76, https://doi.org/10.3991/ijoe.v15i13.10744.

[14] S. De Silva, S. Dayarathna, G. Ariyarathne, D. Meedeniya, S. Jayarathna, A.M.P. Michalek, G. Jayawardena, A rule-based system for ADHD identification using eye movement data. Moratuwa Engineering Research Conference (MERCon), IEEE, 2019, pp. 538–543, https://doi.org/10.1109/MERCon.2019.8818865.

[15] R. Bussing, M. Fernandez, M. Harwood, W. Wei Hou, C.W. Garvan, S.M. Eyberg, J.M. Swanson, Parent and teacher SNAP-IV ratings of attention deficit hyperactivity disorder symptoms: psychometric properties and normative ratings from a school district sample, Assessment 15 (3) 2008 317–328.

[16] C. Salvatore, A. Cerasa, I. Castiglioni, F. Gallivanone, A. Augimeri, M. Lopez, G. Arabia, et al., Machine learning on brain MRI data for differential diagnosis of Parkinson's disease and Progressive Supranuclear Palsy, J. Neurosci. Methods 222 (2014) 230–237.

[17] B. Li, A. Sharma, J. Meng, S. Purushwalkam, E. Gowen, Applying machine learning to identify autistic adults using imitation: an exploratory study, in: M. Sakakibara (Ed.), PLoS One, vol. 12(8), Public Library of Science, 2017, p. e0182652.

[18] M. Case, H. Zhang, J. Mundahl, Y. Datta, S. Nelson, K. Gupta, B. He, Characterization of functional brain activity and connectivity using EEG and fMRI in patients with sickle cell disease, NeuroImage: Clin. 14 (2017) 1–17.

[19] J.B. Poline, M. Brett, The General Linear Model and FMRI: Does Love Last Forever? NeuroImage 62 (2) (2012) 871–880.

[20] Z. Zhang, Q.K. Telesford, C. Giusti, K.O. Lim, D.S. Bassett, Choosing wavelet methods, filters, and lengths for functional brain network construction, in: S. Hayasaka (Ed.), PLoS One, vol. 11(6), Public Library of Science, 2016, p. e0157243.

[21] D. Steyrl, F. Patz, G. Krausz, G. Edlinger, G.R. Muller-Putz, Reduction of EEG artifacts in simultaneous EEG-fMRI: reference layer adaptive filtering (RLAF), in: 2015 37th Annual International Conference of the IEEE Engineering in Medicine and Biology Society (EMBC), vol. 2015, IEEE, 2015, pp. 3803–3806.

[22] L. Khedher, I.A. Illán, J.M. Górriz, J. Ramírez, A. Brahim, A. Meyer-Baese, Independent component analysis-support vector machine-based computer-aided diagnosis system for Alzheimer's with visual support, Int. J. Neural Syst. 27 (03) (2017), 1650050.

[23] M. Sharif, M. Yasmin, S. Masood, M. Raza, S. Mohsin, Brain image enhancement—a survey, World Appl. Sci. J. 17 (9) (2012) 1192–1204.

[24] M.J. Rosa, L. Portugal, T. Hahn, A.J. Fallgatter, M.I. Garrido, J. Shawe-Taylor, J. Mourao-Miranda, Sparse network-based models for patient classification using fMRI, NeuroImage 105 (2015) 493–506.

[25] J.D. Power, M. Plitt, P. Kundu, P.A. Bandettini, A. Martin, Temporal interpolation alters motion in fMRI scans: magnitudes and consequences for artifact detection, in: X.-N. Zuo (Ed.), PLoS One, vol. 12(9), Public Library of Science, 2017, p. e0182939.

[26] X. Guo, K.C. Dominick, A.A. Minai, H. Li, C.A. Erickson, L.J. Lu, Diagnosing autism spectrum disorder from brain resting-state functional connectivity patterns using a deep neural network with a novel feature selection method, Front. Neurosci. 11 (2017) 460.

[27] M. Ahmadi, M. O'Neil, M. Fragala-Pinkham, N. Lennon, S. Trost, Machine learning algorithms for activity recognition in ambulant children and adolescents with cerebral palsy, J. Neuroeng. Rehab. 15 (1) (2018) 105.

[28] J. Coloigner, Y. Kim, A. Bush, S. Choi, M.C. Balderrama, T.D. Coates, S.H. O'Neil, et al., Contrasting resting-state fMRI abnormalities from sickle and non-sickle anemia, in: A. Kassner (Ed.), PLoS One, vol. 12(10), Public Library of Science, 2017, p. e0184860.

[29] R.A. Poldrack, Region of interest analysis for fMRI, Soc. Cognit. Affect. Neurosci. 2 (1) (2007) 67–70.

[30] J. Kim, V.D. Calhoun, E. Shim, J.-H. Lee, Deep neural network with weight sparsity control and pre-training extracts hierarchical features and enhances classification performance: evidence from whole-brain resting-state functional connectivity patterns of schizophrenia, NeuroImage 124 (2016) 127–146.

[31] K.R. Sreenivasan, M. Havlicek, G. Deshpande, Nonparametric hemodynamic deconvolution of fMRI using homomorphic filtering, IEEE Trans. Med. Imag. 34 (5) (2015) 1155–1163.

[32] B. Thyreau, K. Sato, H. Fukuda, Y. Taki, Segmentation of the hippocampus by transferring algorithmic knowledge for large cohort processing, Med. Image Anal. 43 (2018) 214–228.

[33] Y. Jin, C.-Y. Wee, F. Shi, K.-H. Thung, D. Ni, P.-T. Yap, D. Shen, Identification of infants at high-risk for autism spectrum disorder using multiparameter multiscale white matter connectivity networks, Hum. Brain Map. 36 (12) (2015) 4880–4896.

[34] Q. Mahmood, A. Chodorowski, A. Mehnert, J. Gellermann, M. Persson, Unsupervised segmentation of head tissues from multi-modal MR images for EEG source localization, J. Dig. Imag. 28 (4) (2015) 499–514.

[35] G. Deshpande, D. Rangaprakash, L. Oeding, A. Cichocki, X.P. Hu, A new generation of brain-computer interfaces driven by discovery of latent EEG-fMRI linkages using tensor decomposition, Front. Neurosci. 11 (2017) 246.

[36] M. Adib, E. Cretu, Wavelet-based artifact identification and separation technique for EEG signals during galvanic vestibular stimulation, in: Computational and Mathematical Methods in Medicine, vol. 2013, Hindawi Limited, 2013 https://doi.org/10.1155/2013/167069.

[37] F. de Vos, M. Koini, T.M. Schouten, S. Seiler, J. van der Grond, A. Lechner, R. Schmidt, et al., A comprehensive analysis of resting state fMRI measures to classify individual patients with Alzheimer's disease, NeuroImage 167 (2018) 62–72.

[38] L. Schmüser, A. Sebastian, A. Mobascher, K. Lieb, O. Tüscher, B. Feige, Data-driven analysis of simultaneous EEG/fMRI using an ICA approach, Front. Neurosci. 8 (2014) 175.

[39] R. Pang, B.J. Lansdell, A.L. Fairhall, Dimensionality reduction in neuro-science, Curr. Biol. 26 (14) (2016) R656–R660.

[40] K.D. Toennies, Classification and Clustering, Guide to Medical Image Analysis, Springer, 2017, pp. 473–528.

[41] M. Larobina, L. Murino, Medical image file formats, J. Dig. Imag. 27 (2) (2014) 200–206.

[42] A. Ahmadvand, M.R. Daliri, Brain MR image segmentation methods and applications, OMICS J. Radiol. 02 (04) (2014) 1–3.

[43] S.E. Spasov, L. Passamonti, A. Duggento, P. Lio, N. Toschi, A multi-modal convolutional neural network framework for the prediction of Alzheimer's disease. in: 2018 40th Annual International Conference of the IEEE Engineering in Medicine and Biology Society (EMBC), IEEE, 2018, pp. 1271–1274, https://doi.org/10.1109/EMBC.2018.8512468.

[44] L. Billeci, A. Narzisi, A. Tonacci, B. Sbriscia-Fioretti, L. Serasini, F. Fulceri, F. Apicella, F. Sicca, S. Calderoni, F. Muratori, An integrated EEG and eye-tracking approach for the study of responding and initiating joint attention in autism spectrum disorders. Sci. Rep. 7 (1) (2017) 13560, https://doi.org/10.1038/s41598-017-13053-4.

[45] P. Płoński, W. Gradkowski, I. Altarelli, K. Monzalvo, M. van Ermingen-Marbach, M. Grande, S. Heim, et al., Multi-parameter machine learning approach to the neuroanatomical basis of developmental dyslexia, Hum. Brain Map. 38 (2) (2017) 900–908.

[46] B. Keehn, A. Nair, A.J. Lincoln, J. Townsend, R. Müller, Under-reactive but easily distracted: an FMRI investigation of attentional capture in autism spectrum disorder, Dev. Cogn. Neurosci. 17 (February) (2016) 46–56, https://doi.org/10.1016/J.DCN.2015.12.002.

[47] D. Brunet, M.M. Murray, C.M. Michel, Spatiotemporal analysis of multichannel EEG: CARTOOL, Comput. Intell. Neurosci. 2011 (2011) 813870.

[48] Opencv.org, OpenCV, Available at: https://opencv.org/, 2019. Accessed 25 June 2019.

[49] Imagej.net, ImageJ, Available at: https://imagej.net/Welcome, 2019. Accessed 25 June 2019.

[50] A. Tenev, S. Markovska-Simoska, L. Kocarev, J. Pop-Jordanov, A. Müller, G. Candrian, Machine learning approach for classification of ADHD adults, Int. J. Psychophysiol. 93 (1) (2014) 162–166.

[51] M.D. Sacchet, G. Prasad, L.C. Foland-Ross, P.M. Thompson, I.H. Gotlib, Support vector machine classification of major depressive disorder using diffusion-weighted neuroimaging and graph theory, Front. Psychiatr. 6 (2015) 21.

[52] P. Tamboer, H.C.M. Vorst, S. Ghebreab, H.S. Scholte, Machine learning and dyslexia: classification of individual structural neuro-imaging scans of students with and without dyslexia, NeuroImage: Clin. 11 (2016) 508–514.

[53] H. Fan, L. Li, R. Gilbert, F. O'Callaghan, L. Wijlaars, A machine learning approach to identify cases of cerebral palsy using the UK primary care database, Lancet 392 (2018) S33.

[54] T.M. Ball, M.B. Stein, H.J. Ramsawh, L. Campbell-Sills, M.P. Paulus, Single-subject anxiety treatment outcome prediction using functional neuroimaging, Neuropsychopharmacology 43 (4) (2018) 926.

[55] M.J. Patel, C. Andreescu, J.C. Price, K.L. Edelman, C.F. Reynolds, H.J. Aizenstein, Machine learning approaches for integrating clinical and imaging features in late-life depression classification and response prediction, Int. J. Geriatr. Psychiatr. 30 (10) (2015) 1056–1067.

[56] J.A. Ting, A. D'Souza, S. Schaal, Automatic outlier detection: a Bayesian approach, in: Proceedings—IEEE International Conference on Robotics and Automation, 2007, pp. 2489–2494.

[57] J.L. Marcano, M.A. Bell, A.A. Beex (Louis), Classification of ADHD and non-ADHD subjects using a universal background model, Biomed. Signal Process. Control 39 (2018) 204–212.

[58] X. Bi, Y. Wang, Q. Shu, Q. Sun, Q. Xu, Classification of autism spectrum disorder using random support vector machine cluster, Front. Genet. 9 (2018) 18.

[59] H. Chen, X. Duan, F. Liu, F. Lu, X. Ma, Y. Zhang, L.Q. Uddin, et al., Multivariate classification of autism spectrum disorder using frequency-specific resting-state functional connectivity—a multi-center study, Prog. Neuro-Psychopharmacol. Biol. Psychiatr. 64 (2016) 1–9.

[60] V.R. Konda, J.N. Tsitsiklis, Actor-critic algorithms, in: Advances in Neural Information Processing Systems 12, NIPS, Colorado, 2000, pp. 1008–1014.

[61] J. Aslanides, J. Leike, M. Hutter, Universal Reinforcement Learning Algorithms: Survey and Experiments, in: 26th International Joint Conference on Artificial Intelligence (IJCAI-17), 2017, pp. 1403–1410.

[62] P. Bellec, C. Chu, F. Chouinard-Decorte, Y. Benhajali, D.S. Margulies, R.C. Craddock, The Neuro Bureau ADHD-200 preprocessed repository, NeuroImage 144 (Pt B) (2017) 275–286.

[63] C. Cameron, B. Yassine, C. Carlton, C. Francois, E. Alan, J. András, K. Budhachandra, et al., The Neuro Bureau Preprocessing Initiative: open sharing of preprocessed neuroimaging data and derivatives, Front. Neuroinform. 7 (2013) https://doi.org/10.3389/conf.fninf.2013.09.00041.

[64] A. Riaz, M. Asad, E. Alonso, G. Slabaugh, Fusion of fMRI and non-imaging data for ADHD classification, Comput. Med. Imag. Graph. 65 (2018) 115–128.

[65] A. Riaz, M. Asad, S.M.M.R. Al-Arif, E. Alonso, D. Dima, P. Corr, G. Slabaugh, FCNet: A Convolutional Neural Network for Calculating Functional Connectivity from Functional MRI, Springer, Cham, 2017, pp. 70–78.

[66] N.C. Dvornek, P. Ventola, K.A. Pelphrey, J.S. Duncan, Identifying autism from resting-state fMRI using long short-term memory networks, Mach. Learn. Med. Imaging MLMI (Workshop) 10541 (2017) 362–370.

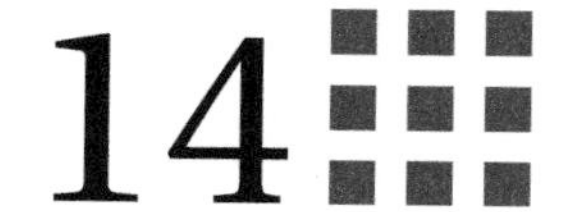

Diabetes prediction using artificial neural network

Nitesh Pradhan[a], Geeta Rani[a], Vijaypal Singh Dhaka[a], Ramesh Chandra Poonia[b]

[a]MANIPAL UNIVERSITY JAIPUR, JAIPUR, INDIA
[b]NORWEGIAN UNIVERSITY OF SCIENCE AND TECHNOLOGY (NTNU), ALESUND, NORWAY

14.1 Introduction

Good physical, mental, and social health of citizens is a reflection for quality of life, progress, and development of a nation. The work efficiency of a healthy people is multifold to that of an unhealthy one. Thus, healthy people play a constructive role in society.

But an increase in environmental pollution, use of fertilizers and pesticides to increase crop yield, an imbalance in physical and mental activities, and an increase in use of packaged food items can lead to overweight and unbeatable health challenges. Fluctuating blood pressure (BP), diabetes, cancer, cardiovascular disorders, and kidney failures are some major threats to life.

Diabetes is one of the most common disorders, affecting a huge number of the population throughout the world. It is a disorder that results in an increase in blood glucose level. It can be categorized into five categories, namely type-1, type-2, gestational diabetes, impaired glucose tolerance (IGT), and impaired fasting glycaemia (IFG) [1].

Type-1 is an innate disorder in which there is very low or no production of insulin. Patients are dependent on insulin for their survival [1]. This is less common category and observed in 5%–10% of the total diabetic patient population.

Type-2 is an adult onset diabetes. It occurs after the age of 40 years when the body ineffectively produces insulin. The patients are noninsulin dependent at an early stage. They use oral medicines to control the condition of hyperglycemia. As per present reports on diabetes [1], the age for onset of type-2 diabetes has reduced. Now, it has been noticed in children. Type-2 is the most commonly detected category. It affects 90% of the total diabetic patient population [1].

Gestational diabetes is a temporary state of hyperglycemia that arises during pregnancy. It disappears after the delivery, but it makes females more prone to type-2 diabetes.

IGT and IFG are transition states from normal to diabetes [1]. IGT and IFG makes the person prone to the type-2 disorder.

Deep Learning Techniques for Biomedical and Health Informatics. https://doi.org/10.1016/B978-0-12-819061-6.00014-8

Overweight, lack of exercise, family history of diabetes, and neural stress increases the possible risk of diabetes [2, 3]. All the diabetic disorders lead to noticeable symptoms, such as frequent urination and thirst, excessive hunger, weight loss, fatigue, etc. The degree of symptoms varies from category to category. These symptoms are more prominent in type-2 diabetes [4].

The condition of hyperglycemia not only self-damaging, it can damage eyes, kidneys, nervous system, and heart. Thus, it is a root cause for other noncommunicable diseases (NCDs) such as retinopathy, nephropathy, and neural and cardiovascular disorders. It is the third NCD responsible for premature deaths after high BP and tobacco use [5].

As per the global report published in 2016 [6], 8.5% of the adult population suffers from diabetes, which is double the number of diabetic patients recorded in 1980. It is spreading at a high rate in developing countries like India, where 49% of all diabetic patients in the world are reported. It is estimated that about 134 million Indians will be diabetic by 2025 [6]. The figure shows a serious health challenge for the country. Thus, there is an urgent need to focus on early prediction of diabetes. This is important for improving public health and preventing premature deaths.

Thus, the government is taking initiative toward dealing with diabetes. India adopted a STEPwise approach to Surveillance (STEPS), an initiative of WHO [1]. STEPS is a systematic way for collection, analysis, and dissemination of data on NCD risk factors. It presents an assessment report on trends of diabetes in different states at different time periods. In India, Tamilnadu is the state most affected by diabetes. The report helps in implementation of a National Multisectoral Action Plan for Prevention and Control of NCDs. Furthermore, it is useful in fixing a benchmark to reduce the burden of diabetes at the state as well as the national level. Mobile Health for Diabetes (mHealth-Diabetes) is another initiative to spread awareness about diabetes by Short Messaging Service (SMS). It also sends information about free counseling sessions and dietary suggestions to prevent diabetes. India annually spends about Rs 27,400 per capita for treatment of diabetes. If it continues to spread at the same rate, then the expenditure will increase up to Rs 1.95 lakh crore per capita by 2025. Therefore, its early detection becomes an urgent requirement for improving quality of life and reducing the economic burden on the nation. To meet the needs of the present scenario, the authors propose an artificial neural network (ANN)-based model for diabetes detection.

14.2 State of art

The study of existing literature in the arena of diabetes prediction reveals the use of machine learning (ML) techniques for early prediction. Researchers in the literature [7–18] applied techniques such as Naïve Bayes (NB) [7–11], support vector machine (SVM) [8–10, 12, 13, 19], decision tree (DT) [7–10, 13, 14], random forest (RF) [9, 11, 13, 14], K nearest neighbor (KNN) [13, 15, 16], K means [16, 17], multilayer neural network (MNN) [11, 18], etc. on the publicly available dataset "Pima Indian Diabetes" [20]. The brief description of these techniques follows.

14.2.1 Machine learning models used for diabetes prediction

Early prediction of diseases/disorders is useful in maintaining good public health. This is made feasible by applying ML technique(s). ML techniques broadly lie in two categories, namely supervised and unsupervised learning [6]. In unsupervised learning, a sample of the input dataset is fed without labeled classes. The algorithms such as K-means, self-organizing map (SOM), etc. are examples of unsupervised learning.

In supervised learning, the system is fed a sample input as well as labeled classes. The model learns from the sample of training dataset and gives a labeled class as output for the testing dataset. Supervised learning includes subcategories such as classification and regression. SVM, NB, KNN, RF, DT, and ANN are examples of supervised learning techniques. Supervised learning is useful for making prediction(s) in the field of healthcare. In the following section, the authors give a brief description of commonly used supervised learning algorithms for diabetes prediction.

Naïve Bayes: This algorithm is purely based on the probabilistic theory. It is simple to implement. It requires a small dataset for training and gives precise estimation [21]. Thus, researchers choose this algorithm for disease prediction. In case of diabetes prediction, it receives symptoms as input and yields the probability that a patient is diabetic.

Support vector machine: This is a nonparametric technique. It requires N number of support vectors to solve a particular problem [22]. Exemplified as: If the range of values to be predicted varies between zero to nine, then SVM requires 10 different support vectors. It considers local as well as global features of a given problem. Thus, it needs high computation time. Use of SVM is common in the prediction of diseases and disorders such diabetes, cardiovascular, brain tumor, etc., because it finds linear boundaries in feature space and clearly discriminates between different classes. It does so by identifying a separation hyperplane for all data items.

K-nearest neighbor: This technique is effective for small dataset. It is useful in resolving both classification as well as regression problems. KNN is nonparametric in nature [19]. This is easy to implement even for multiclass problems. Moreover, it does not require assumptions. Prediction of health disorders may be erroneous if wrong assumptions are fed to an algorithm. To overcome this problem, researchers use KNN for prediction of health disorders. Use of KNN is quite common for diabetes prediction using "Pima Indian Diabetes" dataset.

Decision tree: This technique is a solution for classification as well as regression problems. It effectively manages numerical and categorical data [23]. This is a robust technique and gives precise outputs even in the presence of bias in an actual model [24]. But evaluating the DT on "Pima Indian Diabetes" dataset yields low accuracy. This occurs due to small dataset for training and the problem of overfitting.

Random forest: It is an extension of the DT algorithm. RF deals with categorical as well as continuous variables. While recording the medical history of a patient, it is infeasible to report all the parameters. Thus, there is a fair chance of missing values in the gathered dataset. RF algorithm automatically handles missing values [25]. Thus, researchers prefer this algorithm over DT to achieve better accuracy.

Artificial neural network: ANN consists of three layers, namely input layer, hidden layer, and output layer [13]. The number of neurons in an input layer are dependent on number of features identified for a given problem. For example, "Pima Indian Diabetes" dataset encloses nine different features. Thus, it requires nine neurons in an input layer. The number of hidden layers is problem-specific and also dependent on the size of input data. There is no analytical way to determine the number of hidden layers. The hit and trial experiments help to find a suitable number of hidden layers for a problem. The output layer consists of only one neuron. The ANN considers local features of a given problem for extracting important information for prediction. Thus, in case of diabetes prediction, it receives the symptoms of a patient as an input and predicts whether he/she suffers from diabetes or not. The accuracy of ANN is higher than other previously mentioned techniques. This is due to the fact that it uses activation function and optimizers in its architecture.

Table 14.1 presents a comparative analysis of existing techniques. Column 1 includes the authors' name and year. Column 2 enumerates the technique(s)/algorithm(s) used. Column 3 presents the advantage(s) of the technique. Column 4 highlights the drawbacks of the existing technique, and column 5 shows the accuracy % obtained on performing experiments.

The comparative analysis presented in Table 14.1 shows that the existing techniques face the problems of overfitting and underfitting. These require a large dataset for training the model. Moreover, these techniques are ineffective in handling noisy data. To overcome these challenges, the authors use an ANN-based model for diabetes prediction.

14.3 Designing and developing the ANN-based model

There is a requirement of detection as well as identification of category of diabetes for its prevention and treatment. Thus, the authors designed and developed an ANN-based model. This model performs binary prediction for detection of diabetes. Simultaneously, it performs categorical prediction and assigns the class label for indicating the category of diabetes: type-1, type-2, gestational diabetes, IGT, and IFG.

14.3.1 Artificial neural network

Artificial neural networks are inspired by the biological nervous system. These are computational systems consisting of interconnected neurons. Each neuron can perform a simple task, but a strongly connected network of communicating neurons is effective in performing complex computations. ANN uses an activation function to transform the sum of weighted inputs given at a node into activation of that node or output for the corresponding input.

The activation signal is passed through a transfer function to produce a single output of a neuron. Thus, ANN is widely used ML technique for solving real world problems. In this model, ANN is applied to detection of diabetes and classification of detected cases into different categories of diabetes.

Table 14.1 Comparative analysis of existing techniques on "Pima Indian Diabetes" dataset.

Year	Authors	Technique/classifier used	Advantage	Disadvantage	Accuracy (%)
2009 [18]	Hasan Temurtas, Nejat Yumusak, Feyzullah Temurtas	Multilayer neural networks Probabilistic neural networks	Successful in finding a co-relation between features of the training set parameters	Neural network causes the problem of over-fitting, if training time is more or dataset size is very small	82.37 78.13
2011 [15]	Manaswini Pradhan, Dr. Ranjit Kumar Sahu	Neural network functional link K-nearest neighbor Artificial neural networks	Artificial neural networks are parametric in nature. These store information of entire network. So there is no need to store individual information. Recording individual information is required in SVM. ANN, automatically adjusts the weights of a network by using different optimizers	Artificial neural network covers local minima not the global minima	65.10 69.71 59.80
2014 [16]	Veena Vijayan V, Aswathy Ravikumar	Expectation maximization K-nearest neighbor K-means	KNN algorithm is robust to noisy training data	K-nearest neighbor and K-means takes longer time for large datasets	<70% 73.17 66–77
2015 [7]	Aiswarya Iyer, S. Jeyalatha, Ronak Sumbaly	Decision tree Naïve Bayes	Decision tree and Naïve Bayes are computationally fast because decision tree doesn't require any preprocessing of data and Naïve Bayes are less prone to overfitting	Decision tree is effective for categorical variables but not contiguous target variables	76.95 79.56
2016 [8]	Sanjaya De Silva	Decision tree Naïve Bayes Support vector machine	Less data is required for training the model	Naïve Bayes is based on the concept of probability. Its estimation is zero if there is no occurrence of a class label	84.66 76.66 77.33
2017 [12]	Aakansha Rathore, Sakshi Gujral, Simran Chauhan	Support vector machine	Support vector machine consider local as well as global features that helps to increase accuracy of the model	Support vector machine are nonparametric in nature. It requires N number of support vector for training	82.00
2017 [9]	Md. Aminul Islam, Nusrat Jahan	Naïve Bayes Support vector machine Decision tree Random forest	Decision tree faces the problem of over fitting so random forest overcome this issue	Require large training dataset	75.76 78.01 74.30 74.83
2017 [17]	Han Wu, Shengqi Yang, Zhangqin Huang, Jian He, Xiaoyi Wang	Improved K-means algorithm with logistic regression algorithm	In a hybrid of linear regression with K-means algorithm, linear regression automatically handles nonlinearity effects in a dataset	High execution time	95.42

Continued

Table 14.1 Comparative analysis of existing techniques on "Pima Indian Diabetes" dataset—cont'd

Year	Authors	Technique/classifier used	Advantage	Disadvantage	Accuracy (%)
2017 [21]	Mehrbakhsh Nilashi, Othman Ibrahim, Mohammad Dalvi, Hossein Ahmadi, Leila Shahmoradi	Self-organizing map, principal component analysis, neural network	Use of semisupervised technique for diabetes prediction (a combination of supervised and unsupervised learning) is advantageous as a machine can do prediction for labeled as well as unlabeled data	It is challenging to achieve good accuracy if dataset is noisy	92.28
2017 [22]	Ioannis Kavakiotis, Olga Tsave, Athanasios Salifoglou, Nicos Maglaveras, Ioannis Vlahavas, Ioanna Chouvarda	A combination of supervised and unsupervised algorithms such as support vector machine, clustering algorithms, etc.	SVM is robust and gives good accuracy even though training sample contains bias up to some extent	If the data size is small, these techniques produce low quality results	92.43
2015 [13]	J. Pradeep Kandhasamy, S. Balamurali	Decision tree J48 KNN classifier Random forest Support vector machine	These algorithms have good computation power, so execution time is less	Lower accuracy for noisy data	86.46 82.55 100 77.73
2018 [10]	Deepti Sisodiaa, Dilip Singh Sisodia	Decision tree Support vector machine Naïve Bayes	The outputs of applied algorithms are verified using receiver operating characteristic (ROC). This is successful in determining the probability whether occurrence of diabetes is correctly ranked or randomly chosen. Computational power of applied algorithms is higher than algorithms such as random forest and artificial neural networks	Need a separate technique to validate the results	73.82 65.10 76.30
2018 [14]	Quan Zou, Kaiyang Qu, Yamei Luo, Dehui Yin, Ying Ju, Hua Tang	Decision tree Random forest Neural network	Performed extensive experiments on two different datasets: "Luzhou" and "Pima Indian Diabetes" dataset	The applied algorithms successfully predict occurrence of diabetes in a patient but in successful identifying the type of diabetes	72.75 76.04 78.28
2017 [11]	Dr. Asir Antony Gnana Singh, Dr. E. Jebamalar Leavline, B. Shanawaz Baig	Naïve Bayes Multi-layer perceptron Random forest	Useful in handling noisy data	High computation time as applicable for preprocessed or nonpreprocessed data	78.25 76.96 84.93

It includes three different layers namely input layer, hidden layer, and output layer. Each layer contains a set of neurons. These neurons receive information at the input layer, processes the received information, and passes the processed information to the output layer via hidden layer. The output layer generates the prediction/classification as an output.

It is quite challenging and data dependent to decide the number of hidden layers in ANN. The number of neurons in input layers are dependent on the number of features selected for a given problem. The sum of product of feature values and assigned weights yields the net input of hidden layer neurons. Eq. (14.1) shows the calculation of net input. Here, y_{in} is the net input, b is the bias, n is the number of neurons, x_i is an input to ith neurons, and w_i is a weight assigned to ith neurons between an input and hidden layer.

$$y_{\text{in}} = b + \sum_{i=1}^{n} x_i w_i \tag{14.1}$$

Fig. 14.1 shows the architecture of an ANN-based model used for detection and classification of diabetes. X1, X2, ..., Xn are features given as an input to input layer. HL1, HL2, HL3, and HL4 are hidden layers. Wnj, Wjk, Wkl, and Wlm are weights assigned to neurons at hidden layers. Y is the output generated from the output layer.

An ANN-based model is comprised of four hidden layers. Each hidden layer consists of six neurons. The authors applied Rectified Linear Unit (ReLU) activation function on each hidden layer. It is a piecewise linear function. It gives input as an output if given input is positive or zero. It gives zero as an output for negative input values [15]. The authors use this activation function because it gives a convenient way to train the model. It also achieves better performance for prediction and classification.

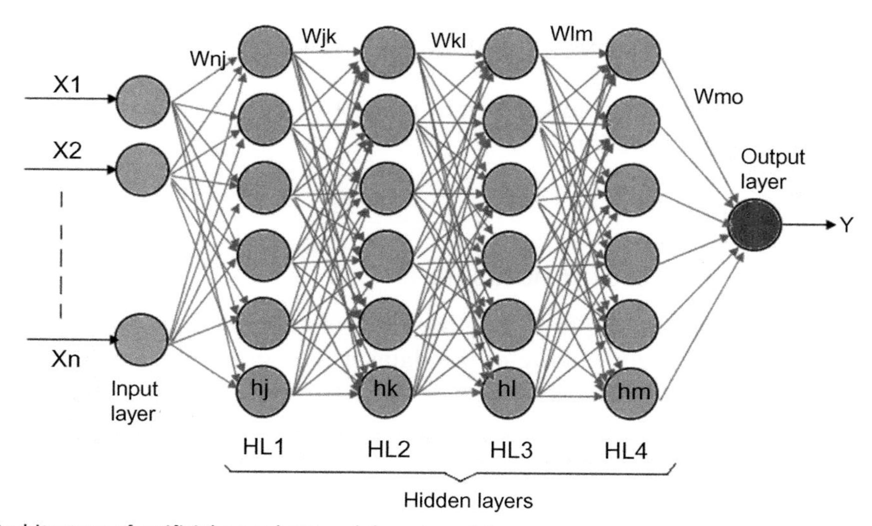

FIG. 14.1 Architecture of artificial neural network-based model.

The model contains only one output layer that contains a single neuron. It applies "sigmoid" activation function on the output layer because sigmoid function is effective in predicting the probability of occurrence in the range of 0 and 1. In this model, it detects the probability whether a patient is diabetic or not. In the case of more than two possible categories, softmax activation function is preferable. The model uses the "Adam" optimizer because it realizes the benefit of Adaptive Gradient algorithm (AdaGrad) and Root Mean Square Propagation (RMSProp) optimizer.

14.4 Dataset

For evaluation of efficacy of the designed model, the authors use the publicly available "Pima Indian Diabetes" dataset (https://www.kaggle.com/uciml/pima-indians-diabetes-database). The authors prefer this dataset because it gives insights about symptoms that can predict the occurrence as well as the category of diabetes. Moreover, the review of related literature shows that this dataset is most widely used for evaluation of different ML models. Thus, it gives an opportunity to compare the designed model with existing models.

The dataset contains the medical history of 768 women patients. All patients lie in the age range of 21 to 81 years. The dataset contains the following nine symptoms as independent variables (medical predictor) and one dependent (target) variable. Each complete presentation of a data point is named as epoch. An epoch is an entity used by machine for learning.

- **(i)** Number of times a woman patient is pregnant
- **(ii)** Plasma glucose concentration a 2h in an oral glucose tolerance test
- **(iii)** Diastolic blood pressure (in mm Hg)
- **(iv)** Triceps skin-fold thickness (in mm)
- **(v)** 2-h serum insulin (lU/mL)
- **(vi)** Body mass index (weight in kg/(height in m^2))
- **(vii)** Diabetes pedigree function
- **(viii)** Age (in years)
- **(ix)** 2-h serum insulin (lU/mL)

14.4.1 Training and testing datasets

In ML models, it is challenging to decide the ratio of training and testing datasets. It is highly dependent on size and nature of the dataset. The presence of noise and missing values need to be considered for dividing a dataset. The incorrect ratio leads to problems of overfitting and underfitting. If the training data contains all cases very close to testing data, then it faces the problem of overfitting. In case the training dataset is very small and considers a lesser number of cases for training, then it faces with the problem of underfitting. To avoid both extreme challenges, the authors performed a set of experiments to decide a ratio. On the basis of experiments, in this model, the authors divided the "Pima Indian Diabetes" dataset into training and testing sets; 70% of the total dataset (records of 538 patients) was used for training the model, and the remaining 30% (records of 230 patients) was used for testing purposes.

14.5 Implementation

The model completes its tasks in a sequence of the following three phases. Fig. 14.2 demonstrates the workflow of the proposed model.

Step 1: Preprocessing of data: Initially, Pima Indian Diabetes Dataset have nine columns. Out of the nine columns, some columns include parameters such as glucose, thickness, insulin, etc. These parameters have a wide range of values that need to be

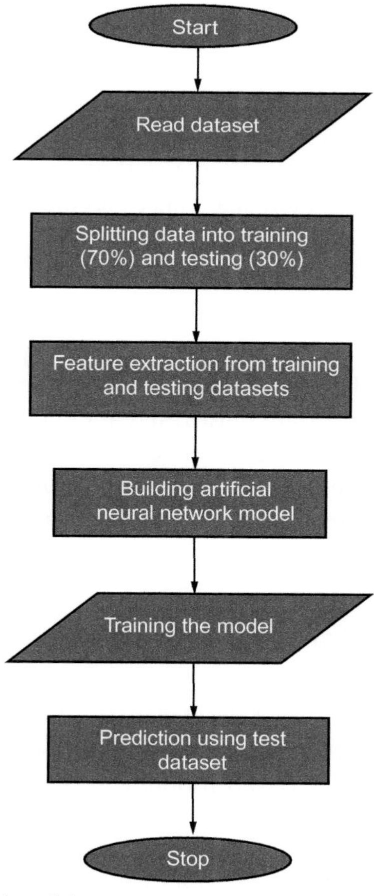

FIG. 14.2 Workflow of the ANN-based model.

preprocessed. The authors performed normalization on these columns to get a range from +1 to −1. In addition, noise removal is applied on the Pima Indian Diabetes Dataset.

Step 2: Training of the ANN-based model: After successful completion of preprocessing of the dataset, there is need of training the ANN-based model of the training dataset (70% of the total dataset). At the time of training, the authors used an input layer with nine neurons. The number of neurons is set at nine on the basis of the number of features in the dataset. The authors used four hidden layers in the ANN model. Each hidden layer contains six neurons. The number of neurons in a hidden layer are set as an average of the number of input and output neurons.

Step 3: Testing of a trained ANN model for multiple epochs: After training the model of the training dataset (70% of the total dataset), the model is run on the testing dataset (30% of total dataset). This step is useful for evaluating the efficacy of the model.

14.6 Experiments

14.6.1 Experimental setup

The model is trained on the Google Colab, a free online training platform. Google Colab uses Tesla K80 GPU and provides a RAM of 12GB. It can be used continuously for 12h. RAM is a limitation here, so the batch size is kept small.

14.6.2 Experimental results

For evaluation of efficacy of the proposed model, the authors performed a set of experiments. They used four hidden layers in their model. The experiments were performed on a range of 0 to 500 epochs. The difference between two values of number of epochs is set as 50, exemplified as the experiments are performed at epoch numbers 0, 50, 100...500. Two different activation functions, namely Sigmoid and ReLU, are applied at hidden layers. On the output layer, the authors used only sigmoid activation function. The results shown in Fig. 14.3 clearly indicate that the model yields the best accuracy of 85.09% for 500 epochs using ReLU activation function.

14.7 Comparative analysis

The study of the literature shows the use of various ML models for prediction of diabetes. In this chapter, the authors compared the accuracy of the proposed ANN-based model with existing models, namely KNN, NB, DT, SVM, and K-means on Pima Indian Diabetes dataset. The accuracy of the ANN-based model is directly dependent on the size of the training dataset. The analysis given in Fig. 14.4 shows K-means yields the lowest accuracy of 72%, and the proposed ANN-based model outperforms the existing models by achieving the highest accuracy of 85.09%. Its accuracy can be further improved by increasing the number of data points.

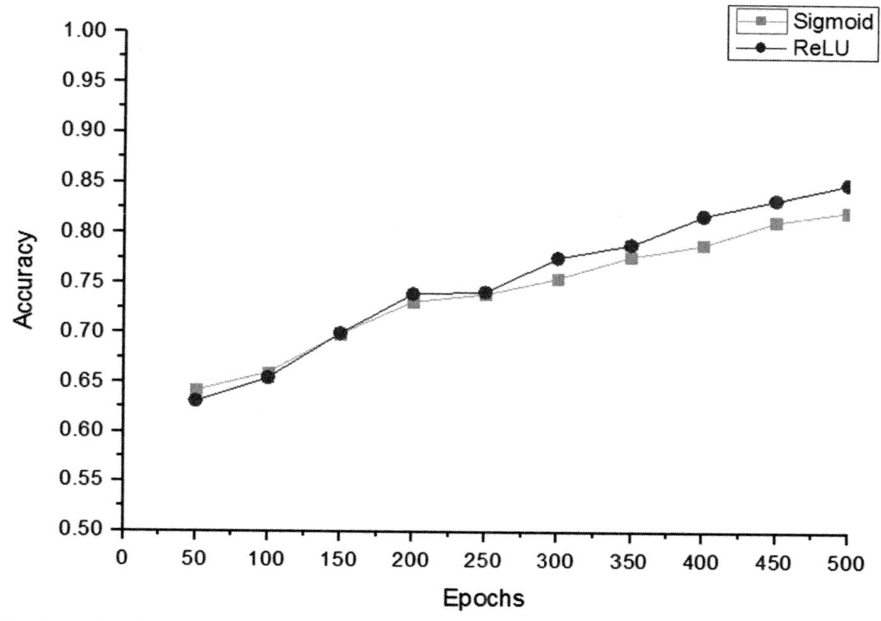

FIG. 14.3 Result analysis based on activation function.

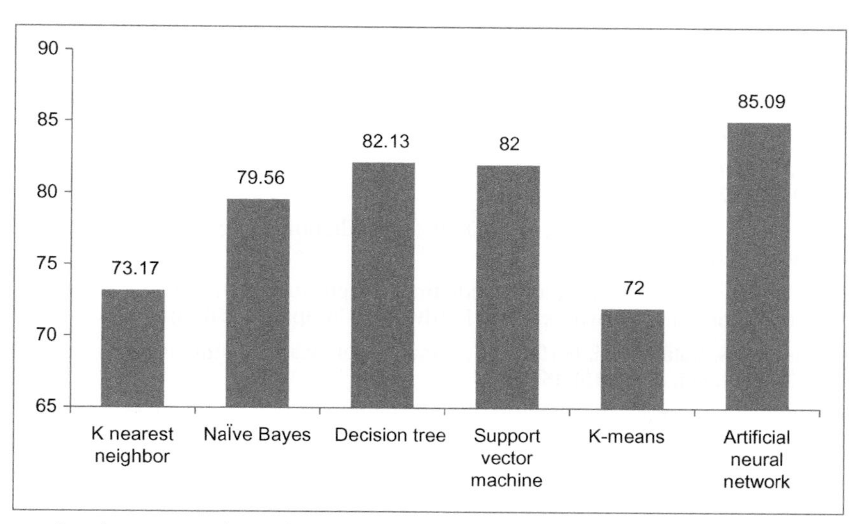

FIG. 14.4 Comparison in accuracy of ANN-based model with existing models.

14.8 Summary

In this chapter, the authors presented the needs of diabetes prediction. It elaborates the statistical scenario of diabetes across the world, as well as in India. The chapter contains the initiatives taken by the government for dealing with an increasing burden on public health and economy of the nation. Then the authors elaborated on various ML models used for prediction of diabetes using the "Pima Indian Diabetes" dataset. They identified the challenges in existing models. To overcome the challenges, they proposed an ANN-based ML model. The chapter gives the detailed description of an ANN-based model in terms of its architecture, experimental setup, Dataset, and evaluation. It also includes the comparison of the ANN-based model with existing models.

References

[1] H. Bekedam, Diabetes Scale Up Prevention, Strengthen Care and Enhance Surveillance, A Report by World Health Organization, 2016.

[2] S. Pobi, A Study of Machine Learning Performance in the Prediction of Juvenile Diabetes From Clinical Test Results, University of South Florida, South Florida, 2006.

[3] M. Davidson, D.L. Schriger, A.L. Peters, An alternative approach to the diagnosis of diabetes with a review of the literature, Diabetes Care 7 (1995) 1065–1071.

[4] O. Chandrakar, J. Kumar, R. Saini, Development of Indian weighted diabetic risk score (IWDRS) using machine learning techniques for type-2 diabetes, in: Proceedings of the 9th Annual ACM India Conference, 2016, , pp. 125–128.

[5] O. Chandrakar, J. Kumar, R. Saini, Identification of parameters impacting diabetes risk score, Int. J. Control Theory Appl. 10 (14) (2017) 1–7.

[6] WHO, Global Report on Diabetes, WHO Press, 2019, pp. 1–86. Available at: http://www.who.int.

[7] A. Iyer, S. Jeyalatha, R. Sumbaly, Diagnosis of diabetes using classification mining techniques, Int. J. Data Min. Knowl. Manag. Process 5 (1) (2015). arXiv: 1502.03774.

[8] S. De Silva, Predicting Diabetes Using a Machine Learning Approach, 2016, pp. 7–11.

[9] A. Islam, N. Jahan, Prediction of onset diabetes using machine learning techniques, Int. J. Comput. Appl. 180 (5) (2017) 7–11.

[10] D. Sisodiaa, D. Sisodia Singh, Prediction of diabetes using classification algorithms, Proc. Comput. Sci. 132 (2018) 1578–1585.

[11] A.A. Gnana Singh, E.J. Leavline, B.S. Baig, Diabetes prediction using medical data, J. Comput. Intell. Bioinforma. 10 (2017) 1–8.

[12] A. Rathore, S. Chauhan, Detecting and predicting diabetes using supervised learning: an approach towards better healthcare for women, Int. J. Adv. Res. Comput. Sci. 8 (5) (2017) 1192–1194.

[13] J.P. Kandhasamy, S. Balamurali, Performance analysis of classifier models to predict diabetes mellitus, Proc. Comput. Sci. 47 (2015) 45–51.

[14] Q. Zou, K. Qu, Y. Luo, et al., Predicting diabetes mellitus with machine learning techniques, Bioinform. Comput. Biol. 9 (515) (2018) 1–10.

[15] M. Pradhan, R.K. Sahu, Predict the onset of diabetes disease using artificial neural network (ANN), Int. J. Comput. Sci. Emerg. Technol. 2 (2) (2011) 303–311.

[16] V. Vijayan, A. Ravikumar, Study of data mining algorithms for prediction and diagnosis of diabetes mellitus, Int. J. Comput. Appl. 95 (17) (2014) 12–16.

[17] H. Wu, S. Yang, Z. Huang, et al., Type 2 diabetes mellitus prediction model based on data mining, Inform. Med. Unlocked 10 (2017) 100–107.

[18] H. Temurtas, N. Yumusak, F. Temurtas, A comparative study on diabetes disease diagnosis using neural networks, Expert Syst. Appl. 36 (4) (2009) 8610–8615.

[19] M. Nirmala Devi, S. Balamurugan, U.V. Swathi, An amalgam KNN to predict diabetes mellitus, in: IEEE International Conference on Emerging Trends in Computing, Communication and Nanotechnology, 2013, pp. 691–695.

[20] D. Dua, C. Graff, UCI Machine Learning Repository, https://www.kaggle.com/uciml/pima-indians-diabetes-database, 2019.

[21] K. Rajesh, V. Sangeetha, Application of data mining methods and techniques for diabetes diagnosis, Int. J. Eng. Innov. Technol. 2 (3) (2012) 224–229.

[22] J. Wu, Y. Diao, M.L. Li, A semi-supervised learning based method: Laplacian support vector machine used in diabetes disease diagnosis, Interdiscip. Sci. Comput. Life Sci. 1 (2009) 151–155.

[23] S. Sankaranarayanan, T. Perumal, A predictive approach for diabetes mellitus disease through data mining technologies, in: World Congress on Computing and Communication Technologies, IEEE, 2014, , pp. 231–233.

[24] R. Manimaran, M. Vanitha, Novel approach to prediction of diabetes using classification mining algorithm, Int. J. Innov. Res. Sci. Eng. Technol. 6 (2017) 14481–14487.

[25] M. Butwall, S. Kumar, A data mining approach for the diagnosis of diabetes mellitus using random forest classifier, Int. J. Comput. Appl. 120 (8) (2015) 36–39.

Further reading

T. Jayalakshmi, A. Santhakumaran, A novel classification method for diagnosis of diabetes mellitus using artificial neural networks, in: International Conference on Data Storage and Data Engineering, IEEE, Bangalore, 2010.

V. Vijayan, C. Anjali, Prediction and diagnosis of diabetes mellitus—a machine learning approach, IEEE Recent Adv. Intell. Comput. Syst. (RAICS) (2015) 10–12.

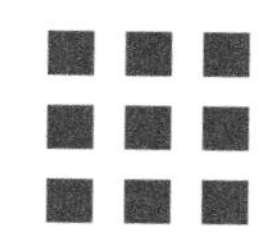

Index

Note: Page numbers followed by *f* indicate figures and *t* indicate tables.

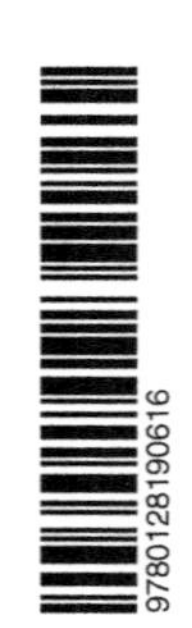